John Lister

Cotton manufacture

A manual of practical instruction in the processes of opening, carding, combing, drawing, doubling and spinning of cotton

John Lister

Cotton manufacture
A manual of practical instruction in the processes of opening, carding, combing, drawing, doubling and spinning of cotton

ISBN/EAN: 9783337264918

Printed in Europe, USA, Canada, Australia, Japan

Cover: Foto ©Andreas Hilbeck / pixelio.de

More available books at **www.hansebooks.com**

Cotton Manufacture

*A MANUAL OF PRACTICAL INSTRUCTION IN THE
PROCESSES OF OPENING, CARDING, COMBING,
DRAWING, DOUBLING, AND SPINNING OF
COTTON, AND THE METHODS OF
DYEING AND PREPARING GOODS
FOR THE MARKET*

For the use of Operatives, Overlookers, & Manufacturers

BY

JOHN LISTER

OF PENDLETON

TECHNICAL INSTRUCTOR TO THE TEXTILE CLASS OF THE NATIONAL ASSOCIATION
OF MANAGERS AND OVERLOOKERS.

WITH NUMEROUS ILLUSTRATIONS

LONDON
CROSBY LOCKWOOD AND SON
7, STATIONERS' HALL COURT, LUDGATE HILL
1894

AUTHOR'S PREFACE.

THE efforts which are now being made for the advancement and extension of Technical Education amongst English workpeople have not been initiated too soon; and as regards the literature of the subject, it may be taken as a fact that in no manufacture is there found a greater need of really practical information, and of practical guides for the successful conduct of a business, than in the different operations connected with our great Cotton Industry. This need is intensified at the present time by the existing keen competition at home and abroad, and by the demands of the operatives for higher wages and shorter hours of labour.

Under these conditions, success cannot be attained by manufacturers unless they have not only the command of the best machinery, but are also able to ensure that it is utilised to the best advantage.

In the present volume, I have endeavoured to supply a work which may assist the intelligent Operative to master the details of the processes in which he is engaged, and at the same time assist the Manufacturer in the conduct of his business and the selection of machinery suitable for his requirements. The volume (it should be explained) is based upon a series of articles written by me at the request of the Editor of *Textile Industries and Journal of Fabrics*,* and published in that periodical.

In preparing those articles, I was able to utilise the experience of over fifty years' active occupation in the Manufacture of Cotton, in the course of which the positions I have filled have ranged from that of an Operative in various branches to the responsible one of Manager in mills, both at home and abroad. And it should be a further qualification for undertaking such a task, that for some time I held the important position of Technical Instructor of Managers and Overlookers.

Throughout the work, I have endeavoured to treat the subject from a thoroughly practical point of view, and to deal clearly but tersely with the various processes in their order of precedence.

To make the matter here presented more intelligible, illustrations of machinery have been freely introduced; and these having been selected with the greatest care, I am confident that they will add very materially to the value of the work as a Manual of Practical Instruction in the Processes of Cotton Manufacture. And while I have found it advantageous, and indeed necessary, to refer to examples of machinery made by specified firms, I must disclaim all intention or desire of giving to any particular firm undue prominence as compared with other makers of similar machines. Amongst the latter are doubtless to be found

* H. & R. T. Lord, Bradford and Manchester.

firms whose machinery and appliances are of equal repute with those of the firms of whom mention is made in this volume.

Whether or not I have been successful in my object, I must leave for others to decide, but I sincerely trust that from my efforts some appreciable benefit may accrue to many of those engaged in the Industry with which I have been so long associated.

My best thanks are due to Mr. R. T. Lord for assisting me in preparing this volume for the press, and for carrying out the arrangements for illustrating it.

JOHN LISTER.

PENDLETON, MANCHESTER.
December 1893.

[PUBLISHERS' NOTE.—Although Mr. Lister took part in the preliminary arrangements for the production of this volume, his death, unhappily, occurred while it was passing through the press. The foregoing Preface was left by him in draft.]

EDITOR'S PREFACE.

THE series of articles on Cotton Manufacture contributed by Mr. Lister to *Textile Industries and Journal of Fabrics* (with which publication I am intimately associated) were so much appreciated by readers of that journal, that their republication in a permanent form was deemed advisable. It is a matter of very deep regret to me, as to all Mr. Lister's acquaintance, that he did not live to see the appearance of this volume.

As his reasons for the production of the work are fully stated in the foregoing Preface, there is no necessity for me to enlarge upon that point. I desire, however, to take this opportunity of thanking the several firms (amongst whom are some of the best Machinists of the country) who have so kindly aided me in the work of editing these pages by placing woodcuts in my hands for use as illustrations. It would have been impossible to properly illustrate a book of this description without the assistance thus given; and I accordingly beg to acknowledge the Author's and my own indebtedness in this respect, as well as the great kindness and courtesy of one and all to whom application was made.

Besides the blocks thus furnished, some additional illustrations—including sectional diagrams of the more important machines—prepared from designs especially drawn and engraved for the work, have been inserted where necessary.

R. T. LORD.

BRADFORD,
April 1894.

CONTENTS.

		PAGE
CHAPTER I.		
INTRODUCTORY	. . . .	1
CHAPTER II.		
COTTON OPENING	. . . . .	7
CHAPTER III.		
CARDING AND COMBING OF COTTON	.	16
CHAPTER IV.		
DRAWING AND DOUBLING	. . . .	30
CHAPTER V.		
THE FLY, OR BOBBIN FRAME	. . . .	40
CHAPTER VI.		
THE MULE	. . . . . .	50
CHAPTER VII.		
MEMORANDA FOR MULE SPINNING	. .	66
CHAPTER VIII.		
THROSTLE AND RING SPINNING	. . . . .	72

CONTENTS.

CHAPTER IX.
Winding and Reeling 82

CHAPTER X.
Cotton Bleaching 94

CHAPTER XI.
Sizing of Cotton 101

CHAPTER XII.
Warping of Cotton 119

CHAPTER XIII.
Colour Dyeing 130

CHAPTER XIV.
Action of Indigo in Dyeing 138

CHAPTER XV.
Calico Printing 145

CHAPTER XVI.
Preparatory Process of Weaving . . . 159

CHAPTER XVII.
The Power Loom 165

CHAPTER XVIII.
Finishing of Cotton Fabrics . . . 196

LIST OF ILLUSTRATIONS.

FIG.		PAGE
1.	HOPPER FEEDER	7
2.	DIAGRAM GIVING SECTIONAL VIEW OF HOPPER FEEDER (IN UPPER ROOM) APPLIED TO OPENER (ON GROUND FLOOR)	8
3.	SYSTEM OF EXHAUST (IN SECTION)	10
4.	COMBINED OPENER AND LAP MACHINE, WITH HOPPER FEEDER (IN SECTION)	12
5.	SCUTCHER LAP MACHINE	14
6.	SECTIONAL VIEW OF SCUTCHER LAP MACHINE	15
7.	REVOLVING FLAT CARDING ENGINE	17
8.	FLEXIBLE BEND	18
9.	DIAGRAM ILLUSTRATING CARD-GRINDING MOTION	19
10.	CARD-GRINDING MACHINE	24
11.	COTTON-COMBING MACHINE	27
12.	SECTIONAL VIEW OF COTTON-COMBING MACHINE	28
13.	DRAWING FRAME	32
14, 15.	SECTIONAL VIEWS OF DRAWING FRAME	35
16.	SLUBBING FRAME	43
17.	SECTIONAL VIEW OF SLUBBING FRAME	45
18.	ROVING FRAME	46
19.	SECTIONAL VIEW OF ROVING FRAME	47
20.	SELF-ACTING TWINER	58
21.	SELF-ACTING MULE HEADSTOCK	60
22.	SECTIONAL VIEW OF SELF-ACTING NOSING MOTION	61
23, 24.	MULE INDICATORS	70
25.	RING SPINNING FRAME	75
26.	SECTIONAL VIEW OF RING SPINNING FRAME	76
27—29.	SPINDLES, IN ELEVATION AND SECTION	77
30, 31.	CONDITIONING BOBBINS	79
32, 33.	SIDE AND FRONT VIEWS OF ANTI-BALLOONING APPLIANCE	80
34.	WINDING-FRAME FOR SHUTTLE BOBBINS	83

LIST OF ILLUSTRATIONS.

FIG.		PAGE
35.	CONICAL DRUM WINDING FRAME	85
36.	DOUBLING-WINDING FRAME	88
37.	DOUBLING-WINDING FRAME IN TRANSVERSE SECTION	89
38.	GASSING FRAME	92
39.	BLEACHING KIER	96
40.	VERTICAL DRYING MACHINE	99
41.	SLASHER SIZING MACHINE	107
42.	HEADSTOCK OF SLASHER SIZING MACHINE	108
43.	DAMPING MACHINE	110
44.	MOTE-CLEARING MACHINE	111
45.	FILLING AND FINISHING MACHINE	113
46.	DOUBLE HANK SIZING MACHINE	117
47.	SECTIONAL WARPING MACHINE	125
48.	BEAM WARPING MACHINE	128
49.	DYE JIGS	132
50.	DYE WENCH, WITH ENGINE	133
51.	COP DYEING MACHINE	135
52.	HANK AND SLUBBING DYEING MACHINE	136
53.	INDIGO DYEING MACHINE	141
54.	EIGHT-COLOUR DUPLEX CALICO-PRINTING MACHINE	146
55.	ANILINE AGEING MACHINE	148
56.	CALICO LOOM	166
57.	TAKING-UP MOTION	167
58.	DROP-BOX LOOM	174
59—62.	TAPPETS	177
63, 64.	LOOM TEMPLES	186
65.	CALICO LOOM, WITH DOBBY	188
66.	DOBBY	190
67.	JACQUARD MACHINE	193
68.	PIN-TENTERING AND STRETCHING MACHINE	197
69.	RIGGING MACHINE	199
70.	GAS SINGEING MACHINE	200
71.	FOUR-BOWL CALENDER	201
72.	COMBINED STARCH MANGLE AND DRYING MACHINE	203
73.	BEETLING MACHINE	205
74.	DOUBLE-TIER DRYING MACHINE	209

COTTON MANUFACTURE.

CHAPTER I.

INTRODUCTORY.

The Cotton Industry.—Numerous writers, from Herodotus to those of our own time, have discussed the history, origin, and nature of cotton. The exact period when it was introduced into England is not known, the earliest record we have being an entry in the books of Bolton Abbey, 1298, showing that cotton was used for candlewicks. In 1760, not more than forty thousand persons were employed in the cotton industry, while, at present, the numbers engaged can be counted by millions. The object of the present volume is to supply, as far as possible, the means of obtaining a practical and intelligent knowledge of every process for the commercial and economical production of yarns and fabrics, and it is scarcely necessary to add that the information to be given of the systems of preparation, carding, etc., will be found of direct utility to all who are interested in the trade.

Varieties of Cotton.—In business, cotton is separated into the following varieties:—*United States* cotton, exported from Charleston, New Orleans, Mobile, and Savannah, and produced in Georgia, South Carolina, Alabama, Mississippi, and Louisiana; but the best in the market is Sea Island, the most perfect form of cotton fibre at present known. It commands a high price, and can be spun into the finest counts. *South' American* cotton comes from the Brazils, Guiana, New Granada, Peru, Venezuela, and Columbia. The *West Indian Islands* produce cotton of a superior quality, which is preferred to the Brazilian. Amongst *African* cottons, the Island of Bourbon sends us a good class of fibre, but Egyptian cotton is the most regular in length of staple and diameter, and can be spun into 150's count, extra care and attention, however, being required in cleaning and carding. Its cost is less than that of Sea Island, but it is improving in value every year, and is a very useful cotton in many respects. Fine-spinning factories in Lancashire use the Sea Island and Egyptian almost exclusively, as what are called Bolton counts, ranging chiefly from 60's to 100's, depend on Egyptian. Other fine long staples can be grown that might be equally serviceable, but the question of price has to be considered. A very good quality of fine long staple is grown in the *Fiji Islands*, and more would be cultivated if required. The best African cotton is grown in Algeria; it is of good colour, as fine as silk, and partakes of the characteristics of the Georgian and Egyptian, and in quality approaches the finest Louisiana variety. *East Indian* cotton is greatly inferior, and the fibre is very short, although, when mixed with American, it is found to possess good spinning properties, and is principally used for low counts. Next to the United States, *British India* furnishes the largest quantity for our home consumption, the best known sorts in the market being Bengal, Madras, Bombay, Surat, Siam, and Manilla. *Levant* cotton comprises cottons which are received from European and Asiatic Turkey, Morea, and the Grecian Archipelago; and the best or principal sorts are the Smyrnian, Syrian, Cyprian, Macedonian, and Persian, all of inferior quality like the East Indian, and used for low counts. A *Chinese* cotton, known as Nankin, of a beautiful gold colour, is a good imitation of camel's hair, and is often used as a substitute for wool.

There are other varieties, at times, offered for sale, but they are not sufficiently distinctive to call for any special remarks. Whilst on the subject, it may not be out of place to observe that samples of cotton grown from New Orleans seed in *Ceylon* are reported to be of high quality, the colour, texture, and length of staple all combining to give a character and value to the fibre, and a good price has been obtained for a quantity sold in the Liverpool market. The climate of Ceylon and other conditions are favourable to an important cotton field; and most certainly the time has now arrived when cotton-spinners should take note of any probable area of production that will render them less dependent than they have hitherto been on

America; and this for many reasons. As the Ceylon cotton is produced from American seed, it is equally suitable for the class of yarns now spun from United States qualities, and the opinions of those who have particular knowledge of cotton cultivation have been expressed from time to time, all encouraging the idea that Ceylon cotton will be even of greater value than that of the States, as the conditions of growth are more favourable.

Test of Value of Cotton.—The value of all cotton is determined by the length and fineness of the staple. American cotton is the standard, and at present its price in the market rules all others. A study of the conditions in which cotton is placed by the demands and requirements of commerce cannot fail to be of practical value to those who may be called upon to fill the all-important and arduous duties of cotton-mill managers, and a thoroughly sound knowledge of every point connected with the material, its qualities, and the characteristics of each sample submitted to his notice, is necessary, whilst a delicate touch and good eyesight are also much-needed physical qualities. Occasional microscopic investigations will be required to determine the produce of a spurious or immature crop, and other useful analyses may come within the scope of this instrument when properly handled. To detect the quality of the mixing, a good specimen of yarn that has been worked with all the inevitable defects would be a capital test.

The advice now given is from the school of experience, dearly bought, and will be found worthy of notice. No concern can be made to pay, in the face of keen competition and depressed prices, unless the most careful and the closest attention is paid to every detail, however trivial and apparently worthless. A calm determination, watchfulness, strict integrity, celerity of mind to conceive and carry out any useful hint or suggestion of practical value, constitute good management in every sense of the word.

Quality of Yarns.—The aim of a cotton spinner is, at all times, to strike a quality of yarn suitable for the market, and to maintain uniformity in so doing; but, in consequence of the want of this uniformity in the quality of the cotton, although of the same class, and of the number of operations through which it passes—in any one of which a defective machine or neglect would have a direful effect —it is no easy matter to obtain more than a comparative uniformity. To obviate, as far as practicable, the difference of quality is the object of the first manipulation to which the cotton is subjected. The most important of all the long and costly series of operations gone through in a spinning mill is to mix or mingle the different qualities of cotton so as to obtain a new quality, which shall be such an average of the qualities of each as may be required for the special kind of yarn ordered. This is really a work demanding the exercise of good judgment, as well as consummate skill in handling and observing the variations of tint and shade, which may diverge from the normal tone or colour of the cotton desired.

Mixings.—Considerable experience is absolutely necessary to obtain comparative perfection as a mixer, for a failure in this first stage cannot be remedied afterwards, and a very heavy loss may be incurred. A mixing of cotton is not always of the same class—the lower counts are, in almost every case, made solely from East Indian, or other lower-class cotton, medium counts from American, and the higher from Egyptian or good American. Intermediate counts are, generally speaking, made from a "blending" of several cottons in varying proportions, according to the judgment of the manager or spinning master, and it will, therefore, be seen that this manipulation requires a considerable amount of care.

The method usually adopted is to have the bags or bales ranged conveniently. The contents of the first bale are spread out in a thin layer, aided by a rake; the second bale is spread on the top of the first; the third, in like manner, and so on, until a great pile is formed, from the side of which the raw material is taken to the first machine. In many cases, two, three, or several, qualities are mixed with a certain weight of waste, produced by the several operations, and care must be taken to thoroughly incorporate this waste with the higher qualities, forming the real basis, the object being to give the most useful and economical averages.

The material is now supposed to be sufficiently analysed in staple, strength, and colour, to be passed through the several stages by which, finally, it is converted into yarn; and here it is necessary to observe that all the care used, and the value of the "mixing," will be lost unless the material is sent to the blower or cotton-opener in the same order as it is spread on the mixing surface. The layers ought to be cautiously handled, and disturbed as little as possible, the mass being presented to the action of the first machine with the greatest amount of uniformity. We have seen many a good mixing spoiled by want of careful management or by recklessness on the part of the operatives.

Preparing of Cotton.—If cotton could be delivered in the condition in which it is left after being dried and picked on the field, spinning would be easy work, and all the preparation before carding might be dispensed with. A large quantity of broken seeds, leaves, sand, and other impurities, however, are left after the first rough cleaning process, called "ginning," has been gone through; and as the difficulties attending carriage cause the cotton to become caked and matted, in consequence of excessive pressure, it is necessary that costly and tedious work be done beforehand to loosen the fibres, to render them soft, and to separate them from foreign or deleterious substances.

Before we proceed to give a description of the machinery employed for this purpose, we may remark that there is an increasing diversity of opinion amongst practical men, whether the present system of preparing cotton for the carder is not unsound in principle. It is contended that "the less you manipulate cotton before carding, the better." Every time it is put through a willow, opener, or double lapper, the vitality is beaten out of it, and the fibres are damaged. No doubt we destroy some staple every time the beater strikes it, but the amount is small in comparison with the quantity dealt with. The object in passing cotton through a preparing machine is to free it from dirt or seeds, and to thoroughly clean it—good yarn is impossible otherwise. If the dirt and seeds are not removed before carding, the former will take the point off the card wire, whilst the latter will be ground up into fine particles and distributed all through the yarn. The card wire will be choked and laid flat, thus reducing the number of carding points, and the particles distributed through the roving will cause it to break down every time one of them is passed through the front rollers, whether on the jack, the roving, or the spinning frames. Every time an end breaks down on any of these machines, there is considerable waste as well as loss of time. This loss is, in our opinion, much greater than that from the destruction of fibre in the opener.

Caution necessary in Use of Machines.—Machines are too expensive to be played with or neglected. No machine ought to be forced, for there is no economy in cramming or gorging beyond the ordinary capacity. No doubt, operatives will be found anxious to show how much can be done in the shortest space of time, and quality is sacrificed for quantity; but quality is a point which ought not to be lost sight of by a manager.

In opening cotton, a certain period of time must be allowed for the proper working out of the process: this is the main consideration. If a greater weight of material must be opened to suit the whim or wish of an overlooker, this extra quantity is only obtainable at the expense of the quality, and, as we have shown, such a policy is eventually disastrous. If really necessary, two machines would be true economy, doing lighter work, having more time to open out the matting thoroughly, and giving the best quality of work.

This advice we give where it is found that there is positively more work than one machine can effect, though driven at a ruinous speed. " Starting fair " is of the utmost importance in spinning, and this may be secured by having the first work done thoroughly.

A machine is as good as new as long as it will produce an equal quality of work and as much quantity, however old it may be. A new machine may turn out to be wrong in principle, or it may be superseded by something better. As one machine after another is constructed and applied, improvements are suggested, which are continued and kept up, every day bringing something new. Enthusiastic inventors have an ambition with which their knowledge does not keep pace; hence the quantity of impracticable machinery which has to be broken up and thrown into the scrap heap. Accuracy of construction is necessary in all cotton machinery, but above all in the carding department. The best information attainable in this direction is to observe the action of a machine in motion, and our object in these pages is not so much to describe machines, as to show how they should be managed with advantage in economy and production, without particular reference to the merits of any make.

CHAPTER II.

COTTON OPENING.

The Hopper Feeder.—This machine has not been in use in this country very long, but, from its high value, it has become an essential part of the plant of all well-equipped cotton mills, and now gives the

Fig. 1.—Hopper Feeder.

initial process, after bale-breaking, in the cleansing and opening of every class of cotton fibre. Our first illustration, therefore, is that of the hopper feeder (Fig. 1), the second being an elevation showing one of the many methods adopted in applying the feeder to cotton openers (Fig. 2). The opener is here shown upon the ground floor and the feeder in the room above. (Later on, on p. 12, we show another type of hopper feeder, in close connection with an opener.)

A few words may be here said upon the action of the machine. The hopper is loosely filled with cotton, which is subjected to gentle treatment by a series of spikes, which have the effect of combing out all lumps in the cotton. It will be obvious that less damage is done to the staple than by the action of beaters, besides which, much dirt is extracted. The cotton is finally delivered in loose, fleecy layers, which are passed on to the opener in such a condition that they may be more easily beaten and evener laps obtained.

The Opener.—Originally, cotton openers were introduced for the purpose of loosening the hard masses and flakes, but they are

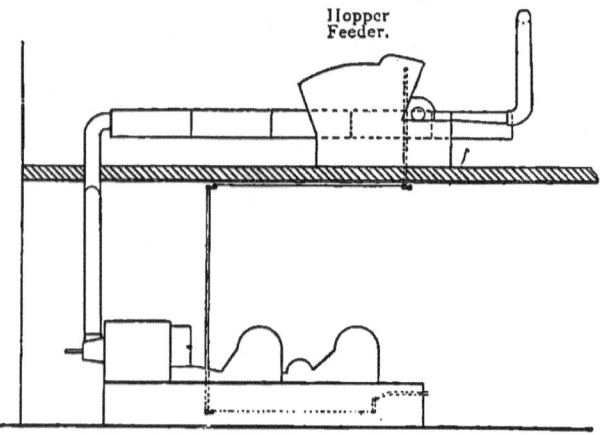

Fig. 2.—Hopper Feeder (in upper room) applied to Opener (on ground floor).

now adapted to assist in cleaning the material. It is no new thing for those in the trade to be reminded that the present machines are, with few exceptions, of the same type as those which were in use nearly a century ago, the greater speed giving precisely the character of the modifications which have taken place. Whatever may be the variation in construction, there is one common principle running through them all—namely, while the cotton passes slowly into the machine, it is struck rapid blows by revolving beaters, and the dirt is thus driven out of the loosened mass. While in this condition, a current of air, induced by a fan, is made to pass through the cotton, which current carries away, not only light particles of dust, etc., through what is known as the dust cages, but also the cotton from one part of the machine to another, for the purpose of being shaken about and subjected to another beating operation. The use of this air current is known as the pneumatic principle of openers and scutchers.

The Crighton opener has been so often described and illustrated, that it will be unnecessary to give a very elaborate account of its construction and capabilities. The principle of this opener is as follows:—The cotton is fed at the bottom of a revolving vertical cone, and is carried round and upwards, until it reaches the top of the machine, by the current of air created by the rapid revolutions. The dirt drops through the grids by which the cone is surrounded, and, as dirty, heavy material is kept down by its own weight longer than clean, fleecy cotton, it, of course, is longer exposed to the cleaning process: at the top is the outlet to the cage.

In this opener, the operation of the "beater" being upon the grids only, it is, to all intents and purposes, similar in action to the old method of "batting" with a stick, therefore reducing to a minimum the damage to the fibres of the finest cottons, and the delivery has no tendency to stringing or curling, and is in a perfectly open condition. It has been found suitable for all kinds of cotton, from the lowest class of Surats to fine Sea Island, and has made its way into cotton factories all over the world. There are already as many produced as, if worked to their full extent of power, would treat every bale of cotton grown. The principle of the machine has remained the same from the first, but has been modified from time to time to meet different requirements. Some are made single, some double, and some with a feeding table.

A very important feature is the "porcupine" cylinder, revolving at a high speed, armed with strong steel teeth, set at different angles. The cotton is struck by this cylinder and dashed against a grid, so that it is scarcely possible for any foreign matter to escape being acted upon. A fan is always applied, the object of which is to draw through the perforated cylinder cage the air displaced by the revolving beater, thereby separating the air from the cotton at the cage. If the draught of this fan be made sufficiently powerful, the cotton can be drawn from a considerable distance through a tube, but, at the same time, the knives on the beater shaft will not have sufficient opportunity of acting upon it. The cotton, in fact, will have the

COTTON MANUFACTURE.

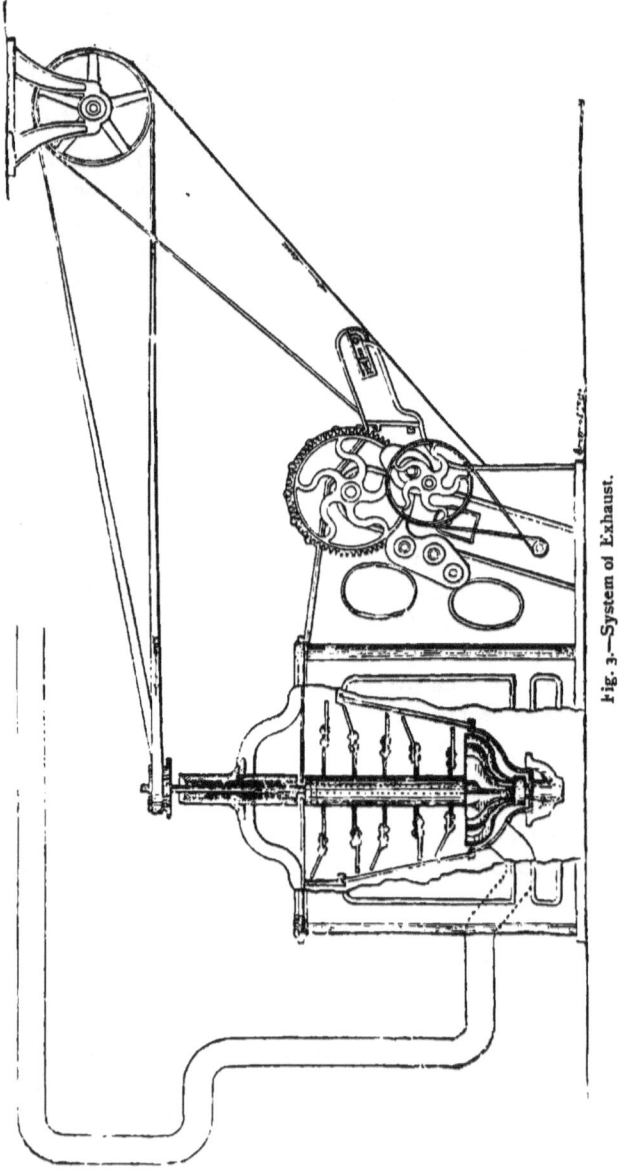

Fig. 3.—System of Exhaust.

same force impelling it through the opener proper as it has for impelling it through the tube, and the opener will not be able to act upon it in its normal and proper manner. The opener, as ordinarily made, has a good draught at the feeding orifice—a sufficient draught, in fact, to lift the cotton through a tube from one floor to another; but, by the application of the exhaust to the foot of the beater, as illustrated in Fig. 3, the cotton can be drawn any reasonable distance and around a considerable number of bends, the influence exerted by the exhaust upon the cotton ceasing immediately upon the latter entering the opener, thereby leaving the beater to operate upon it as it ought to do.

The combined opener, with a single beater lap machine, shown in Fig. 4, obviates the necessity of feeding two machines by hand—viz., the opener to deliver cotton loose, and the first scutcher lap machine—and the laps from it have been found sufficiently even and clean to be taken from it direct to the cards in fine spinning mills. The cotton is placed in the automatic hopper feeder, and, instead of being delivered loose, is made into laps. The leaf extractor, between the scutcher beater and the cages, consists of a travelling table with an air-tight bottom, upon which is placed a series of bars, the full width of the machine, driven by a chain from the upper cage, which run in an opposite direction to the cotton, thus catching the leaf and dirt, and depositing these impurities on the floor under the machine. The bars are made with much wider spaces than usual, leaving room for the extraction of large pieces of leaf.

This machine, followed by a single beater lap machine as an intermediate, and by another one as a finisher, is sufficient to clean the lowest class of cotton; and, if followed by a finisher only, is equally efficient for high-class materials.

The single beater scutcher lap machines are made with the feeder sides to put up three or four laps, or with plain sides for feeding loose cotton. Sometimes a cone feed regulator, with piano-pedal motion, is applied, as shown in Figs. 5 and 6, and the machines have sometimes only two calendering rollers at the lap end, whilst at other times they have four. These variations are to meet the requirements of different classes of cotton. In all the machines, the risk of fire is reduced to the minimum point.

Besides the machines illustrated above, we may say that there are other machines, by different makers, that are perfectly efficient and

12 COTTON MANUFACTURE.

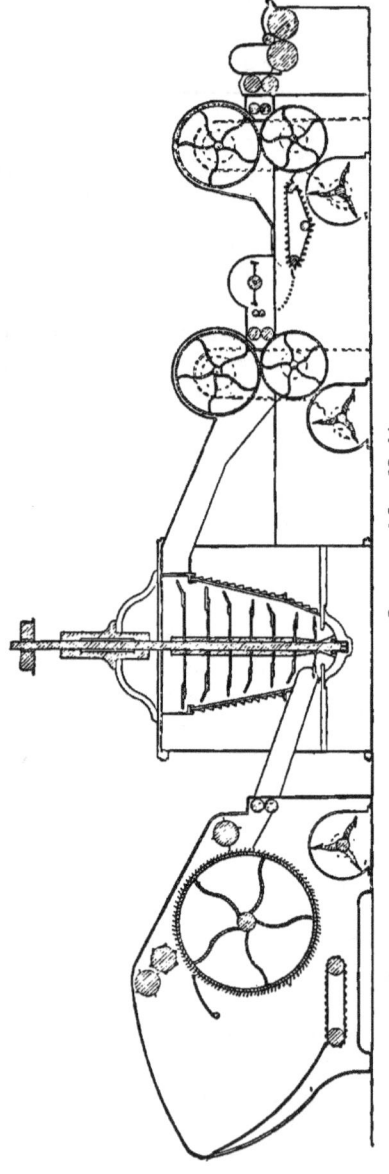

Fig. 4.—Combined Opener and Lap Machine with Hopper Feeder.

deserving of notice, but as they are more or less upon the same lines as these, this remark will be sufficient testimony to their merits.

Scutching.—We need say little upon the operation of scutching. The loose opened cotton is taken up, as it issues from the cage of the second beater, by rollers, and from thence round the lap rollers, and, in winding round it, it increases in diameter, and is then pressed by two fluted rollers in order to compress it into a flat, adhesive mass. This lap is then taken to the next machine, the feeding of which requires both skill and care, so that the same thickness will be obtained throughout the working. As a test, a length is taken, weighed, and compared with another length, taken after an interval of time, when an estimate is formed of average thickness. If this process is properly conducted, the more regular will be the counts of yarn. All the working the cotton needs is simply enough to clean and straighten the fibre—more than this is injurious; and we may here say that, as new machinery is of so much importance, it is necessary to know when a machine becomes old; but this is not a question that can be decided off hand.

Weight of the Mixings.—A very commendable practice obtains in well-regulated cotton mills—viz., a small quantity of the raw material, after having been weighed, is exposed for some hours, spread open in a room at something like the heat it will have to pass through in the after-process; then it is again weighed, and the loss is evaporated water. To find the average of the varieties composing the mixing in use, a small quantity, say three or four pounds, is taken from each bale, and passed through the scutcher and carding engine; it is then weighed correctly, and the particulars entered into a loss book, which is valuable as a reference.

The mixing book ought to show dates of cotton purchased, quantity, marks, number of bales, gross, tare, net weight, and price per pound of each quality, so as to ascertain the value of the mixing —any waste also must be entered, and the quantity of each sort noted, with the prices attached, and the totals divided by the number of pounds the mixing contains. The result will be the average cost, and the test sample will show the percentage of loss. The exigencies of competition demand the utmost care and economy of material.

Natural Twist of Cotton.—Cotton differs in one particular

Fig. 5.—Scutcher Lap Machine.

from all other vegetable fibres, each fibre being an independent twisted thread, so close that the twist cannot be seen except by the aid of a microscope. This property gives to the fibre great adhesiveness, rendering it capable of a tenuity far beyond any other fibre. The fibre is considerably affected by atmospheric changes, and often absorbs a goodly percentage of its own weight in dampness; and, although a non-conductor, it excites electricity more or

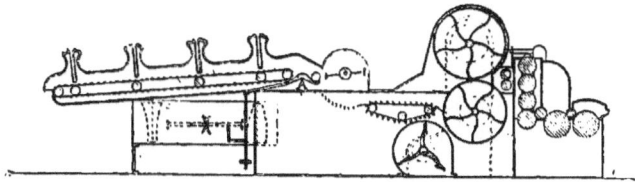

Fig. 6.—Scutcher Lap Machine.

less by the machinery through which it passes in its manufacture, and frequently causes trouble in carding. When the air is dry, it is full of electricity, attracting the cotton fibres in all directions—drawings roll up, rovings will not draw properly nor twist smooth; but a remedy for this will be duly considered in its proper place.

These remarks the reader will find are of sufficient importance to be carefully thought over. The most of everything in cotton, as in other trades, must be obtained at the least expense. There is a best way to do it, and the very best way is to work it as little as possible.

CHAPTER III.

CARDING AND COMBING OF COTTON.

Object of Carding.—However well cotton may be opened and prepared in laps, the fibres are not parallel; they lie across each other at every imaginable angle, and any attempt to combine them in this state would be fruitless. They must be rendered parallel, and to effect this is the object of the beautiful operation of carding, one of those operations which has exercised such a large amount of inventive ingenuity.

The main point in carding is to free the cotton from every impurity, and this most desirable result requires the carding engines to be thoroughly efficient in every respect. Out of the many excellent machines, with every modern improvement, we give as an illustration (Fig. 7) the "Simplex" revolving flat carder, which is peculiarly well adapted for fine counts of yarns. It is not necessary for us to enter into lengthy details—we can only deal with the salient features of this new carding engine. One of the most important points in connection with its construction is the almost imperceptible fraction of any deviation; when one inch is divisible by 1000, this represents such a fraction. The flexible bend is supported at five points, and the portion over the doffer is held by a link which swivels on its centre. At the side over the licker-in we have the setting point—an engine-marked indication, with full explanation, shows the amount of setting. The dial is in divisions of the fiftieths of an inch, and each of these again represents $\frac{2}{1000}$ parts of an inch of the flat's downward movement. (Truly, carding must be comparatively easy in its operation compared with the troubles of the past.)

This measurement really means, in plain language, that each turn of the setting angle only moves the flat across $\frac{1}{5000}$ part of one

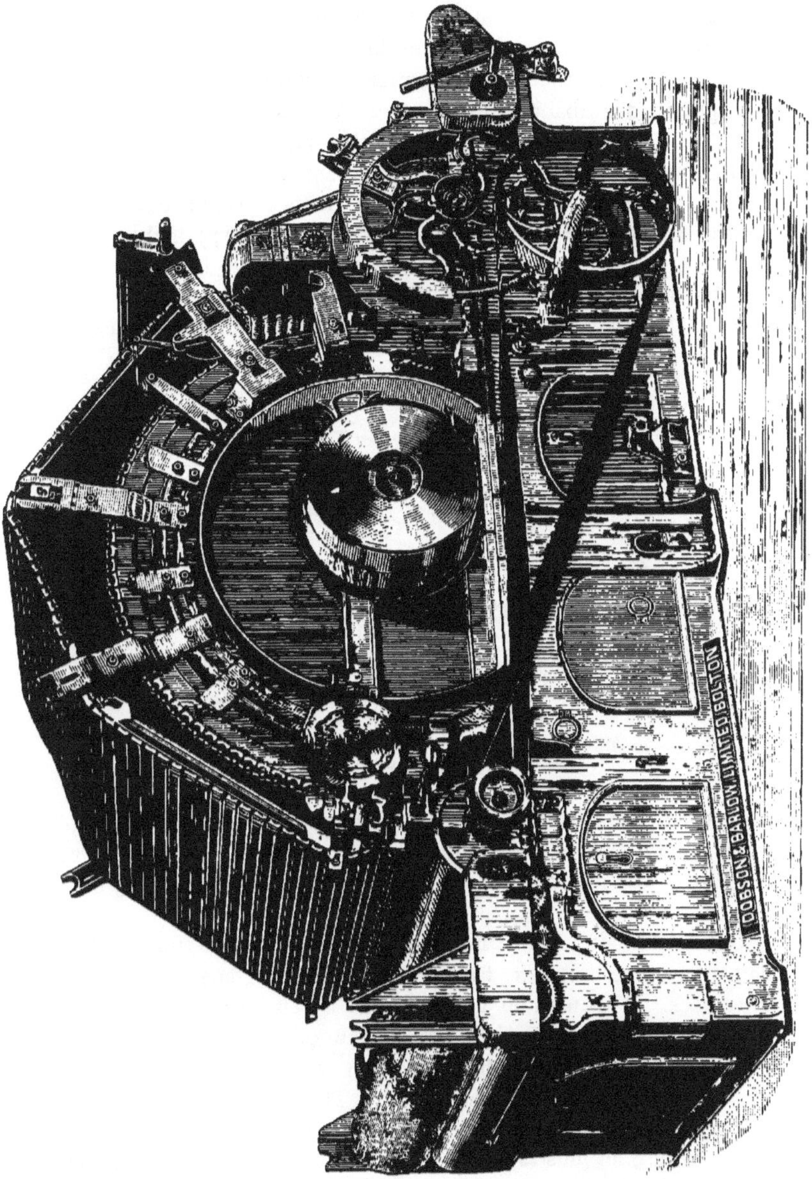

Fig. 7.—Revolving Flat Carding Engine.

inch, which seems a figure almost too small for comprehension, but so it is—a marvellous piece of mechanism! The carding on this machine ensures the most positive uniformity in yarns. No atmospheric disturbance can interfere with the laps, which can be carded the full width with unexceptional selvages, and special kinds of clothing are not necessary; the flats are always fully employed. Cylinder diameter, 50 inches and 45 inches wide; doffers, 44 inches; taker-in, 9·5 inches diameter; 110 flats, 1·375 inches wide.

Upon reference to the drawing here given (Fig. 8) it will be seen that the flexible bend shown at D, D is supported by the steel pins, E, E. This figure gives the principle upon which the curves are

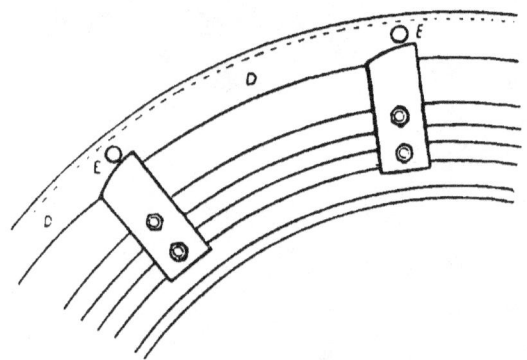

Fig. 8.—Flexible Bend.

worked out. We have referred in the above description to the index wheel. The segment of a circle is that formed by the bend; the advance of the index wheel will raise or lower this bend $\frac{1}{2000}$ of an inch.

Card-Grinding Motion.—In ordinary grinding of the revolving flats of carding engines, flats are supported on special facings at their back. This method gives various bevels and length of wires —a very defective principle, which can only be avoided by supporting flats during grinding on their working surface, each flat receiving its angle by the one wedge. This would save the labour of preparing special grinding surfaces at the back of the flats. We give in Fig. 9 an illustration of one of the latest methods of grinding revolving flats.

This arrangement will readily be understood by means of the drawing.

To the grinding bracket, A, a grooved guide, B, is fixed, in which a toothed bar, C, can slide. To the bottom of this bar is attached a wedge curved to the radius of the flexible bend. The flats, D, are pressed with their working facings against this wedge, by means of the lever, E, and slide, E¹, the other end, F, being loaded by a weight or spring. As the flats revolve, each of them seizes the projection, G, of the wedge and carries it along until the wires have

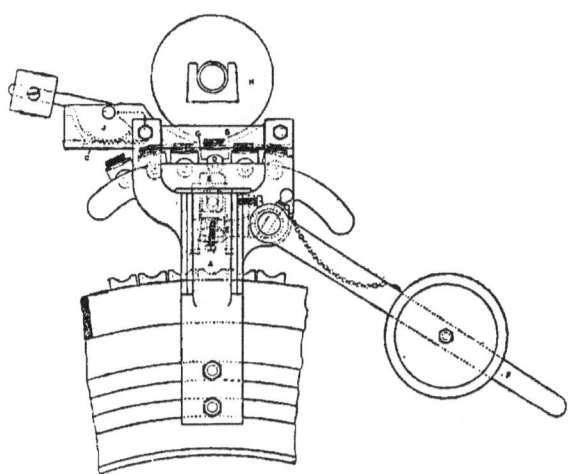

Fig. 9.—Card-Grinding Motion.

passed under the grinding roller, H. The flat, in its forward traverse, comes into contact with an incline, attached to the back of the guide, B, which presses the flat down and so releases the wedge, and, by means of the wedged toothed segment, J, it is returned to its original position, in readiness for the next flat. The accumulation of fluff or dirt on the working surfaces is entirely avoided by these surfaces being on the underside of the guide, B. By changing or altering the wedge, the bevel of the wire can easily be altered if required.

Clothing of Cards.—It is of more importance to have clothing properly drawn than it is to have a straight edge merely to look at, and this obtains with all clothing on main cylinder, tops, or strip cards. These cards require less drawing than tops, because the leather is thinner, the teeth are not so closely set, and they do the stripping better by being a little loose. All classes of cards and every card should be carefully set up after the clothing is finished, whether it is to be ground or not, as teeth out of place damage a card. These remarks apply to leather clothing. Many other materials are used, which vary a great deal in results, so that no general rules apply.

Setting the Card.—In setting a card ready for work, see that it is thoroughly clean, and set the doffer as near as it will run, without actually coming into contact with the main cylinder. Having got it into this close position, fasten it at once securely, so that it will be unable to get out of place when working, then set the tops, beginning with those nearest the doffer; set the front twice as far from the cylinder as the back is set—this will bring the nearest point about two-thirds from front to back, and if it is found that they load too heavily on the front, raise them a little, and if the fronts fill too scantily, lower them. Let the tops be set square on all the screws. If workers and stripper are used, set them as near as they will run without touching. Do the same with lickers-in. A finishing card top-feed-roller should be set as close to the cylinder as possible, on breakers with lickers-in—set the bottom roller the nearest. When all has been got ready, turn the cylinder cautiously backwards to find out if it comes in contact with anything, and thus cards will not be damaged, as the combs ought just to clear the doffer.

The Carder's Duties.—The carder can always find plenty to do; there is ever something to engage attention. The machines must be kept in good point, strippers smooth and well cleaned from time to time, cylinder and doffer in good point, to prevent excess of waste—the machines are, or ought to be, kept for the purpose of combing out the cotton and freeing it from dirt, and not for grinding it up—and each card made to do its particular share of work. If cotton is dirty, it is wise to reduce the speed—this will allow the cylinder time to free the material from impurities. Feed rollers become greasy and must be kept clean, or the drawings will snatch and make uneven work. The rest of the machine may be in capital order; but, if the feed rollers are neglected, bad work will ensue.

Remedies.—If the licker-in becomes greasy, the cotton will pass in lumps to the cylinder. The stripper belts should be regularly examined and new laces put in. Stripper belts, if neglected, break, get into the card, and a smash may take place, which, with care, might have been easily prevented. A slightly moist, clean, oily cotton can be made with less twist than when in the dry state, and the percentage of loss is less. When in a dry state, the fibre shrinks and curls up, and an extra amount of twist has to be put in, whatever may be the quality of the cotton in use. A card ought to be so arranged that it will catch quickly, and speedily let go the cotton; this requires a fine needle point on the teeth. Cotton may be, and is, often forced through wire-edged cards, but it is not carded, it is simply ground through. A well-regulated carding engine takes the material in slowly, in the shape of a lap or sheet, and the filaments of the cotton, as it comes from the card, ought to be as nearly as possible as nature made them—light, lively, flexible, elastic, strong, adhesive, and capable of great attenuation. It is often said that the cotton is carded too much—the remedy is to card less. The difference between single and double carding is easily seen in the yarn and cloth—single carding may give a stronger yarn; but it is dirtier and rougher, the spinning is indifferent, and the cloth not so smooth and clean as in double carding, because, in this process, nearly all impurities are removed. Again, if goods are required to be very fine and light, double carding is a necessity; if, on the contrary, coarse, heavy goods are needed, single carding cannot be objected to.

Stripping.—If tops were stripped only half as much as they are at present, good cotton would be saved, waste percentage would be smaller, and the wear and tear of machinery less. The stripping of the lower tops oftener than those higher up is a very erroneous practice. If one half of the cylinder could be stripped in the forenoon, and the other in the afternoon—every other top on a card and every other cylinder in the same row—the work would be lighter and less affected than when stripped all at one time. Of course, where the work is light and clean, this would, perhaps, be too often, and where it is dirty and heavy, not often enough; but whether stripping takes place many times or otherwise, one-half should be done at a time.

Speeds of Cylinders.—The speed of main cylinders varies—

if run too fast, more waste is thrown off; if too slow, the cotton wil not be properly cleaned. The draft of cards varies as much as the speed—a thin lap requires a low draft and a slow speed of the doffer, and the cylinder in this case will not clear well enough; but if a thick heavy lap is given, a high draft is needed, the result being that the cotton is not held firmly enough between the rollers to be well carded, and it will draw off in flakes; but it is good policy to have a medium draft. It is impossible to make good yarn of cotton which has not been properly carded, as no after process will remedy the evil. The speed of the different parts, as far as rollers and doffers are concerned, depends upon the draft—there should be barely enough between the lap and feed rollers to keep the lap straight, the same between the comb and calender rollers, if such are used, or between the comb and apron, where there are no calender rollers. One very important matter in carding and preparing cotton for the spinner is to handle it carefully.

Condition of Rovings.— Rovings are in the best condition for immediate use when new, as they do not improve with age like spun yarns, and, if there is a necessity for making stock, let them be carefully covered and left undisturbed until required. Carders who have been successful and have attained perfection in producing good, smooth, even, clear, and clean rovings, attend to every detail, and believe in prevention being better than cure. Hard, steady, and untiring vigilance covers all the ground in every man's experience for all time; and it is not enough to order something to be done, but, at the same time, it is necessary to see that it is properly executed. All we have stated in connection with carding, etc., is not theory only, but is the best possible practice obtaining in properly conducted cotton mills.

Carding Calculations.—Before entering upon the process of combing cotton, it may be useful to give necessary calculations relative to carding machinery. The word "draft" means to draw out, stretch, or elongate a sliver of cotton, and for every inch a frame receives, it delivers 3, 4, 5, 6, or 7 inches, according to circumstances. If the machine delivered just what it received, there could not be any draft or stretch, and if all the rollers were alike in diameter, the drivers and driven the same figure, the draft would be 1, because the drivers contain the driven once only, therefore, this unit is, in all cases, deducted from any draft shown in figures, the remainder being the real or actual draft or stretch. When a draft is called 1·125, 1·25 or 1·5 it is really a draft of ·125, ·25 or .5.

In calculating the draft of any machine, begin to count the teeth at its delivery and work back to its receiving part. The middle rollers of a drawing, roving, spinning, doffer, or feed rollers of a card, are not required in the calculation, except in a card where there are no calender rollers—the doffer in this case becomes the delivering roller, and must be reckoned with. Take nothing for granted, because, in many cases, the machinery may not be perfect in all its parts—the pulleys, rollers, and cylinders may not measure exactly what is stated. Go over every one carefully and note the figures; then, to find the draft of a carding engine, the following particulars are required:—The wheel on the front roller—say, 25 teeth, working a wheel on the second roller, 28 teeth; the wheel on the front roller shaft, 30 teeth, worked from a wheel on the end of the doffer, 150 teeth; the wheel on the other end of the doffer, 36 teeth, working into a wheel on the side shaft, 36 teeth; the wheel on the other end of the side shaft, 16 teeth, working into a wheel on the end of the feed rollers, 130 teeth; the diameter of the feed rollers, 1·75 inches; the diameter of the delivery rollers, 2 inches, what ought to be the draft?

$$= \frac{28 \times 150 \times 130 \times 2 = \text{dividend.}}{25 \times 30 \times 12 \times 1·75 = \text{divisor.}}$$

The draft is 69·33. Having thus determined the draft of the engine, all other calculations are comparatively easy. In many carding engines at present in use, there is a wheel on the front roller shaft that drives both the first and second rollers in the draw-box. In these machines, the wheel on the second roller must be omitted on taking the draft of the engine, and the wheel on the front roller shaft, driving first and second rollers, taken instead. To find the number of feet of fillet required to cover a cylinder—take the diameter of the cylinder in inches, its breadth in inches, and 3·1416 for a dividend, and multiply the breadth of the fillet by 12 for a divisor, and the quotient will give the number of feet required. The number of carding engines for a given number of deliveries at the drawing frame may be ascertained by multiplying the number of deliveries, the number of ends put up at the back of each delivery, and the speed and diameter of the back roller together for a dividend; then multiply the speed and diameter of the delivering roller at the carding engines together for a divisor, and the quotient will be the number of engines to supply the drawing, both running the same length of time. A side shaft wheel, to give a required draft, is found by wheel on front roller, 24 teeth, working into a wheel on second roller, 32 teeth; wheel on front roller shaft, 34 teeth, worked from wheel on end of doffer, 154 teeth; wheel on other end of

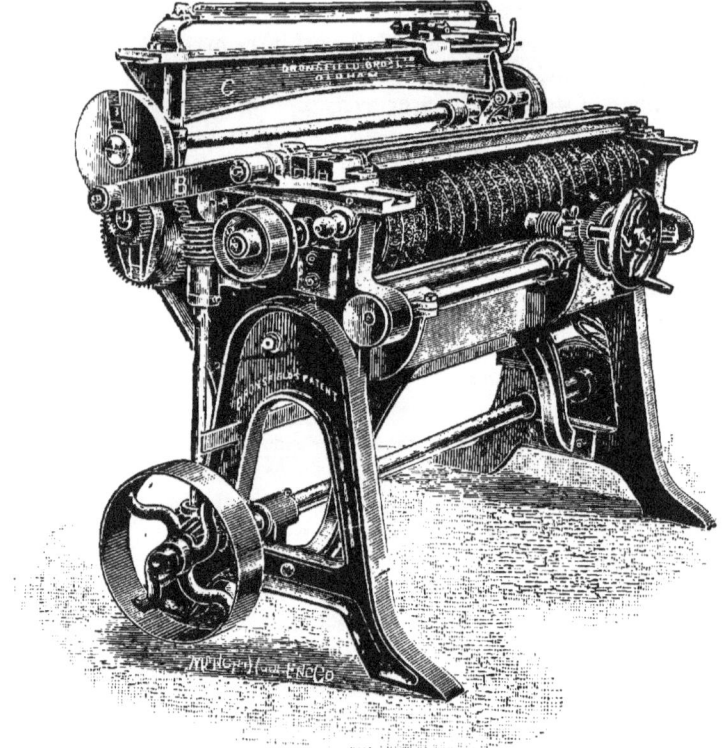

Fig. 10.—Card-Grinding Machine.

doffer, 38 teeth, working into wheel on side shaft, 38 teeth; wheel on other end of side shaft required to produce an 80·5 draft, wheel on end of feed rollers, 134 teeth; diameter of feed rollers, 1·75 inches; diameter of delivering rollers, 2 inches:

$$= \frac{32 \times 154 \times 134 \times 2}{24 \times 34 \times 80\cdot5 \times 1\cdot75} \frac{\text{dividend}}{\text{divisor}}$$ quotient, gives, omitting

fractions, a side shaft wheel of 12 teeth. To alter weight of carding by side shaft wheel is found as follows :—Suppose a length of carding weighs 90 grains, with a 12-tooth side shaft wheel on, what wheel will make the same length weigh 100 grains? Multiply wheel that is on by the grains the carding requires, and divide by weight the carding is making, and quotient side shaft wheel required = $\frac{100 \times 12}{90}$ = 14 nearly; but this full figure can be used, as there are no fractional parts in the teeth of wheels.

If the draft of a carding engine is required to produce a given weight of carding from a given weight of lap, the following example will show the method :—2 yards of a lap weigh 20 ounces, and 1 ounce is allowed for waste in passing through the cards, what draft in the carding engine would cause 2 yards to weigh 100 grains? Find number of grains in 19 ounces,—437·5 grains in 1 ounce × 19 = 8312·5; this product, divided by 100 grains, gives 83·125 draft required. We could go on multiplying examples, but the exigencies of space will not permit, and we must conclude our remarks on the process of carding by saying that the greatest advantages may be obtained by care and skill in the management of this important branch of cotton manufacture. To sum up, there ought to be in every carding room a pair of straight edges, full width of the cylinder, or as long as the emery roller, and 3 inches broad by $\frac{3}{4}$ inch thick, to be used for trying the cards, card cylinder, lickers-in, and rollers; a tape accurately marked in feet and inches, 50 or 75 feet long, also a spirit level, 2 feet long by 3 inches in width, having both horizontal and perpendicular spirit glasses inserted, as the age for rule of thumb has departed. Should the cards work badly after being set carefully, all the cylinders should be stripped and brushed, the straight-edge applied on the surface of the teeth, holding it parallel and looking under it against the light. By this means it will be easy to discover whether grinding will be required to make them true. In this way the doffers, under similar circumstances, ought to be examined, and the emeries made straight; then put one on for such period of time as is required to produce the desired effect.

Card Grinding.—There are opinions without end as to the best method of grinding. All kinds of grinders are driven more slowly than formerly, and with success; but they are far from perfection. The cylinder might be driven slower, and the grinder retain its present speed. In adjusting a grinder to a card, every nut and bolt should be screwed up to the greatest possible degree of tightness, as, if the grinder gets loose, it will, in a very short time, cause serious trouble. In covering a grinder, no delay should take place—everything ought to be prepared beforehand—glue must be spread evenly over the surface, plenty of emery used, the grinder all the while moving with a slow, intermittent motion, even after the clothing is put on, and then a day or two should be allowed for drying. When cylinders are not true, or out of balance, it takes up more time to grind a new card, and one side of the cylinder is ground far more than it ought to be, and teeth will rise if the leather is loose. It is far better to grind cards often than to grind them long at a time. This is absolutely necessary to produce good work. More cotton is spoiled by improper grinding, or unskilful setting, than by all other causes combined. Some carders are so apprehensive lest the teeth of the cards should be cut down too soon that they will not allow grinding to be done except at the last pinch, perhaps once in two or three weeks. It would, therefore, be equally as correct not to sharpen a knife, razor, or chisel, lest they should be worn out too soon. An illustration of a card-grinding machine is given in Fig. 10.

Cotton Combing.—Combing is used mainly for the production of fine yarns, or of those of a high quality, uniformity of length in the fibre being the main object, and to effect which all fibres shorter than the required standard are combed away and rejected. It is not a necessary step in the preparation of cotton, although, for fine counts, it is an improvement on carding, being very efficient in removing the short staple; yet it can only operate on carded cotton, and it is, to all intents and purposes, a substitute for the second carding engine. One particular point in connection with this process is that the material must be acted upon by the combs lengthwise. The feed rollers have an intermittent motion, and only allow a little less of the carding to pass at a time than is equal to the length of fibre which is to be combed and retained. On the circumference of a roller there may be from 20 to 30 combs; each of these, in regular order, being finer than those before, there being in some machines 30 or more in the length of one inch, and, perhaps, 100 in the last. The short fibre rejected by the combs is stripped off them by a revolving brush, and is used for coarse yarns.

CARDING AND COMBING OF COTTON. 27

Fig. 11.—Cotton-Combing Machine.

The combs are very liable to injury, but can be easily replaced. The parts are adjustable, and are so regulated that a great amount of wear is obviated. A standard type of this machine is shown in Fig. 11, which cannot fail to be appreciated by fine spinners. Although the details of this machine may appear difficult to under-

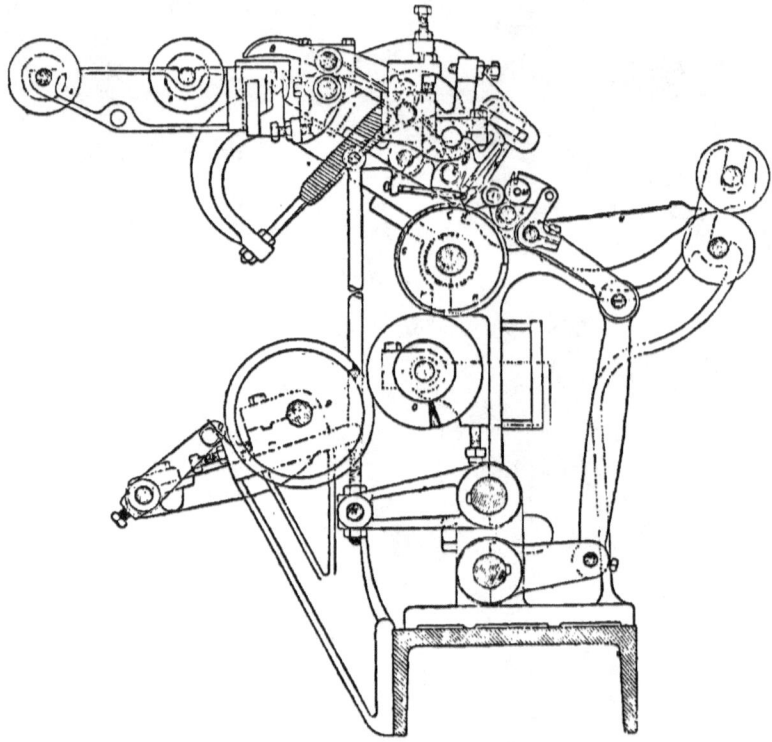

Fig. 12.—Cotton-Combing Machine.

stand, a close study of the sectional drawing will show that it is simple enough.

The operation of combing is, no doubt, a delicate one. Some of these machines are made with six and eight heads ; but, as the work of each head is similar, one only need be described. A narrow lap, about 7·5 inches wide, is passed into the head by means of a pair of fluted feeding rollers, and the action that takes place may be briefly described as follows :—When the "stroke" is finished, the lap is seized by a pair of nippers, leaving a small portion of cotton projecting. This is combed by a revolving cylinder, in part clothed with comb teeth. When the projected material is combed, a straight comb drops into the nippers, which open, and the cotton is then drawn through this straight comb, and the fibres are pieced or laid on to those previously combed by a stroke, and in this way a connected or entire fleece of combed cotton is obtained. The cotton during the combing is actually separated from the narrow lap, carried over and really pieced to the end of the combed fleece. For a brief moment only it is without connection, and that is during the carrying over.

The improvements that have taken place have brought the comber to as near perfection as possible. One of the best types of these improvements is shown in Fig. 12 ; and, by this particular make of a combing machine, 9-inch laps with perfect selvages can be made, in place of the usual 7·5-inch lap, whilst the production is materially increased, without any increase in the size of the machine. Taking the drawing, which is a transverse section through one head, A, A show the lap rollers, B the trough which carries the lap along unrolled to the feed rollers, C, D, which deliver the cotton to the nippers placed in front of the rollers. The nippers have a cushion plate, E, and knife, F. The lower jaw of the nippers is covered with leather, and the upper one is fluted, so that the nip of the fibres is formed at the point of the flutes. G is the combing cylinder, and on the opposite side of this cylinder is H, which is a fluted segment. I is the detaching roller for forming the cotton after being combed into a sliver. K is the top comb, above the cylinder, G. The delivery rollers, L, M, pass the sliver over the guide, N, to the trumpet tube, and thence to the condensing or calender rollers, shown on the right-hand side of the figure. From these rollers the fleece is placed on a plate, and from thence to a draw box, in which are drawing rollers ; and, lastly, it passes into coiler cans. O is a cylinder brush, P a clearer, and Q the doffing comb. There are three distinct operations—feeding, combing, and detaching.

CHAPTER IV.

DRAWING AND DOUBLING.

Object of Doubling.—We will now enter into the process of drawing or doubling, which succeeds carding. The object is twofold —to straighten, and to lay the fibres of cotton parallel. The slivers as they come from the cards are exceedingly tender and loose, and the different slivers vary in weight, because all cards do not clear equally, or there may be a variation in the feed rolls, doffer, doffer pulleys, or calender rollers, from all or any of which causes arises the necessity for doubling. One result of the drawing process, if properly conducted, is that the drawing is perfectly equal in thickness in every part, and is formed of parallel fibres; and, in order to secure this, the operation is repeated more than once, each sliver being doubled with others before each successive drawing. This is, perhaps, the most important principle in the whole range of the cotton manipulation. To equalise the slivers, and also to straighten the fibres, bringing them closer together, is the main object of the drawing frame.

The rollers in this frame are generally so adjusted that the drawing is done between the first and third rollers, the middle roller having little or no influence so far as stretching is concerned. Where there are more rollers, the drawing is performed twice, each pair drawing a certain amount; and the distance between the rollers is so regulated that the longest fibre of the cotton does not reach from the centre of one roller to the centre of the other. This prevents the rollers from tearing the fibres; because the first pair of rollers pulls the fibres, while the second set holds them fast. It is better to have the rollers too close together, than too far apart, provided they are always so far distant as not to injure the staple. The lower rollers are fluted; the top ones, with the exception of the first, which is also fluted, are clothed in flannel and leather.

The Sliver.—The slivers of a section of cards, varying in number from six to twelve, pass between several pairs of rollers, and, if all these rollers revolved at the same speed, no change would take place; but, allow the second set to revolve one-half faster than the first, or in the proportion of $1\frac{1}{2}$ to 1, and the sliver will be gradually drawn out, the fibres clinging to each other, until it is one-half longer; the same in the third and fourth rollers.

Short-stapled cotton requires less drawing than long-stapled—the long fibres are liable to double up in carding. It is of importance to have the rollers adjusted to suit the class of cotton. To prevent unevenness or clouding, the upper and lower rollers must press firmly together, so that the cotton may not be allowed to slip; and the slivers ought to be as light as possible, for, if bulky, the edges are apt to draw too freely, more so than the centre which bears the greatest weight.

Object in Drawing.—The object to be obtained in drawing the slivers is to reduce their thickness after they have been doubled. The more a sliver is doubled and elongated, the more perfect should be the yarn spun from it. If a drawing frame of three heads and four deliveries each draws eight slivers, combined into the first head, and again drawn into the second, and redrawn from the second at the third head, the sliver will have passed through $8 \times 8 \times 8 = 512$ doublings and drawings; but this process may be carried too far, as excessive drawing, as well as excessive clearing and carding, tends to weaken the fibres—they separate, will not interlock, they all fall away from each other, and become brittle and useless. The sliver from the last drawing head should be of a silky lustre, and its component fibres perfectly parallel with the sliver and each other. Very little waste occurs in this operation, and that only through inadvertence or negligence. The leather upon the rollers is liable to be affected by damp, therefore it is necessary to have a dry atmosphere in the rooms, as the adhesiveness caused by dampness laps the cotton round the rollers, creating a great deal of trouble and causing loss of time. Drawing frames are of various makes and sizes, but the principle is the same throughout. An approved example is shown in Fig. 13.

A railway head is a most convenient and economical method

Fig 13.—Drawing Frame.

of gathering a number of card slivers together in shape. To receive the first drawing, it is necessary to deliver these slivers in a smooth even sheet. The edges should not be allowed to rub against the railway box in their passage through it; if they do they will be fretted, and a smooth, even-edged sliver cannot be produced. It is at this point that the cotton receives its first doubling and drawing proper, and where it begins to tell on the evenness of roving and yarn. A railway generally has, or ought to have, four sets of rollers, because the weight needed to hold the sliver while it is drawn through two or three sets of rollers can be better divided with four sets—each set will be comparatively light, and the top rollers will be kept in better condition, with the advantage of lasting longer. All the sets of back rollers hold, whether two or more, and the front set does the drawing. As railways take the slivers of a section of cards varying in number from six upwards, the weight to apply to the top rollers of a railway, or indeed of a drawing frame of any kind, should be only enough to hold, and no more. In some railways there is a rack by which the weight of all rollers is connected, and the power is applied with one long lever and one weight—a capital arrangement. As we have previously stated, the front fluted roller is usually $1\frac{1}{4}$ inch in diameter, and the calender roller, $2\frac{1}{2}$ inches. If a driver of 36 teeth is put on the front roller, and a 72-teeth on the calender, there will be no necessity for an allowance of draft between the front set of rollers and the calender rollers; but as this idea may not be entertained, the calculations for drafts will still continue in practice, and we shall give a few rules and examples before concluding the process of drawing.

Electricity by Friction.—All railway heads and drawing frames excite electricity by friction. Many remedies are tried for this evil—steam pipes are opened, which causes a dampness in the air—but prevention is better than cure; therefore, have no more weight on the rollers than is absolutely needed; keep them well oiled; avoid all unnecessary friction everywhere; varnish all rollers that are old, rough, and dry; set buckets of hot water about the railways and drawing frames, and the air will soon get moist and drive the troublesome guest away. To produce good drawing, the top rollers ought to be constantly examined, both in railways and drawing frames. Where solid rollers are in use, the front ones must be oiled frequently, owing to the heavy weight; for if they get warm and dry, they will make heavy work, if they do not cut. It is sheer waste and recklessness in oiling to put too much on at a time—little and often is better and, in the end, more economical; for, if the leather gets saturated with oil splashes, the rollers soon become worthless. Doffers and feed of cards, driven by the railway, stop when the railway stops, as the main cylinder throws some extra cotton on to the doffers, and, when they start again, are apt to break down more or fewer ends, causing an irregularity in the sheet as it goes to the railway. It would be desirable to run the railways as regularly, and to stop as seldom, as possible, and only for a very short time.

Drawing Frame.—We will continue these remarks by a few particulars of one of the latest improved drawing frames, of which we give two views in section. It would be impossible by reference letters to convey anything like an adequate description of the motion, and, therefore, we must content ourselves with saying that, upon a stand is a series of rollers, four in number. The lower series is fluted; the top ones are generally covered with flannel or leather. These are driven at different speeds to produce a continuation of the sliver's decreasing size as it passes, and this decrease depends upon this variation of speed. The size of the machines is generally determined by the number of head deliveries—in fact, one of these heads is a machine in itself, though several are combined in one frame. There may be three heads of five deliveries. One important benefit, in connection with any modern drawing frame, is the stop-motion, preventing the overwinding of the bobbins; and, further, the building up of a bobbin can be stopped at any point. This arrest was generally made at the top of a bobbin, but later improvements obviate the awkward position in which the end had to be broken. The variation of the bobbin's speed, whatever may be the improvements, must take place with almost positive accuracy, because the attenuation must be in accordance with the weight desired. In the frame under notice, one good feature is the prevention of roller leathers being spoiled by reason of undue pressure. A weight-relieving motion takes off all pressure when the machine is not in motion. This in itself is sufficient, as far as economy is concerned, to render this drawing frame, with all the other improvements attached, worth recognition in the spinning industry.

Speed of Rollers.—The number of times the slivers should be doubled and drawn is dependent on the class of cotton. When any

DRAWING AND DOUBLING.

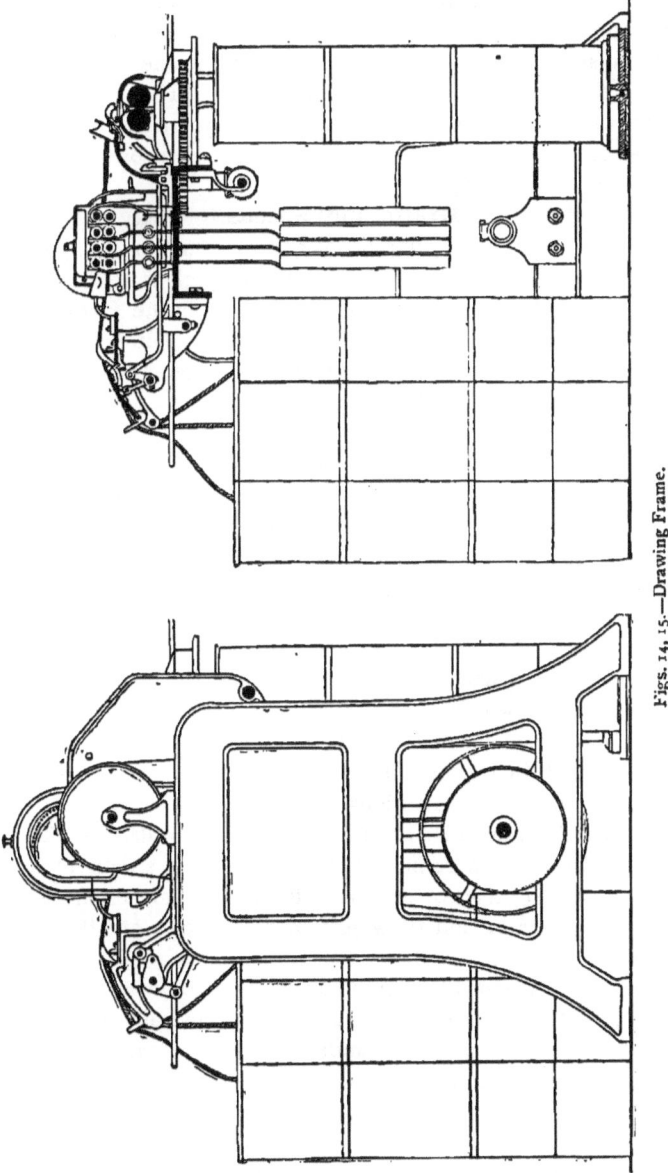

Figs. 14, 15.—Drawing Frame.

number of slivers is drawn into one, if one or more be broken or run out without stopping, the sliver will be reduced in substance and strength, so that, to prevent waste and injury in the after manipulations, great care and attention are necessary. The usual speed of rollers is, as a rule :—taker in, $1\cdot00$; second, $1\cdot50$; third, $5\cdot75$; fourth, $6\cdot00$—these will differ with the draft, being more or less, but the proportion is maintained.

Calculations for Drafts and Speeds.—A very easy method of taking the draft of drawing heads and speeders (or roving frames) without counting the teeth of the various pinions, or taking the diameter of rollers, is to break off the ends from the first can close to the calender rollers; then, with a foot-rule laid to the cotton close to the back rollers, at their pinch or bite, measure off 6 inches; roll the sliver at the end of the 6 inches, with finger and thumb, and make a mark—this measure taken at one end of the back rollers is enough, there is no necessity for measuring along the whole head—turn the head slowly, until the length (6 inches) measured off is fairly run through up to the mark, break off close to the calender rollers without stretching it, lay it on a board or a clean place on the floor, where its length in inches $\div$ by 6 will give the draft of the frame. Suppose the sliver measures 30 inches, $\div$ by 6 gives 5, that is, a draft of 1 into 5. This calculation can be made sooner than written about, and is very desirable when there is not sufficient time to count the wheels.

A draft between front and calender rollers has a bad effect, and too great a draft in any one head will injure the staple. The principal part of drawing, in all kinds of drawing, roving, and spinning frames, should be done between the first pair of rollers, or the two first sets. Draw no more in any one place than is strictly required, increase as it progresses or advances, from one frame to another on its way towards yarn, and draw just enough to take out the curl of the fibre.

When any of the small rollers in a railway or drawing frame need to be changed, always work the old ones back and put the new one in next to the front, but never, under any circumstances, put a new roll in at the back; it can and should be better employed, while an old one will do very well at the back—this is economy. Cans ought not to be too full, because the drawing becomes snarly.

Doubling and drawing are so closely connected that if we have one, we must have the other—they must go together. If we undertake to draw the sliver as delivered from the card without any doubling, it would become so small that nothing could be done with it; but, by doubling, the different strands are equalised, and are nearly all alike in weight per yard. Top rollers must be perfectly round, straight, and both ends of a size. Cover the front and back ones with good calf skin—it is more reliable and does not bed down so much as sheep skin. The driving belts should be allowed to run as slack as they will drive—they stop better than tight belts; or, rather, the stop motion works easier, and fewer ends are run through. Doubling in the railway head will be according to the number of cards that furnish it.

Rules for Drawing Frames.—A few rules for drawing frames may be given here :—

To find what the carded hank will be after passing through the frame, × draft and carded hank together, ÷ by number of ends put up. If hank carding be ·185, number of ends 5, draft 8, then ·185 × 8 ÷ 5 = ·296 hanks.

To find draft that gives required hank from a given carded hank, × ends by hank required, ÷ by carding hank. Thus, if carding hank be ·185, number of ends 5, the drawing hank ·296, then ·296 × 5 ÷ ·185 = 8 draft.

To find the weight of the drawing from weight of carding, × number of ends put up by weight of carding, ÷ by draft. Carding, 2 yards, weighs 90 grains, ends 8, draft 6; thus 90 × 8 ÷ 6 = 120 grains.

Number of ends put up, × by weight of carding, ÷ by weight of drawing, will give draft. Without crowding space by further examples, which may easily be determined by the ordinary arithmetical rules, we give the method of finding draft for any drawing frame :—
× the crown wheel, the back roller wheel, and diameter of front roller wheel for a dividend; then the front roller wheel, the change pinion, and diameter of back roller together for a divisor; and the quotient = draft.

To find change pinion for draft, × front roller wheel and draft together for divisor, and crown wheel by back roller wheel for dividend, and the quotient = change pinion.

A little consideration will point out how to obtain any other wheel or draft required.

Slubbing.—The drawing and doubling having equalised the slivers, the straightening and laying parallel of the fibres, another process, called slubbing, is called into requisition. The sliver made by the drawing frame is as fine as it is capable of being made, consistent with the strain in being drawn from the cans, but further attenuation is needed, which is strengthened by a slight twist to hold its shape, and to render it more convenient to handle, as ends increase in number.

Trial of Rovings.—It may not be amiss to say here that the last head of a drawing frame is where a trial of the rovings ought to be made. Break off all cans at the front of the last heads close to the calender rollers, and run the head so that the ends just touch the floor, or the surface of pulleys on which the receiving cans belonging to the last head revolve—repeat this twice, and put all the ends together. The distance from the calender rollers to the surface of the pulleys may be taken at 3 ft. $4\frac{1}{2}$ in., double of this, 6 ft. 9 in., or if there are, say, four heads, 27 feet in all. If a proper standard is fixed upon, the weight, in grains, of the rovings should be ascertained by correct scales or yarn quadrant; if it becomes too light, or too heavy, owing to change of cotton, or weather, or a defect at the drawing heads, a pinion one tooth less will decrease the weight 6 or 7 grains, if required; one tooth more will increase it in the same proportion—this change, of course, depends upon the fixed standard, size of roving, and counts of yarn being spun.

The rovings ought to be tried every two or three hours. If any change has been made in the weight of the lap, or in the card pinions, repeated trials must be made until all is right. The 27 feet would weigh from 32 to 34, on the small scales, and would make in roving 2·42, or nearly $2\frac{1}{2}$-hank roving, and the yarn spun from it would average from 18 to 25 hanks per pound. A great deal of waste is made, or yarn is spoiled, by being overdrawn—the fibre becomes broken, and will not produce good twist. With three heads, the cotton is only doubled 216 times, too little for warp or weft twist. If doubling is continued through four heads, one can to every roller, all running into one can in front, the doubling would be 320, or if two cans could be used, running up to each roller, at the fourth head, without increasing the draft beyond 1 into $6\frac{1}{2}$, the doubling could be carried to 640; if a doubler is used previous to putting it up to the first head, the slivers will be prevented from running single, and a saving of waste will be made by using four cans at the back of all the heads, instead of 12.

A change in the roving for the purpose of making finer yarn without adding to the spinning draft may be made as follows:—20's yarn is being spun, the roving to be altered for 25's, and the weight at the last head 214 grains; weight required for this change—as 20 : 214 :: 25 to 171$\frac{1}{5}$; therefore, to alter it, five numbers require 43 grains lighter rovings; then, if the pinion on the last head contains 28 teeth, what must be the number for the alteration—as 214 : 28 :: 171 to 22 teeth; this would be too much on any one head, but a difference of two teeth may be made on the first heads, and the rest on the last. When a change of this kind is made, the cans should be run out of the frames.

We have been very minute in these details, because, after the carding, and before spinning, all the deficiencies are accentuated, and any remedy to avoid such must be of special interest.

Final Preparation for Doubling.—In the final preparation for the mule or throstle, let it be clearly understood that, if a card sliver is $\frac{3}{4}$ hanks to a pound, after passing through drawing, slubbing, intermediate, roving, second roving, or jack-frame, it may be, according to orders, 5, 7$\frac{1}{2}$, or 12 hanks to the pound, and, particularly if the jack-frame is used for fine counts, it may reach 40 hanks. As we have previously remarked, these frames are simply duplicates of each other, in which there are two motions, one absolute and regular, the other relative and varying. The relative and varying motion is, in all cases, produced and regulated by the use of cones. The relative power is applied to the spindles and rail, the absolute to the flyers, while in the slubber and fly frame, it is applied to the bobbins on the spindle and to the rail, and the absolute to spindles, and to these the flyers are attached. In many cases, they are complex and intricate, requiring skilful management to secure good work, and, despite all the care in cleaning the cotton, carding, drawing, all may be spoiled in these frames. In winding on the bobbin, regulate the speed of the rail, so that the roving, as it is wound round the circumference of the bobbin, will exactly cover—neither pile up nor show spaces between the coils, but make a smooth layer.

CHAPTER V.

THE FLY, OR BOBBIN FRAME.

The Flyer Frame.—In order that the utility of the flyer may be better understood, it will be necessary to give as clear an explanation as possible of the fly or bobbin frame, which is essentially a machine, like the drawing frame, for drawing the sliver, but the difference consists in obtaining greater tenuity; but on no consideration should the amount of twist given to the roving be more than sufficient to enable it to be wound on or taken off the bobbin—a greater amount would materially interfere with subsequent processes. The sliver is supplied from a can, and, after going over a guiding rod, is passed between a series of drawing rollers, thence to the flyer, which winds it upon the bobbin. The machinery to accomplish the various movements in this frame, and in this branch of cotton manufacturing, is very ingenious and complicated—so much is this the case that diagrams, if given, would convey no practical information unless motion could be imparted to them; we must, therefore, endeavour to be clear and concise as far as the nature of the subject will permit. The rollers are similar in principle and arrangement to those of the drawing frame, and are usually in three pairs, the lower ones fluted, and the upper ones, with the exception of the first, are also fluted.

In order to take a better grip of the cotton, they are covered with flannel and leather, and their circumferences are tightly compressed by means of weights. The draft is principally between the middle and front rollers. Cleaners, as in the drawing frame, rest upon the top rollers. Now, on leaving the front roller, the roving enters the tube of the flyer, passing out by an orifice near the top of it, and down one hollow arm, after which it is turned once or twice round a "finger," or rather, presser, and deposited by it upon the bobbin.

The centrifugal presser consolidates the rovings upon the bobbin, and, by this means, a larger quantity is wound on. This "finger" is hinged upon the hollow arm or leg of the flyer, so that it can swing with perfect freedom. Between those two points lies its greatest weight, and, as the spindle revolves, it acquires great speed, which, being communicated to the "finger," causes it to press with corresponding force on the layers put upon the bobbin. The twining round of the roving on the "finger" is to keep it tight while being wound on.

The ends of the spindles are slightly tapered, so as to fit into the tube of the flyer, until a slot in the end receives a pin, which crosses the tube, making a thorough connection between the two. The flyer is movable, so that when the bobbins are full they may be taken off to be replaced by empty ones. The flyer is attached to this spindle, which is vertical, and receives motion from a pulley. At the top are the two arms of the flyer, one of which, as already stated, is hollow, and through this the roving passes to the bobbin, the motion of which is independent of the spindle. If the spindle and bobbin revolved with the same speed, the roving would certainly receive a twist, but no winding on the bobbin could possibly take place. If, however, the bobbin is given a smaller degree of velocity than the spindle, then the roving not only receives the twist required, but will be wrapped round the bobbin at a rate due to the difference. As the bobbin revolves, the coils of roving increase in diameter, and, if this continued uniformly, the roving would be stretched and torn from the bobbin; its velocity has, therefore, to receive a diminution at the completion of each layer. Again, the bobbin must receive a motion up and down the spindle, so that the roving will be uniformly wound along the length as well as the circumference.

Combination of Movements of the Flyer.—Let us carefully understand this singular combination of movements, intricate in conception, but, seen in motion, mechanically simple. The speed of the spindle is unvarying, or, rather, fixed and uniform; so is the quantity of roving when once adjusted to the amount of necessary twist. The flyers making a fixed number of revolutions in a given time, and the front roller a certain quantity of roving, it is clear that the bobbin requires a varying motion, corresponding to the constantly increasing bulk of roving layers wound upon it. The mechanism by which this seeming paradox is accomplished is one of the finest geometrical problems that has ever been applied to practical use. Nothing can exceed the ingenuity of what operatives know as "Jack-i'th'-box," or, properly, the differential motion, which effects all the desired movements by a curious combination of wheels, now so thoroughly well known that a description of its operations would be superfluous. Improvements, however, are still necessary, and, assuming that thorough efficiency and economy are the order of the day, it is desirable, in writing on the subject of cotton manufacturing, to notice every attempt to facilitate this object. Notwithstanding the fact that regularity of movement is a necessary condition in spinning, it has been impossible to attain this in a perfect manner. The primary cause of this, in roving and other frames, is the train of wheels that communicates motion to the shaft or shafts driving the bobbins, and better known as the swing frame. In going up, the shafts move at a different rate from that when the rail is going down, because the wheels of the swing frame move bodily in the same direction as the driving wheel in one instance, and, therefore, retard movement; and, in the other case, they go in an opposite direction, and, therefore, at an increased speed. The difference amounts to nearly one revolution of the bobbins in each up or down movement of the rail, and this really means from seven to nine inches of sliver, with a three-inch bobbin, so to that extent the sliver is irregular. We have seen in operation a remedy for this defect, which is a substitute for the swing frame—an arrangement of bevel-wheels centred on a swinging spindle, one sliding loose on the lengthened bars of the other, and made practically into one, as regards rotary motion, by a key. The bevel-wheel, on the lengthened boss of the differential motion bevel, drives a bevel on the swing spindle, which is carried over the boss mentioned, having the other end of it free to rise and fall with the bobbin rail. The two bevel-wheels on the swing spindle, although like one, are really separate—the lower one drawing out when the box is going down. The driving and driven wheels, being on opposite sides, compensate the irregularity arising from the motion of the ordinary swing frame from going against or with the direction. The truth of this is easily realised, and is quite conclusive to a practical spinner. If the rail is raised and lowered without turning the bobbins, it will be found that the relative positions of the bobbin and shaft are undisturbed. In the swing frame, the bobbins will move a little one way during the upward movement of the rail,

THE FLY, OR BOBBIN FRAME. 43

Fig. 16.—Slubbing Frame.

and in the opposite while going down. It is obvious that this device for driving the bobbin is more regular, yarn being better spun, or produced, from the sliver than by the ordinary process in roving and fly frames.

The Slubbing Frame.—Slubbing is somewhat similar in action to drawing—a further combination of the slivers and the drawing, carried so far as to require a twist, which is all the distinction between the two operations. The slubber, intermediate, and rover, are used in succession for fine counts of spun yarns. The mechanism is, however, the same, or nearly so, in each.

The winding motion, as improved in the latest machine, of which we give a full illustration (Fig. 16) and also a sectional view drawing (Fig. 17), supersedes the "sun and planet" arrangement in older machines, spur wheels being used in place of bevels, and thus reducing the amount of power required for driving. The underlying principle, in connection with the modern improvements in these frames, is to ease the drive of the cone belts and regulate the minor portion of the draft. The continuous changing number of the revolutions of the lower belt cone is given by spur wheels. The proportion of all the wheels is such that there is no motion in the box at the commencement of the winding, and every change of speed is given with more regularity and certainty. This new differential gear works with less friction, because the spur wheels, loose upon the driving shaft, turn in the same direction as the shaft, but, on the ordinary principle, they turned opposite to the shaft, causing greater friction, with a relative velocity. The transmission of speed by cones is more certain and effective by the use of very narrow belting—the less the difference in the diameters of cone surface covered by a belt and the more positive is the speed.

The Roving Frame.—We give a description of a roving frame (shown in Figs. 18 and 19), which is certainly worthy of attention.

The driving shaft is shown at A, upon which is fixed a pair of fast-and-loose-pulleys, giving motion to two spindle-driving shafts, B, the flyers being attached to the top of the spindles. In some arrangements, the bobbin leads the flyer, and in others the flyer leads the bobbin. The centre wheel, C, carries a pair of bevel-wheels, which are put into gear by a long bush; motion is in this way communicated to the bobbins, H. The lifting rail is seen at G, and the racks at J. The centre wheel, C, can be made to revolve in a reverse

THE FLY, OR BOBBIN FRAME. 45

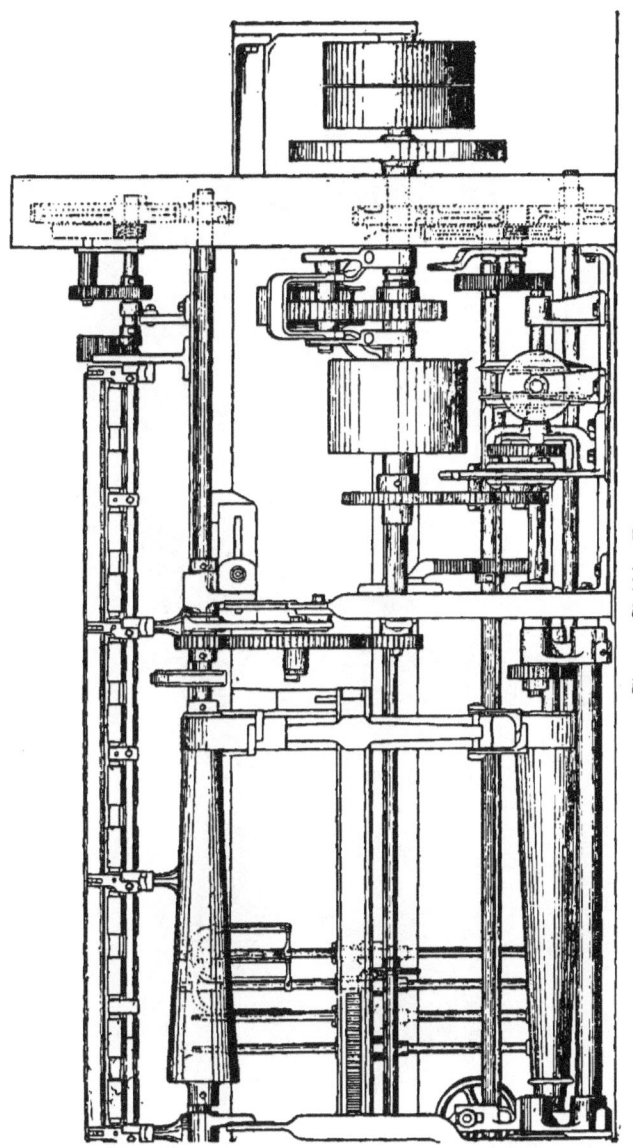

Fig. 17.—Slubbing Frame.

way by the small pinion, K, and in the opposite direction to the bevel and spur wheels; the speed is in this way gradually increased,

Fig. 18.—Roving Frame.

or decreased, according to the size of the rovings wound on the bobbins. This peculiar motion is fully explained earlier in this chapter. L is the reversing bevels; the roving rods are given at R; the transverse guide to the three pairs of rollers at S; and the spindles, with their flyers, at T. The bottom rollers in this machine

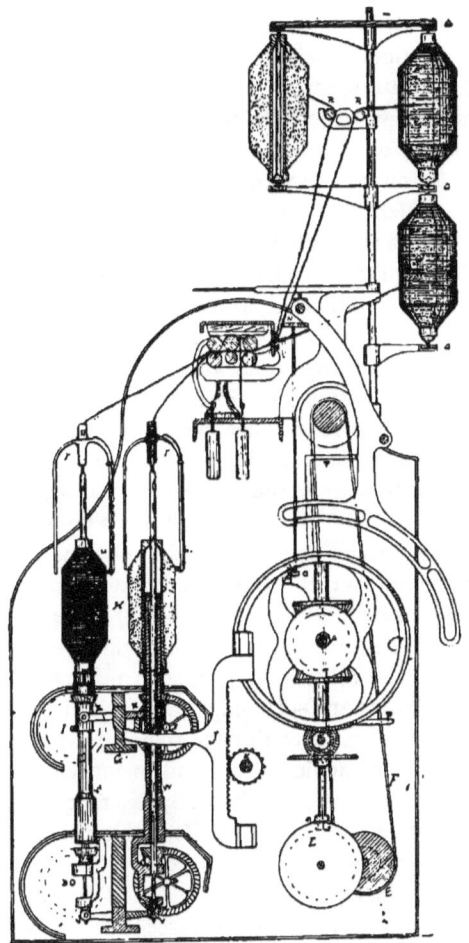

Fig. 19.—Roving Frame.

are fluted, and the top ones plain and covered with leather. The middle pair run more quickly than the back ones, and the front more quickly than the middle pair. From these rollers the rovings pass into the neck of the flyer, D, and down the hollow by U, and out at the bottom through an eye on to the bobbin. The foot of the spindle itself rests upon a swivelled footstep, V. By a careful inspection of the illustration the nature of the various movements may be ascertained by the reference letters. A few necessary calculations are given, from which others may be easily obtained.

Working Calculations.—To find draft of slubbing or roving frame:—Front roller wheel 22, crown wheel 90, pinion 41, back roller 50, diameter of back roller 1 in., diameter of front roller $1\frac{1}{8}$ in. $= 22 \times 41 \times \frac{8}{8}$ for divisor, $90 \times 50 \times \frac{9}{8}$ dividend, quotient 5·6 draft. If the rollers are different in size, multiply the pinion by the diameter of the front roller, and divide by the diameter of the back roller.

Twist wheel required in changing from one hank to another:— Suppose $1\frac{1}{2}$ hank is produced with a 23-twist wheel, on what twist for a 2-hank? $= 23 \times 23 \times 1\frac{1}{2} = 793 \div 2 = 396.75$; the square root of this is 20 for teeth of change twist wheel.

Finally, to find the number of times the spindle of the frame must go round for the front roller once, with a fair sample of cotton:— The square root of the hank roving going to be produced, multiplied by 3·33, the number of times ordered for a hank, the roller being one inch in diameter, will give the number of times the spindle should go round for the front roller once; but if the roller be thicker than one inch, a different multiplier will be required, that will stand as follows:— multipliers 3·33 = 1-inch, 3·74 = $1\frac{1}{8}$-, 4·16 = $1\frac{1}{4}$-, 4·57 = $1\frac{3}{8}$-, 5·00 = $1\frac{1}{2}$-, 5·41 = $1\frac{5}{8}$-inch rollers; in fact, proportion will give all these calculations, thus—as 1-inch roller is to 3·33, so is 1-$\frac{5}{8}$inch roller to 5·41.

Advice to Managers.—Having now arrived at the last of all processes, previous to mule or throstle spinning, we feel it incumbent to give a few words of advice to all mill managers and superintendents who wish to fearlessly and honestly discharge their duties.

In the late evening of an active life, half a century of which has been spent amongst textiles of every description, we may safely say that spinning machinery consists very much of drawing and retaining rollers, and it is necessary, in most cases, that the relative surface speed or draft of these rollers should be known. This is arrived at by multiplying the diameter of the drawing roller and the number of teeth in each of the driving wheels of the gearing together, and dividing this sum by the product of the diameter of the retaining, multiplied by the number of teeth in each of the driving wheels, the result being the draft required. It is not, however, necessary, in every case, to repeat this calculation, as a constant number can be found by leaving out the change wheel, or varying element, in the calculation, and this constant number can be operated on by the given draft, if its corresponding wheel is required, or by the given wheel, if its corresponding draft is required.

Now, these have frequently to be ascertained by those in charge of the operatives and machinery, amid, perhaps, all the noises and commotion of a mill, and are, therefore, liable to error. In many cases, the drawing roller is subject to wear, and has occasionally to be turned up. This reduces the diameter—a fact scarcely, if ever, noticed. Errors may arise from a broken wheel being replaced by one a tooth more or less, or different in size, from not having one just to hand, or from the happy-go-lucky system which often prevails; but, besides these errors, which may occur inadvertently, others can be, and often are, made by overlookers to save themselves trouble. For instance, if it is found that several spinning frames, which are spinning the same sort of yarn, are producing it too light or too heavy, the overlooker, instead of changing the draft of all frames, may only change one or two, to such an extent as to counterbalance the error on the other frames; and as the average weight of yarn would thus be correct, the inequalities in the weight would not be detected without considerable trouble.

It is very well known by spinners that when the amount of draft exceeds a certain limit, uneven yarn must be produced, and when it falls short of a certain limit, the preparing machinery cannot supply the spinning machinery. The proper amount of draft is, therefore, usually determined by these limits, and it is important that the manager or really responsible official should know that the draft is such as these conditions require. Unless, however, he is prepared to stop a machine, and ascertain the draft for himself, he must rely for his information on the overlooker, who may, as he is only human, be tempted to conceal an error that may have been made in the weight of a rove. If the trouble of calculating is too much, a draft and speed indicator would obviate these deficiencies, which also create errors when estimating the weights of woven fabrics, causing perplexities that are attributed to waste in winding, warping, and weaving.

CHAPTER VI.

THE MULE.

The Mule and its Structure.—The mule may be considered one of the greatest triumphs of mechanical skill: each separate movement is a study in itself, and yet is perfectly reliable in its action and co-operation with all other parts of the machine. No sooner was the mule found to be capable of spinning than its advantages were at once acknowledged; and, when the point was reached at which a much finer count could be obtained than on the throstle frame, it may be said to have become the principal source of supply for spun yarns—both warp and weft. In all frame spinning recourse is obliged to be had to the fly, or something equivalent to it, for the continuous action of the frame does not admit of its being evaded—this is an inherent defect in all frame spinning, when applied to fine counts.

In the structure of the mule, the part in front of the rollers, where the fly is used in the frame, is composed of the best part of Hargreaves' jenny, and so the necessity of the fly is avoided; and, instead of the sliver as it issues from the front roller being burdened with the strain and pressure of this fly, it is left perfectly free from all strain or external pressure, having only to sustain its own weight. This is one of the great advantages in mule spinning, as this object is attained without losing the benefits of roller drawing—the mule is, in fact, the "excelsior" of all spinning machinery inside a cotton mill. There may be instances in coarse counts where the strain and gird of the fly may be of use in producing a bare and wiry thread, but in spinning as an art, especially in the production of very fine counts, the fly and all its allied appliances will always be in the rear.

In all textiles where finish, fineness, and extreme delicacy of manipulation are requisite, the mule is our most advanced machine, and neither the throstle nor ring frame is able to compete with it. Briefly stated, as a thread approaches the extreme verge of tenuity, it must be relieved of all the burdens of rings and flies, of all weight and trammel, restraint and pressure, for, as it enters into the gossamer lightness almost invisible to the human eye, surely it needs no great amount of thought to understand that its own weight, which it cannot cast off, is a sufficient burden to its already reduced power of tension, and, as the finished stages of spinning are approached, airy lightness, delicacy of hand, together with gentle equability, should characterise every movement in fine spinning.

Economy is another consideration—in fact, the prime one. We do not think it would require a savant to divine which yarn, mule, throstle, or ring frame, is in the most convenient form for ready use: practical people can soon show the wide difference, in point of labour and waste, attending the working up of the one, in comparison with the other two. On the one hand, cops can be packed by hundreds in a very small space and carried to any part of the world; but bobbins cannot be so packed—the bulk will tell. The carriage cost of mule cops may be figured at 10 per cent., but the lumbering addition of bobbins would reach 200 per cent.; this can scarcely be considered a favourable feature for either throstle or ring spinning, especially when the yarns are not required in the hank.

Since the days when the mule and flyer throstle were first introduced, no distinct advance has been made beyond improvements in points of detail, arising from experience, and each style of machine still retains the respective original working and, therefore, commercial defect. Originality is very rare, and when it does occur, attention is directed to it, and ordinary minds, which can follow but never lead, are at once exercised to find out modifications, which may or may not be improvements. We have, in our time, seen so many ruinously costly experiments made in changing machinery, that we strongly advise manufacturers, desirous of alterations, to consult only the best makers, who will give an honest verdict.

Description of the Mule.—To write anything like an adequate description of the self-acting mule would fill a goodly sized volume. The multiplicity and complication of wheels and levers are so great, and the arrangement and construction so varied, by the different makers, that all we can do within the limits of these pages is to describe the general working, the principle of the important movements, the part each has to perform, the necessary calculations, the defects, and the possible remedies.

Since the mule was invented, above a century ago, up to the present time, there has been a constant endeavour to increase the number of spindles. Fifty years ago, each mule contained as many as from three to four hundred spindles, filling in width the narrow mills of that time. The limits of the walls being reached, an attempt was made to couple the two corresponding mules of two pairs together, doing away with one pair of headstocks—this was called "double-decked," but it has not been so common, as an increase in spindles has been preferred. Mills are now built sufficiently wide to hold more than 1,200 across the room, and these are at distances varying from 1 to $1\frac{1}{2}$ inch upon the "carriage," or movable part of the mule, in one long row, which is interrupted in the centre for about thirty inches by the "headstock." The principal machinery is here located, and from this, as a centre, the different movements are communicated. A beam stretches right and left from the headstock, supporting the drawing rollers; behind are the creels holding the bobbins of roving from the fly frames; within the carriage there is a tin cylinder, with bands or cords to drive the spindles.

The carriage was originally moved by the left hand of the spinner, but is now self-acting—the motion being made by machinery. Along the top of the spindles stretch two wires—the fallers—one of which guides the yarn upon the cop. The movement of the carriage is effected by means of a back shaft, extending to the end of the mule, and on this shaft are fixed pulleys. A cord is fastened to the carriage which, passing round a fixed pulley upon the end piece of the mule, at the extremity of the "draw," takes the carriage out. This back shaft is driven by the main shaft, which is placed on the upper part of the headstock, through a series of connecting wheels. At the end of the main shaft, there is the "rim"—really the fly-wheel—which drives the tin cylinder by means of an endless double band, passing round carrier pulleys.

It may be as well to note here that the later improved mules have the main driving pulleys, main band rim, cylinder band pulleys, and all band carrier pulleys of much larger diameter. Less power, therefore, is required, and the movements run much lighter than in older mules with small pulleys. Roving bobbins are set in the creels upon pegs, the lower portion of the peg being tapered to a fine point, so that the friction may be reduced to the smallest degree possible. The half-spun roving, being very tender, is easily broken by the slightest strain in unwinding. This roving passes to the drawing rollers, guided by steel plates with slots, and between the first roller and the second is another row of guiding plates. Both rows are moved backwards and forwards, lengthway of the frame, by an eccentric—this is to obviate the running of the roving too much in one place and the cutting up of the leather covering.

The rollers are the same as those which we have described in the intermediary machines, but not quite so large, being mostly an inch in diameter. There are three sets; the first and second take in the cotton, the third draws it to the size required. Of course, they are always adjusted to suit the different qualities of cotton. The first roller speed is fixed, and from it the other rollers are driven by wheels and pinions, so that the spinner can, at any time, alter the counts of yarn. The rollers are equally weighted by means of saddles.

Gaining of Carriage.—The cotton having passed the rollers, reduced to the size ordered, is twisted by the spindles, which move slowly backwards from the beam—this movement is in excess of the velocity or speed of the rollers, or, in plainer language, the carriage moves perceptibly faster than the cotton is delivered from the front rollers. This is called by the spinners "gaining," and the object is to prevent snarling, which would take place if it were not for this stretching.

When the carriage has gone to the end of the "draw" or "stretch," it stops, but the spindles continue to revolve until the exact turns of twist are put into the yarn. Then a few revolutions are made backwards, until the corresponding number of turns of yarn which run up to the point have been unwound, and at the same time the front faller wire drops down upon the ends and guides them upon the spindle or cop, as the carriage returns to the beam. The faller then resumes its position, and the carriage prepares for another stretch. The "counter-faller" keeps the yarn at one certain degree of tension whilst the winding on is taking place. While the carriage is going from the beam, and during the time spinning is being done, the front faller guide is stationary at a little distance above the ends, and close by the point of the spindle; but the position of the counter-faller is under the ends and a little farther off the spindle. When the spinning is finished, and the front faller guides the yarn upon the cop, it slackens a chain and sets the other faller free to rise against the ends. The main shaft drives the spindle by the rim wheel, and it also drives the taking-out motion of the carriage and the drawing rollers.

The Hand Mule.—The hand mule is much simpler than the self-actor, several of these movements, such as bringing the carriage back to the beam and the building of the yarn upon the spindle, being imparted by the spinner. These mules are principally confined to the spinning of the very finest counts. A self-acting mule of the most improved make, for spinning from 150's to 300's, has the usual appliances, which include a motion to drive the spindles at two different speeds, and a motion to stretch the yarn slower when the rollers are thrown out of gear. There is also a motion causing the rollers to turn slowly while the spindles are winding—this is not exclusively confined to fine spinning—a slow roller motion that can be regulated to different speeds, and is in action while the twist is being put in at the head, and a snarl motion to guide the thread round the spindle when the faller begins to rise, working in conjunction with other arrangements to take the yarn softly from the spindles at the moment of backing off and winding. The exactness with which the under faller is balanced may be judged when 96 ends of 150's are able to move it without damage to the yarn. The accuracy of the adjustments and this careful balancing of the under faller causes this self-actor to be considered a very near approach to the final extinction of the hand mule in fine spinning.

Warp and Weft Yarns.—We have twist mules with 1000 spindles, 19-inch rims, running at 10,000 revolutions per minute, averaging 30 hanks of 32's twist per spindle per week, and weft mules with 1,200 spindles, 18-inch rims, running proportionately slower. The front rollers in these mules have three ends to a boss, and are made with them loose. The taking-in shaft is driven by a band from the counter shaft, instead of by wheels from the rim shaft, and there is no rope-tightening motion.

There is a very simple contrivance by which the motion of the carriage is made to guide the strap gradually from the fast pulley near the end of the " draw," so that it is almost off before the cam changes, and to guide it on again in the same way as the carriage approaches the roller beam in putting up—this is a most excellent device. In the weft mules there is a movement by which the rollers are caused to make a revolution as the carriage puts up. This increases the delivery of the rollers, and, at the same time, eases any strain in winding the yarn. The quality of the yarn produced is unequalled, and the cost of production is at the lowest possible figure. The spindles on weft mules are usually not more than $1\frac{1}{8}$ inch apart, as that is sufficient space to admit an ordinary-sized cop.

Weft for reeling—that is, to be put up in bundles of hanks—and warp yarns are spun on spindles about $1\frac{1}{2}$ inch apart. The spindles are made stronger to prevent vibration. A cop, in this case, may contain four or five times greater length than when the spindles are nearer. It will easily be seen that there is also a great advantage in the less frequent stoppages for doffing.

Lubricating the Mule.—In all mules a great loss occurs in the oil reserve—the footsteps of the spindles occupied by the spindle feet will only allow a minimum of room for oil—and the only reserve there can be is in the countersunk dish at the top. The spindle being taper, and revolving rapidly, the oil rises up along its surface, and is thrown off anywhere by centrifugal force, reducing the given turns of the spindle. A loss of oil also takes place by small pieces of loose fibre, that are always flying about the spinning rooms, adhering to the oily spindles, and being afterwards thrown off in a manner that may be easily noticed in some machines where the spindles remain in one place, instead of moving in a mule carriage.

In a winding frame, for instance, where the speed is much less, this fibre, after accumulating for a while at the top of the step, often begins to wind round with the spindle in a ring, at first by starts, and then more regularly. After a while, the accumulation breaks loose, and begins to revolve itself in the dish round the spindle until it is thrown out. As the capillary space between the spindle and the footstep and the motion of the spindle bring the oil to the surface, the waste sucks it up like a sponge, and thus all surplus is lost.

The waste of oil in connection with mule spindles can be easily prevented by a simple contrivance. A brass flange or disc, three-eighths of an inch in diameter, flat on the under and convex on the upper side, has a hole through its centre of the required size, and is driven on the lower end of the spindle until it fastens on the taper part, about five-eighths of an inch higher up. A metallic dome, about half-an-inch in diameter and rather less in depth, with an aperture at the top of five-sixteenths of an inch, is placed on the spindle before the flange is attached, and its top covers the top of the flange, while its edge or lip fits loosely in the dish of the footstep. By this means, waste cannot accumulate about the footsteps of the spindles, and the oil that rises up them is thrown from the edge of the disc, and runs down within the dome to the footstep again, thus effecting a saving of oil, banding, and labour.

Economy in Mule Spinning.—The chief consideration in the economy of mule spinning is the supply of piecers; and the inducement for a youth to serve an apprenticeship to spinning is to become a spinner, and not a big piecer only. There are thus too many of the one and too few of the other; and hence the number of improvements or attempts made by inventors to reduce the cost of mule spinning by automatic movements, which in some few instances have been successful. Double-decking, to which we have already alluded, is not entirely lost sight of, and it would probably have been more generally adopted but for imperfections in carrying it out. On account of these imperfections the cops spun on the coupled mules—or in the "back wheel gate," as it is often expressed—were frequently inferior, the soft scarcely ever corresponding exactly with those produced by the headstock mules, and consequently not marketable except at a loss.

The common plan of conveying the motions of the carriages by bands is open to the danger of slipping, and an improvement in this direction for producing yarns and cops of perfectly uniform character on the headstock mule and the one coupled with it is an arrangement of bevel-wheels and a shaft, the motions being by this means identical in the two carriages. It has been, and is, usual to convey the motion from the fallers of the headstock mules to those of the off mules by wires, which method has always proved unsatisfactory and liable to produce bad work; but in the improved arrangement each carriage has an independent copping motion, so that the one mule has the same power of making perfect cops as the other. This improvement is of importance in many ways; 120's can be spun without any difficulty, and each spinner can attend to 3000 spindles. This result is highly satisfactory in the production of fine yarns.

It may be observed that there are many branches of spinning.

Oldham has, practically, a monopoly of 32's twist and weft; but the system which obtains in that town is fast spreading throughout Lancashire, on account of its great economy in production, some mills having as many as 80,000 spindles. The mule seems to be a machine that every machinist believes can be improved without limit, and a very ingenious individual has spent time and money without stint in moving the roller-beam—the practical spinner follows the carriage to get to the roller-beam, and would certainly object to it slipping away from him as soon as he reached it. For piecing ends, taking off roller laps, stripping the clearers, changing a roller, supplying an occasional drop of oil, regulating the rovings, and for general working, the roller-beam is much better as it is—fixed—than moving; therefore this so-called improvement is a step backward, and a great deal of trouble has been expended which, it is to be feared, will not be adequately rewarded.

An Improved Mule.—Those who have watched the development of spinning machinery during the past few years, must be surprised at the enormous improvements which have been made in this class of textile machinery. We give an illustration (Fig. 20) of an excellent improved mule now in operation, which is very much commended. It contains all the latest advantages, especially in the driving arrangements being placed over the headstock, with a belt running parallel to the sides. For mills of a narrow width, this type of mule is extremely suitable. The self-acting headstock is shown in Fig. 21.

The cam shaft, being placed along the frame side of the headstock, allows a long lever to act direct with the revolving stops, independent of extra levers. The working of this cam shaft is without any noise, giving a great amount of power, with very little friction. A pulley, with three grooves, drives the backing-off, taking-in, and cam shaft; the rope transmitting the power is kept at an equal tension by an ingenious tension frame, the tightening pulley being always between the bands, whatever the angle may be. The taking-in friction has sufficient power to actuate long mules equally as well as shorter ones.

A further improvement is the fitting of a safety latch to the friction lever, so that it is impossible for the lever to get into gear before the carriage completes its outward run and backing-off.

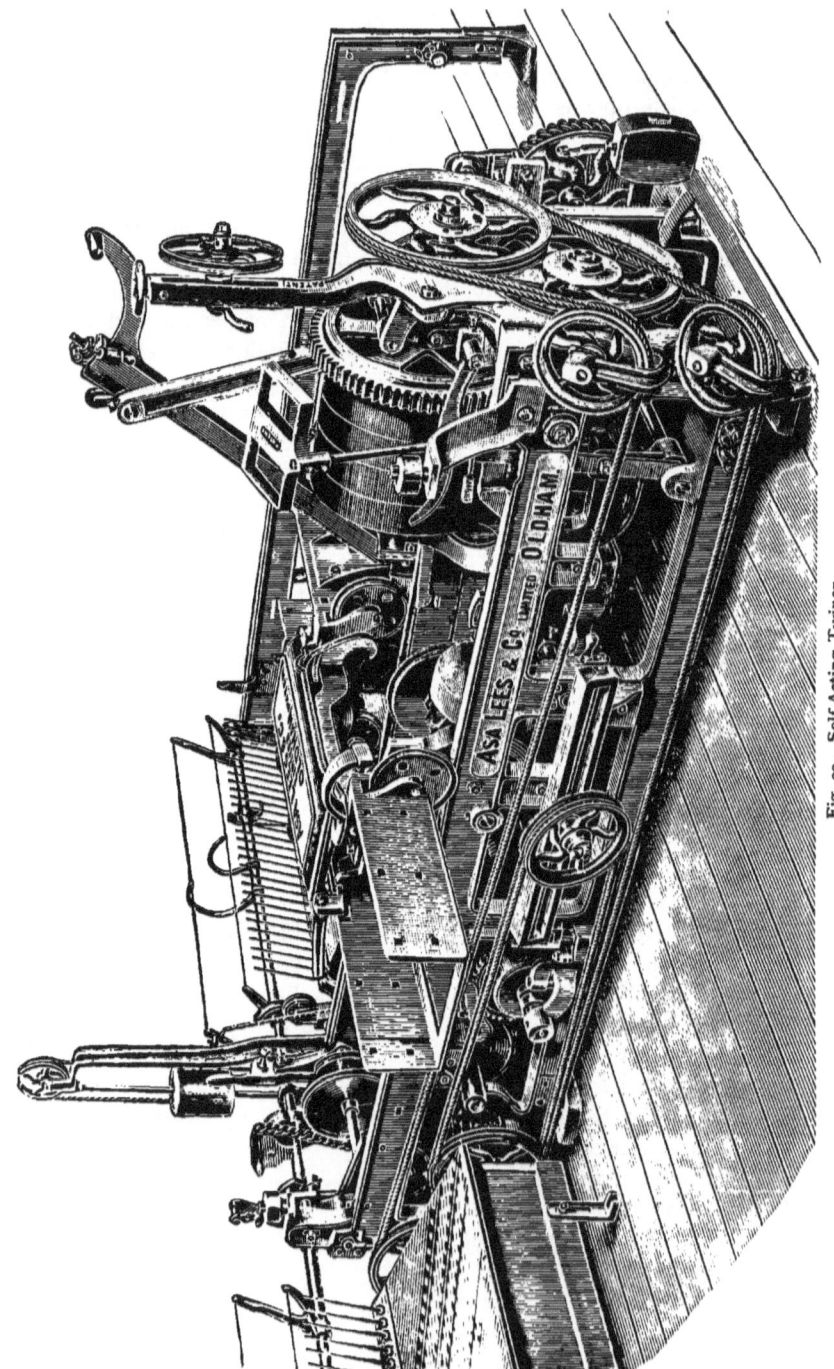

Fig. 20.—Self-Acting Twiner.

This is most valuable as a preventative of serious accidents to minders or others. The rim shaft is case hardened.

For a 64-inch traverse, the drag wheel teeth each represents one-half of yarn. The drag lever has a locking movement, especially suitable for high speeds. This checks all jerking as the carriage begins to run out, and, of course, cutting the yarn is avoided. When a change of the cam lever takes place, the drag lever is at once locked, and so remains until the carriage moves a few inches, and the latch is relieved automatically, and the lever is prepared to lift, if any obstacle should occur. A check scroll band is secured by a clip, and one of the best arrangements is the builder of the cop bottom, the motion requiring no attention when set properly, any after setting, if required for another doffing, being at once effected by turning back the shaper screw. The yarns are rendered free from snarls—an immense boon to the operative spinner. In the inward run of the carriage at the finish, the spindles are increased in speed, before the change of the cam, and thus the nose of the cop is better shaped than is usual and is more solid. The winding-click, as the fallers lock, is put into gear, just before the tin roller commences to revolve. When a full set of cops is spun, the mule is at once stopped, in a proper position for doffing, and it is claimed that every set of cops will give the same weight of yarn.

Self-acting Nosing Motion.—This mule is fitted with a patent self-acting nosing motion (Fig. 22) which we believe is the only one not dependent upon a toothed ratchet wheel, which hitherto has been located between the arm, A, and the shaper, F, shown in the illustration. This improved motion being worked direct from the shaper, F, is considered sufficient to operate, thread by thread, in exact proportions to the cop as it is being built. The loose arm, A, is affixed to the quadrant arm, G, and, upon the side of the head-stock frame, is attached a pendant lever, C. These parts are all connected by B, which can be increased or diminished in its length as required. At the foot of the frame is the hooked lever, D, with the fulcrum in the centre, the hooked end encircling the lower end of the pendant lever, C, and the other end of D overlapping the shaper screw, F. The shaper plate, E, in its inward motion, pushes D before it; and the lever, C, being forced forwards, also moves the connector rod, B, and brings down the arm, A, to establish a gradually increasing drag on the quadrant chain, I—the spindles are

60 COTTON MANUFACTURE.

Fig. 21.—Self-Acting Mule Headstock.

thus increased in speed to meet the change consequent upon their diminished diameter. H is the quadrant. The illustration is very simple and easily understood without further description.

In mules there is no lack of really first-class makes, and it would be merely a repetition to give further details of the construction of different machines. Each has its special features, of the merits of which experience alone is a trustworthy test.

The Mule Carriage.—When the carriage of a mule has reached the extremity of the stretch, it comes in contact with a projection upon a lever or rod—we are here alluding to the ordinary mule. This lever, either directly or through a second lever or cam-shaft,

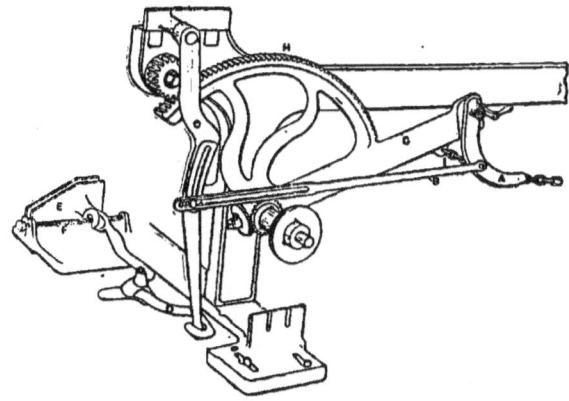

Fig. 22.—Self-Acting Nosing Motion.

disengages the clutch boxes of the back shaft and rollers, and, at the same time, the action of a powerful spring is brought to bear upon the swinging belt-guide, tending to throw it off the pulley which drives the spindles. This force is held in check by a small catch, until the yarn has received the proper amount of twist, when the catch is lifted by a finger. Instantly, the strap passes on to the other pulley, and the spindle ceases to turn. A worm upon the end of the main shaft, and pinions which are calculated to give the twist for the particular yarn which is being spun, work the finger. As a rule, the quantity of twist to be given increases with the fineness of the yarn, so that the twist pinion requires changing when an alteration is necessary in the counts. The same motion which has thrown the strap upon the second pulley brings a friction box into conjunction with a corresponding friction cone fixed upon the main shaft, and by this means the backing-off motion is communicated to the spindle, until the turns of thread between the cop and the point of the spindle have been unwound.

The length of yarn to be wound upon the spindle is always the same, as it equals the traverse of the carriage, the speed of which is the same during the period of building the cop; but the quantity of yarn wound up by one revolution of the spindle keeps increasing until the full diameter of the cop is reached—that is, until the abrupt taper is finished.

A very ingenious and simple device is in use for cleansing the top of the mule carriage at each return. A curtain or piece of cloth is attached to the rail below the rollers, the length and width of the box on which it rests, and, as the carriage recedes, the cloth falls, brushing away the dust and flowings to the floor.

The Quadrant.—As in the case of the fly-frame, where the differential motion is used to compensate the constantly increasing size of the bobbin, so, in the mule, a delicate, ingenious, though simple, device is adopted to regulate the speed of the spindles, when winding up the yarn. The quadrant, as this device is called, though a very important adjunct, is not sufficiently understood by operatives. There is more difficulty attached to keeping it in order than obtains with the differential motion. The calculation through the different wheels will give positive results in the one case, but with the quadrant an approximation is often only attainable. It swings upon a stud, and has this oscillation given to it by a small wheel, and with every stretch it ought to describe an arc of $90°$. Within its long arms, and extending its entire length, there is a screw which works a sliding block. To this the winding chain is fixed, and as the long arm, from being nearly upright at the beginning of the winding, falls into a position almost forming a straight line, it will be easy to perceive that the amount of chain unwound, will, in some degree, depend upon the position of the sliding block. If the chain be held at the lowest point, then the amount of unwinding is at its greatest, as the effect of the yielding by the quadrant is at a minimum; but when the chain is fixed at the extremity of the arm, there will be less of it unwound, by the distance between the two extreme points.

When a spinner commences a set of cops, he turns the handle at the top of the long arm, until the sliding block has reached the lowest point, if the yarn has to be wound on the bare spindle; but if tubes are used, there is an increased winding surface, and this is allowed for by not going quite to the lowest point with the sliding block. It is the great object with inventors to apply contrivances for adjusting the chain as the spinning proceeds, but nothing, so far, is equal to watchfulness on the part of the operative, who may have, now and again, to shift the screw by hand.

The Counter-Faller.—All motions, having for their object the regulating of the degree of tension to which the yarn is subjected, are controlled by the counter-faller; therefore, in mules of the most improved type, on the long arm of the quadrant there swings a sort of pendulum, the lower end of which is the segment of a circle. This is toothed, and works into a wheel which communicates its motion to the quadrant screws, but only when the pendulum moves backwards from the carriage. From the counter-faller, a long arm is suspended. When the winding is taking place, and as the carriage reaches the beam, this arm touches a jointed step fixed to the floor, which throws it back, and, being adjusted with such exactness, the counter-faller wire, if in the slightest degree depressed lower, on account of the tightness of the yarn, than its usual position, a shoe, sliding loosely and hanging by a short chain, takes hold of a toothed rack and keeps the arm from falling back against the carriage. Considering the great loss in waste, bad and uneven cops, etc., this arrangement, in connection with the quadrant, is worthy of consideration.

On account of the rack being nearly straight, and that part of it which is next to the carriage being nearest to the centre of the arc, described by the loose shoe, its tendency to catch decreases the farther it goes back, so that, just in proportion to the depression of the faller, is the point at which the arm is held back and, consequently, the amount of traverse it gives the pendulum when it comes against it through a projecting stud or casting. The movement of the pendulum is communicated to the screw, and the sliding block to which the chain is attached moves towards the extremity of the arm. When the carriage is up to the beam, and the quadrant straight out, or horizontal, the pendulum, owing to its weight, swings forward, but is stopped by a projection from the sliding block. It is from the shape of this pendulum that the greatest traverse is given to it at the beginning of a cop, and the position of the chain alters slowly, until the full diameter is attained, when it ceases entirely.

Building of Mule Cops.—The building, then, of the yarn, is uniform in the shape of a cone, the quadrant always giving way to suit this, so that the spindles revolve more rapidly as the faller guides the yarn from the greater to the lesser diameter. If fine yarns are spun, a slight stretch is given after the rollers have stopped—this makes the yarn more even—an allowance of from one to six inches, according to the counts, being the rule, and, whilst this is being given, the carriage is made to recede with diminished speed. On the under side of the carriage are wheels in a horizontal position, round which are bands fixed at four points, at the limit of the stretch, in front and behind: this keeps the longest mules free from vibration, and parallel with the beam. Stops or buffers are placed at both extremes of the stretch, so that a positive stoppage takes place when the carriage reaches these points.

We have given the salient features in connection with an ordinary mule, but, if we had unlimited space at our command, life would be too short to enter into full details of the many and varied modern improvements at present in existence. The pessimist view—that the mule has seen its best days—is not yet within measurable distance, and more than this generation will have passed away before this form of spinning becomes obsolete.

Principal Points in Mule Spinning.—These may be stated as follows:—Rovings must have the proper amount of twist—we have shown, in former papers, how this proper twist is obtained; if there is not sufficient, the bobbins will not run in the creels, or the roving will be unduly stretched, causing not only trouble but waste; if there is too much twist it will not draw perfectly in the rollers, but will form a snarly, uneven thread; if the hank roving is too fine, the yarn will not have sufficient twist in it, and will be constantly breaking; when too coarse, the yarn will have too much twist and be full of snarls. The roller, creels, and all the movements for drawing, are the cause of a great deal of bad spinning, if not carefully watched and kept in order. If the creel pins become blunt at the bottom, they ought to be replaced, and the footsteps cleansed of all dirt and flowings, for if this is not attended to, the drag is increased, so that the roving becomes stretched until it breaks away from the creel. This may be considered of very little consequence, but it is the cause of much mischief, and is easily avoided, except by the very careless. When creeling, no portion of a roving ought to be left hanging about.

Roller Covering.—If a good, level yarn is to be obtained, the covering of the rollers must be kept in good order, and, to begin with, covered, in the first instance, with thin leather of the very best quality, soft and pliable: good roller coverings, without undue piecings or joinings, are of the utmost importance in successful spinning of fine yarns, but bad joinings will cut the rovings like a knife.

Rollers ought not to be worked too long before being changed. When the rollers are too dry, a coarse count is produced. Many spinners are quite indifferent about the state of their rollers, which is the cause, in many cases, of unpleasantness, as fault is found with the material, and many excuses are made, and it is only when the roller leather is thoroughly worn out that a change is consented to. This is really the poorest of all poor economy, both for the employer and the employed, as rollers ought to be taken out the moment a defect is observed. When we consider that the motion of these rollers is only obtained by friction, we must see that they will slip if in any way worn too much, and this slippage is further increased by dirt and want of proper oiling.

The roller traverse motion guides ought not to be allowed to wear too much nor to become dirty, as this is also a fruitful source of breakage, of stretching, and, consequently, of bad spinning. If the bottom rollers become strained, weak and stretched yarns are produced; if pinions are too deep in gear, or the teeth worn, or wheels loose, the yarn is cut. The draft ought to be in proper proportion—neither too much, nor too little, between any pair of rollers. In making out this draft for mules, the gain or loss in the carriage of the rollers is the first thing to take into consideration; if the carriage gains on the rollers, a less draft will be required in the rollers; if the rollers gain on the carriage, a greater draft will be required. In all coarse counts of short stapled-cotton, the rollers gain on the carriage: in fine counts from long-stapled cotton, the carriage gains on the rollers.

CHAPTER VII.

MEMORANDA FOR MULE SPINNING.

Length of Stretch.—The length of stretch is the number of inches the carriage runs out or puts in.

To find the draft between the rollers when there is a gain or draft in the carriage.—Example :—What draft is required to spin 36's from a 5-hank roving, the rollers to deliver 60 inches, length of stretch 63 inches = 5 × 63 = 315 divisor, 36 × 60 = 2160 dividend, 2160 ÷ 315 = 6·857 draft required. The rule, then, is multiply the hank roving by the length of stretch for divisor, and the counts by rollers' delivery for a dividend, and this will give the draft. When the gain is in the rollers of the carriage this rule will give the draft.

To find the draft between rollers when there is neither gain nor loss in the carriage.—Divide the counts by the hank roving, thus :—36 ÷ 5 = 7·2 is the draft required. Suppose counts spinning is 36 and draft in rollers is 6, what is the hank roving? 36 ÷ 6 = 6 hank roving, and if the hank roving is 6, the draft multiplied by it will give the counts = 6 × 6 = 36.

To find what the rollers deliver.—Multiply the hank roving, the draft between the rollers, and the length of stretch, for a dividend, and divide by the counts being spun, and the quotient will give the length delivered from the rollers.

To find counts spinning when carriage gains on the rollers.—The hank roving is 5, the draft in the rollers 6·857, stretch 63 inches, rollers deliver 60 inches, thus 5 × 6·857 × 63 for dividend ÷ by 60 = counts. The calculations are very simple, and are in proportion—either direct or inverse.

To find pinion wheel to produce any given counts from a hank roving.—If counts wanted be 40's, the hank roving 5, the front roller wheel 20 teeth, crown wheel 120 teeth, back roller wheel 54 teeth, diameter of back roller $\frac{7}{8}$ inch, diameter of front roller 1 inch or $\frac{8}{8}$, then what is the pinion? = 20 × 40 × 7 for divisor, 120 × 54 × 5 × 8 for dividend = 46-tooth pinion. Let it be thoroughly understood that if the carriage gains on the rollers, multiply the length turned out by the divisor and the length put up by the dividend.

To change from one count to another.—Multiply counts spinning by the pinion that is on for a dividend, and divide by the counts required, and the quotient will give pinion wheel for the change.

To find a back roller wheel for any draft.—Front roller wheel 20, crown wheel 120, pinion 40, diameter of front roller $\frac{8}{8}$ inch, diameter of back roller $\frac{7}{8}$ inch, what back roller wheel will give $9\frac{1}{1}$ of a draft? = $\frac{20 \times 40 \times 7 \times 9\frac{1}{4}}{120 \times 8}$ = 54 back roller wheel, and if the middle roller is required to draw from the back roller 8 into 9.

The wheel to put on the middle roller to give the proper draft is found as follows:—Multiply wheel on back roller by the diameter of the middle roller, and the figure 8 for dividend, and, for the divisor, the figure 9 multiplied by diameter of back roller, and the result is a 16 wheel.

Finding the Twist for a Mule Yarn.—The most important matter is to find the twist for a mule yarn, which, on account of the various qualities of cotton, requires some previous experience and consideration. The section of a thread is that of all cylinders being a circle, and the contents of circles are in proportion to their diameter. The thickness of a 36's thread is not, as many suppose, as 16 : 36, but as $\sqrt{16} : \sqrt{36}$, or as 4 to 6; a thread of 16's has double the diameter of a 64's; the latter has, however, with the same angle of the screw line, double the number of turns of this screw compared with the 16's.

The proportion of twist which is necessary for the different yarns requires the square root to be multiplied by some constant, in order to find the actual twist. Let us suppose the constant to be a, and its value for a certain twist to be 4; then we have the necessary twist for each number of the same class of yarn,—4 $\sqrt{n}$—in this formula n represents the number. For 16's we have 4 $\sqrt{16}$ = 4 × 4 = 16 turns per inch, while 64, in like manner, 4 $\sqrt{64}$ = 4 × 8 = 32 turns per inch, and so on—the smaller a really is, the larger the angle of the screw and the fewer turns per inch. Yarn may be unduly twisted, through want of proper treatment, until it is weakened and strained, and too little twist leaves no cohesion. Warp yarns have more twist than weft; for a, the figures are generally from 3·7 to 4·8, and for weft yarn, from 3 to 4; in high numbers, for which long-stapled cotton is used, the figures range from 3·2 to 3·75.

In order to obtain an even twist, the delivery of the rollers should, by all means, be in proportion to the revolutions of the spindle. Thus, if a delivery of 200 inches of yarn correspond with 3000 turns of the spindle, then 3000 ÷ 200 = 15 twists per inch. If the finished yarn is 16's, then this yarn would be spun with the figures 3·75, for a : $\sqrt{\frac{15}{16}} = \frac{15}{4} = 3·75$, so that the principle which we have given will be found to hold good and give positive results, for every count, when the quality of cotton to be used is duly considered. The want of a thoroughly sound system in spinning cotton yarn is the cause of such a discrepancy in weights and lengths of the same counts of yarn, that variations have been often found of one, two, and three numbers in one pound of the same class of yarns and counts. Of course, it is necessary to have more twist in the yarn made from a very short staple or from an inferior grade of cotton. If spinning from Surat cotton, the square root of the count, multiplied by 4, will be found a reliable guide; then, if we wish to spin 36's twist, the square root is 6 × 4 = 24 turns of the spindle for each inch of yarn.

The question might be asked, What difference will one turn of the front roller, put on slow motion, make in the yarn twist? The answer is, Find the number of revolutions the spindle would make for the front roller once, if on quick motion; then subtract the difference, which will be the extra turns of the spindle in the stretch; and this, divided by the length of the stretch put up, will give the extra turns per inch. Many queries crop up, from time to time, which require all the skill and experience of a good spinner to grapple with—we have indicated a few, and might go on *ad infinitum*.

Working Capacity of Mules.—It is very necessary that managers and spinners in every cotton mill should have some means of ascertaining whether or not mules are operated to the full extent of their capacity; as a guide, we give the following formula in as simple a manner as it is possible to convey it in writing:—Production of any number of yarns per spindle per hour, in decimals of a pound, at any spindle speed:—

Let SR = revolutions of spindle per minute.

N = number of yarn.

M = multiplier of square root of number, to determine the twists per inch of yarn.

Then we have $\sqrt{N} \times M$ = number of twists per inch, and $\dfrac{60\,SR}{\sqrt{N} \times M}$ = number of inches twisted per hour. The hank measures 840 yards × 36 inches = 30,240 inches. In a pound of any number of yarn, indicated by N, there are N hanks; hence, 30,240 × N = number of inches of yarn per pound; then we shall have

$$30{,}240\,N : 1\text{ lb.} :: \dfrac{60\,SR}{\sqrt{N} \times M} : \dfrac{60\,SR}{30{,}240 \times \sqrt{N} \times M} = \dfrac{SR}{504\,N \times \sqrt{N} \times M} =$$

decimal of 1 lb. per spindle per hour, of N yarn; for hosiery yarn, $M = 2\cdot 5$; for doubling, $M = 2\cdot 75$; for weft, $M = 3\cdot 25$; common mule twist, $M = 3\cdot 75$; extra, $M = 4$; super extra, $M = > 4$. We will, by way of example, give this formula, as it may apply, say, to 14's yarn; then the $\sqrt{14} = 3\cdot 741$ with $SR = 6000$ revolutions of spindle per minute, and $M = 4$ extra twist : $\dfrac{6000}{504 \times 14 \times 3\cdot 741 \times 4}$ = ·00568 lbs. of No. 14's per spindle per hour, and for 10 hours' efficient work, ·0568 lb., with $M = 6$ instead of 4, the production would be ·0378 lb. per spindle, the difference being ·0568 — ·0378 = ·019 lbs. per spindle, showing a loss daily, from 10,000 spindles, of 190 lbs. of 14's, from excessive twist, and by comparing with practical results in any mill, the heads of the spinning department are at once advised by this formula about the maximum output.

Mule Indicators.—The modern mule is now furnished with an indicator, the *bête noir* of both employer and operative, as one contends that it registers against him, the other that it is not for him in quantity. Two illustrations of this contrivance (Figs. 23 and 24) are given herewith. The fault of this contrivance, if fault there be, is not in the indicator, if placed in a proper position on the back shaft, where it is almost impossible to tamper with it. When it was first adopted in its present form, it was fixed on the cam-shaft, but the excessive vibration at every change of the cam caused it to be more or less out of order. As usually constructed, it is arranged to register 20,000 hanks, and, though a mule may, and does, spin 28,000, this creates no difficulty in the reading.

The arrangement of the inside wheels on the cam-shaft, for a mule of, say, 942 spindles, with a 64-inch draw, is $\frac{23 \times 56 \times 80 \times 942 \times 64}{10 \times 840 \times 36}$ = 20,542 hanks, and allowing $2\frac{1}{2}$ per cent. for breakage = 514, the total is 20,028 hanks. If the indicator is placed on the back shaft in a mule, with 900 spindles and 64-inch draw, then the arrangement is $\frac{3 \times 6 \times 20 \times 30 \times 900 \times 64}{840 \times 36}$ = 20,571 hanks, and the $2\frac{1}{2}$ per cent. allowance = 514, deducted, gives 20,057 hanks. Indicators are calculated to allow this $2\frac{1}{2}$ per cent. for breakage, but practical and experienced cotton spinners

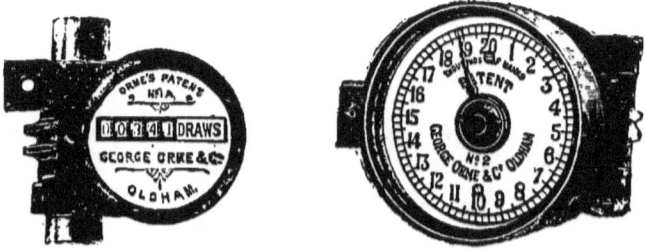

Figs. 23, 24.—Mule Indicators.

consider that 4 per cent. of an allowance in the construction of an indicator would be better, afterwards allowing the $2\frac{1}{2}$ per cent. reduction when arranging prices. The reason for this is stated very simply and to the point. Thus, the yarn wound on a cop is always tighter than on a wrap reel, and the indicated hanks are always greater than the calculated; hence the unpleasantness. The allowance of $2\frac{1}{2}$ per cent. is insufficient to make the difference; therefore, the disparity between the registered hanks, and the hanks calculated from weight and counts would be avoided, together with the grievance, by the 4 per cent. allowance. In any case, for all practical purposes, the weight spun may, with a little care, be obtained exact enough by dividing the total hanks by the average counts, and by timing the mules; the length each minder should turn off in a week may be determined, and a glance at the indicator will go far to prove whether this has been done or not.

Hints to Minders.—To sum up:—the minder ought, every morning, to oil spindle tops, shafts, and pulleys in the headstock, which run at the greatest speed and bear the most friction, and spindles should always be kept clean. Should the top rollers be those which have three ends to a boss, or six ends to each roller, these must be kept clean and oiled during the week, especially so where the shaft comes in contact with the saddle upon which the pressure is laid. The under iron rollers ought to be cleaned at least one-half per week—this is a work of some anxiety—in order to prevent slight accidents during the operation. End bands, jack bands, scroll bands, and rim bands require constant attention, and spindle bands must be kept in good order. A sharp look-out is necessary with regard to driving and counter-straps, upon which there is great pressure in large mules. In spinning ordinary coarse counts, at least eight doffings ought to be made per week, amounting to 17,280 cops.

Great care is needed in keeping all the moving parts of mules well lubricated and cool, or the high speeds will generate a great amount of electricity, particularly at the spindle ends, and this is dangerous in the charged atmosphere of a cotton mill, and, if precautions are neglected, may be the means of originating fire and serious inconvenience, if not destruction.

CHAPTER VIII.

THROSTLE AND RING SPINNING.

Throstle Spinning.—Throstle spinning differs from the mule, being continuous, and is never used for weft, and the warp counts rarely exceed 36's. This machine has changed but little in form or principle since the days of Arkwright's water frame.

If we bear in mind the true nature of the process of spinning, we can easily observe that all the beautiful machines invented within the century for spinning cotton are merely contrivances for effecting two objects—the elongating of the roving till it contains in thickness as many fibres as are necessary to produce required size of yarn, and the twisting of these fibres into a compact thread. The throstle, in appearance, resembles the fly-frame; the bobbins are, however, much smaller and more numerous, and the highest speed may be taken at 7000 revolutions per minute. The flyers build up the yarn at about the same rate as the front rollers deliver it, and the bobbin is dragged round by the flyer at its own speed, less the quantity of yarn wound up, whilst the yarn is kept tight by the bobbin being placed upon flannel washers, which tend to retard it, without breaking the thread. The strain of pulling the bobbin round prevents very fine counts being spun, and the building of the yarn is caused, as in the case of the fly-frame, by the rising and falling of the rail.

Throstle yarn is said to be the best that can be brought into the market, being rounder than either mule or ring yarn, whilst it tensions better, and the fibres are laid more in one direction, and it is not so oozy as other yarns, because it is calendered by lapping it three or four times round the fly leg. The spindles cannot, however, produce the same weight in proportion to the mule spindles, the speed depending very much on the quality of cotton and twist required. In practice, on the ordinary throstle, 40's can only be obtained from combed cotton with extra drawing, with second intermediate preparation, with a very light bobbin, and with the minimum amount of drag.

The throstle is too slow in speed for modern requirements; its power is excessive, and waste of oil and flannel out of proportion to the amount of production. All improvements tend towards an increase of the spindle speed, but it would be of no great interest to our readers to give a description of their merits, when we find it improved, and in its modified form, with the new name of ring frame, coming to the front with a rush; a description of which will suffice to terminate our remarks on the spinning process in cotton manufacturing.

Power-spinning machines consist of throstle spinning, using a flyer and bobbin; mule spinning, in which a certain length of sliver is given out by rollers, and stretched, spun, and wound on spindles; ring spinning, in which a spindle revolving rapidly turns a bobbin with it inside a ring having a rim or lip, around which a small loop of metal called a traveller turns and guides the yarn, as it is twisted from a sliver or roving, which is steadily stretched and given out by a series of rollers in the upper part of the machine. The frictional resistance of the traveller upon the ring causes it to turn more slowly than the spindle and bobbin, and, as a consequence, the yarn winds on the bobbin. The rate of winding is regulated by the weight or size of the traveller. The ring and traveller have a slow, up-and-down motion during this operation, which allows the thread to be wound evenly upon the bobbin. The spindle and bobbin must have a common centre with the ring, or a tight and slack effect will be the result at every revolution, which would prevent the equal twisting of the yarn and its regular winding.

Ring Spinning.—The ring-spinning frame has had almost a century of improvements. Amongst a large number of early ones the Booth-Sawyer was the first success, to be left behind by the Rabbeth spindle, capable of 8000 revolutions per minute in the production of the lower counts of yarn; but with high speeds the wear and tear was great, owing to imperfectly balanced bobbins. This, however, was obviated by the introduction of the gravity spindle, which allowed an unbalanced bobbin to find its own centre;

nevertheless, high speeds had a tendency to cause the bobbins to jump or rise off their seat on the spindle.

To overcome this difficulty has engaged the attention of many inventors, and we believe the problem has been solved by the construction of a flanged bobbin, with a projecting foot at the base, which enters a circular cup. It would be almost impossible to follow, step by step, the many devices that have been, and are now, taking place for the perfecting of ring spinning.

The great difficulty in spinning weft on this machine is that it is required to be spun softer than twist. A lighter spindle is being made, and the rings are more accurate, and, in many cases, along with guides, are made from porcelain, which is always clean and smooth, does not corrode, is harder and stronger than glass, and will outlast hardened steel. In contrasting the two systems of spinning, the rollers of a ring frame are always running and giving out yarn, whilst, in the mule, they have to stop when backing off and winding the yarn on the spindles. The ring frame will not be as liable to cut the yarn as the mule, where the rollers have to stop and start every 64 inches of yarn, with a lot of backlash, particularly when the rollers are geared up at the out ends of the mule. Again, the ring takes less power, banding, and strapping; there is much less breakage, and less cost for attendance or operating.

All persons connected with the cotton trade know that there is a fierce contention as to the merits of the mule *versus* the ring, and, in dealing impartially with this vexed question, there seems to us good reason to believe that, in spinning coarse counts or twist, the ring-frame will equal, if it does not actually advance beyond, the mule. As the counts rise above 24's the mule is the favourite, because it beats the ring in speed; moreover, if there were positive marked advantages in the ring principle for 32's twist, spinners in Oldham and elsewhere would be most ready to take the advantage. Experts say that 32's ring twist, of a quality equal to the mule, requires a finer roving and a more expensive cotton; and allowing 8000 spindle turns in the ring-frame against 10,000 of the mule, then we can see at once where the balance lies. The 32's Oldham mule twist is quite good enough for ordinary printing cloths and other fabrics of that kind, and to use a more expensive yarn would be to throw money away.

We give two illustrations of the ring frame (Figs. 25, 26), which

Fig. 25.—Ring Spinning Frame.

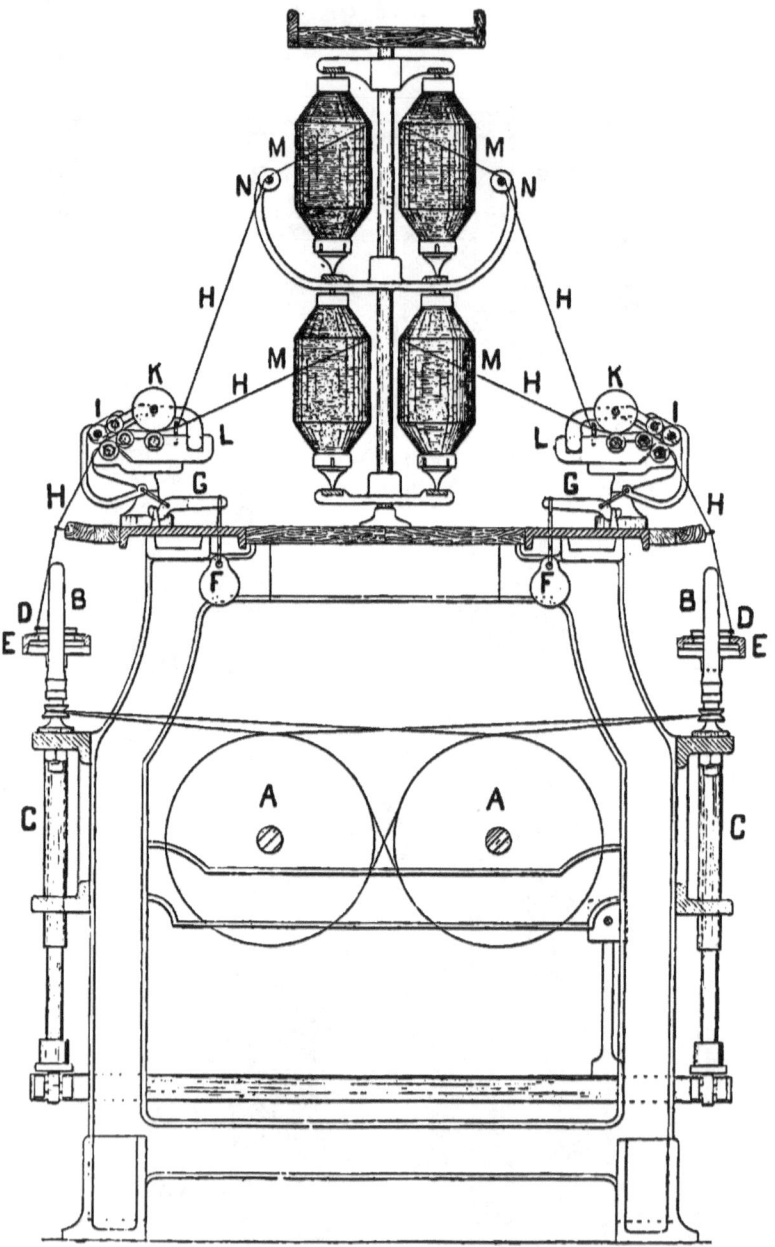

Fig. 26.—Ring Spinning Frame.

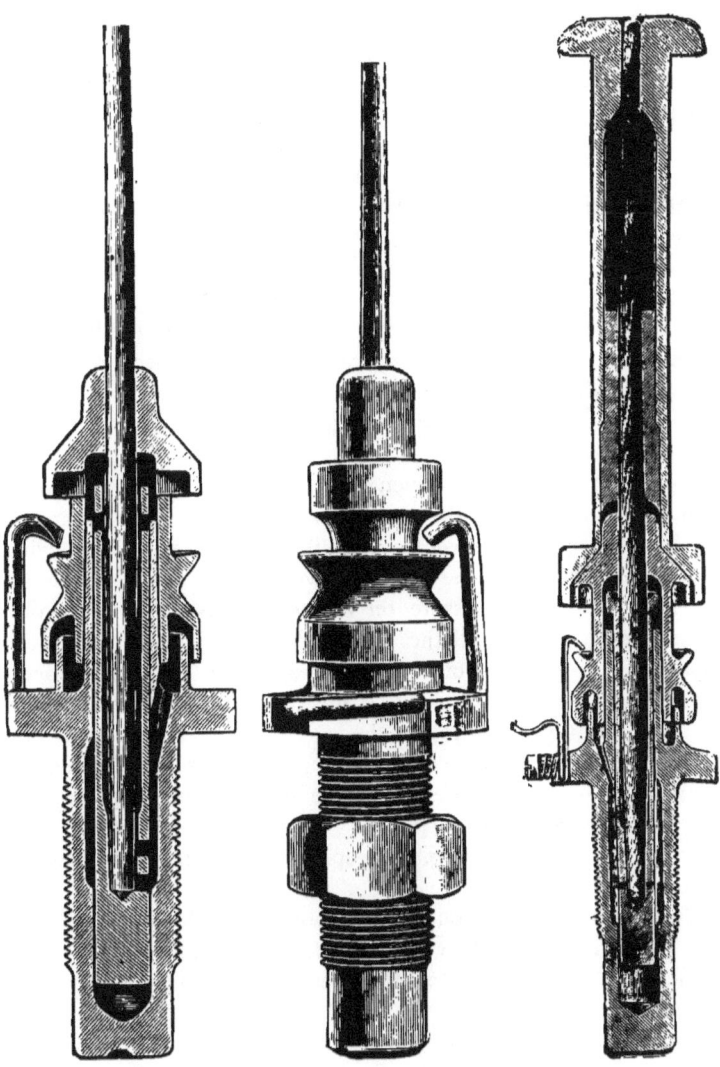

Figs. 27, 28, and 29.—Spindles.

show all the latest improvements. The makers of this machine are noted for their clever inventions in the cotton-spinning industry. The difficulties of constructing an efficient ring-spinning frame are well known to practical spinners. A great amount of time and expense has been devoted to bring ring spinning to some degree of perfection. In this frame—which has the "Within" improved gravity spindle with solid steel rings—steadiness of motion, with no perceptible vibration, is obtained. Further, one decided advantage for weft spinning is that a greater angle can be produced by the special arrangement of the rollers, the back and middle one being almost on a plane, whilst the front in its relation to the middle roller gives an angle of nearly 18 degrees. The full benefit of the horizontal and inclined stand are in this way gained in a most effectual manner, and the spindles run at a more than ordinary speed, with satisfactory results. A very few details will suffice to thoroughly understand the sectional illustrations.

A, A are the tin rollers driven by bands, communicating motion to the spindles, B, B, and the cops formed upon them. C, C, are the spindle rails, D, D, the travellers upon the rings, E, E. The weights, F, F, are attached to the stirrup levers, G, G, which cause the necessary weight or pressure on the top rollers, I, I. The roller clearers, K, K, remove all the fly and dust from accumulating and impeding the progress of the travellers, D, D. The stand containing the rollers is shown at L, L, the roving bobbins at M, M, with the yarn, H, H, going through the guides, N, N. Such is a brief description of a very near attempt at supplanting mule spinning.

Spindles.—Of these we give three illustrations. Fig. 28 shows a perspective view of a spindle; Fig. 27, a view of the same spindle in section; and Fig. 29, a spindle, in section, with a bobbin upon it.

Conditioning Bobbins.—The bobbins here illustrated, (Figs. 30 and 31) undoubtedly show the best form of bobbins in existence, as they may be subjected to either steaming or damping without the slightest injury resulting.

Ballooning.—Considering the great competition in cotton spinning, two points must be considered—quality and quantity. In ring spinning the most vital part of the machine is the ring, whilst the spindle is also an object demanding great care and attention, owing to the high speed at which it runs.

THROSTLE AND RING SPINNING.

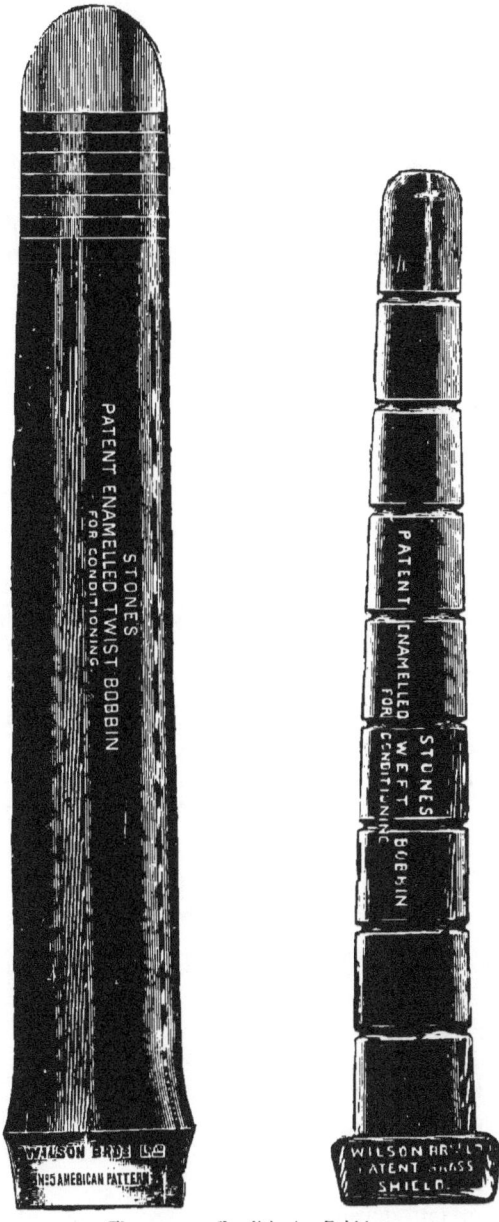

Figs. 30, 31.—Conditioning Bobbins.

The ring throstle is apparently a very simple machine, yet many improvements have been found necessary to remove defects, one of the most essential being some device for dealing with the tendency of the threads to *balloon*, caused by high speeds, the centrifugal motion throwing them outward, in the form of an "inverted balloon," and hence the name. At the lower part of each set, when the lifting rail is removed farthest from the guide wires, through the loop of which the threads pass after leaving the rollers, these threads fly or bulge out and are carried round, and the balloons, in like

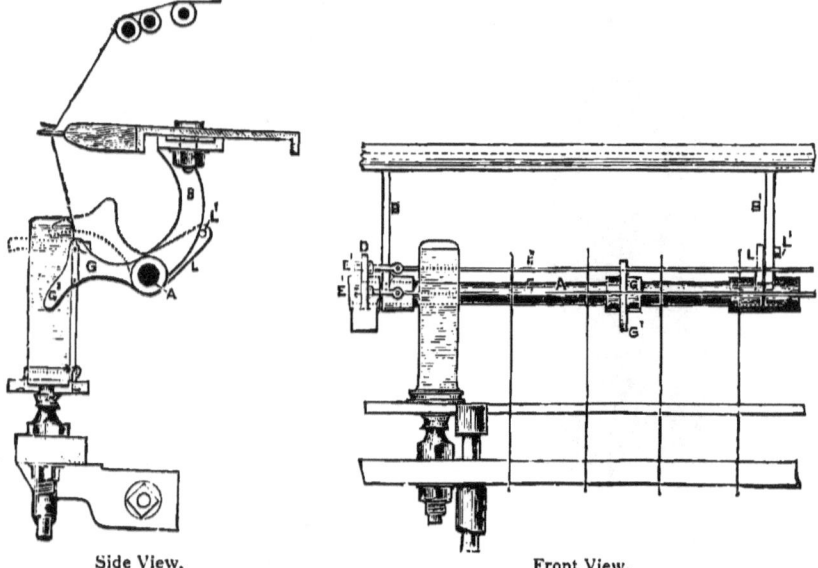

Side View.　　　　　　　　　　　　　　　Front View.
Figs. 32, 33.—Anti-Ballooning Appliance.

manner, formed by adjacent threads, come into contact with each other, and one, or more, are broken down.

Two ideas have been mooted for the removal of this defect—viz., reducing the speed of the spindles, which would, of course, reduce the production, and using a heavier traveller, which causes an additional strain on the yarn. The remedy really required should be free from these drawbacks, and should in itself be simple and efficacious. An apparatus which (it is believed) possesses these essentials has been introduced to the trade, by a gentleman who is at the head of one of the largest firms in the cotton industry, and is thus described:—Fig. 32 is a cross section of a portion of a ring frame, the invention being shown and lettered A to L. Fig. 33 shows a portion of the front of a ring frame. From these illustrations it will be noticed that a short shaft, A, is supported from the roller beam by brackets, B and B 1. On each end of the shaft, A, there is a lever marked D. This lever has two holes for the admission of bolts, E and E 1. The wires, F and F 1, are fixed to these bolts, the former extending quite across the frame, where they are fixed in precisely the same manner. There is a finger, G, on the shaft, A, which is made with two holes, through which the wires pass. The finger, G, is formed with a projecting tail, extending over the ring rail. In order to perfectly support the wires, these fingers are arranged at regular intervals across the frame, and, so that they may be easily adjustable, they are attached to spring pieces of the frame by studs, mounted loosely in brackets. Upon the shaft, A, there is another lever, L, having a projecting peg, L 1, which, coming into contact with the bracket, B 1, holds the finger, G, upon the shaft, exactly in the same position as that of the wires. The wires being arranged midway between the ring rail and the thread guides, the ballooning of the yarn at the lower stages of the winding is prevented, and, as the tendency to balloon diminishes through the ring rising, the wires are turned back and are, therefore, out of the way of the spindles. The motion is repeated each time the ring rail rises and falls, and thus the ballooning of yarn is effectually prevented.

CHAPTER IX.

WINDING AND REELING.

The Winding Frame.—The winding and reeling of the yarns from mule cops and throstle bobbins is the simplest of all the processes in cotton manufacturing. In ordinary single-thread winding, a very simple machine, with the commonest care on the part of the attendant, amply suffices for all requirements. The correct reeling of yarns has yielded to the efforts of inventors by the adoption of a suitable stop-motion, which is now applied to all reeling machines where economy is desirable. The yarn is arranged in the most convenient way, and each thread is passed through a thread wire, over the guide, and upon the reel. On the breaking of a thread, or the finish of a cop, the drop-wire, which is held in position by the tension of the thread, falls down and checks the movement of the rocking-bar plate, carried upon the extremity of levers, which extend the entire length of the swift. The stop-motion is simple and not liable to get out of order, and, however careless the reeler may be, the reel is stopped instantly: it can be driven by hand or by power, and is adapted to reel from cops or bobbins.

Reeling is a process in preparing yarn for bleaching, dyeing, printing, or bundling. In the old machines, which were very much smaller than the present ones in use, the hanks (each 840 yards in length = 560 threads 7 leas, the circumference of the reel being 54 inches or one thread), when finished, were doffed by lifting one end of the swift out of its bearings, and passing them over the end, thus involving loss of time and causing oil stains. To obviate this, the wheel doffing motion was invented. It was improved by "gate" doffing, and since improved by the "bridge" doffing arrangement. The end of the frame at which the doffing takes place is open, and

WINDING AND REELING.

Fig. 34.—Shuttle Bobbin Winder.

carries two standards—the back one supporting the reel swift, when in its working position.

When the doffing takes place, the swift is closed and the hanks are gathered to the end and placed in a recess, when, by a slight push, the end of the swift shaft is shifted to the front standard, by means of the "bridge," and the hanks are then lifted away. To recommence the work, the swift is slided to its original position with the greatest ease. The end of the swift is encased in a sleeve containing the oil, and, therefore, any risk of oil stains is obviated. Further description of this very simple process is unnecessary, single-thread winding being the next step.

Many types of machines are used for this purpose, but the details which we give of one (Fig. 34) will convey sufficient information of the process of winding on bobbins for warping purposes. The spindles are arranged horizontally, and extend across the frame, having their wharve in the middle, and driving four bobbins—two on each side of the machine—and all the length of the spindle is used (the old machines would only allow the upper half). With these improved spindles, each will hold four 5-inch bobbins, driven by a tin cylinder extending the length of the frame. Each has four friction rollers, one for each bobbin. The yarn is taken up at a uniform speed, and the bobbins are supplied with a very light steel thread guide, a little over 5 inches in length, and pivoted $1\frac{1}{4}$ inch from its bottom end. The guide is really a double lever, the short end of which, being attached to, and actuated by, a cam arrangement, has a movement of about an inch, which causes the long end to make a 5-inch movement, corresponding in its proportion to the small end of the lever which moves it. When a thread breaks or a spool has been emptied, and it is required to renew the connection with the end of the thread upon the bobbin, the latter is lifted out and placed in the bracket, when the end having been found and pieced, it is replaced. One girl can attend to a considerable number.

The advantages of this almost automatic machine will be obvious to manufacturers, as frequent breakage of bobbins is dispensed with, Ring-frame yarns need one process less than when delivered in the cop or hank, and the bobbins are wound firm, are not liable to ruffle, and can be used in any warping or beaming machine. In the form of these bobbins, the yarn can, if required, be bleached or dyed, because the cross winding will readily admit liquid penetration to the

Fig. 35.—Conical Drum Winding Machine.

centre. Winding for shuttle bobbins is another purely mechanical process.

Conical Drum Winding.—The method of building the cone of the bobbin most commonly employed is by winding the yarn upon it, whilst it is placed in a metal cup or hollow cone. This arrangement is, no doubt, far before the old system of the presser frame, which is nearly obsolete; but the bobbin being made to revolve in a stationary cup, a great amount of friction is created, which is very detrimental when harsh dyes are wound, as black, blue, orange, etc., as the cup becomes too hot and the yarn is stained with splashes of oil.

To facilitate the winding, tallow is applied to the material, which does not improve the yarn, but, on the contrary, spoils it. The friction in the cup shades very delicate colours, glazes the yarn, and chafes it. There are machines made to obviate this friction, in which there are no cups nor stationary parts. The driving is by contact with a revolving disc of polished iron, bevelled at the outer edge to the angle of the cone of the bobbin. The size of the disc is so adjusted that, in revolving, it passes through equal spaces with the revolving bobbin in contact with it on all parts of the cone; therefore, no friction can occur, no grease is required however harsh the dye, glazing is avoided, more tender yarn can be wound, and no spindle banding is needed, which, becoming slack or broken, is a source of annoyance and loss in many machines at present in use. Jerks and irregularities are also prevented, and the uniform tension builds the bobbin firmly at the point and prevents slubbing, and the high speed of sixty revolutions of the hank per minute is obtained.

The differential motion is simple, and consists of a leather friction driving-bowl, made to slide to and from the centre of a driven friction plate or disc, the motion being so arranged that the driving-bowl is nearest the centre of the disc when the yarn is being wound on the point of the bobbin. The ends can be pieced without removing the spindle, and instantaneous throwing off when bobbins are full is assured. Tin drums and their frequent repairing is dispensed with, and less power in driving is required. The machine here illustrated (Fig. 35) is one out of many improvements which claim attention for efficiency and economy, and is deserving of notice.

Another form of winding cotton yarns, called "doubling," required, on the part of the operative, considerable care and attention, before the introduction of the present improved winding and doubling machinery. The process is a very important branch of the cotton industry, and requires from us more than a few passing remarks. A given number of threads are twisted together, called two-fold, three-fold, and so on, but the proper amount of twist is the principal factor. A few examples will furnish a guide, so that waste may be prevented. If a 2/50's is required, two single threads, each a little finer in counts than 50's, would be needed to allow for the take-up of the twist, say 51's, to be exact; then the two threads combined would be 25's. The square root of this is $5 \times 3.75 = 18.75$, or $18\frac{3}{4}$ turns per inch; but if a softer yarn is required without curling, the multiplier would be 3.25. In many cases, where a good deal of curling is needed, the multiplier may be 4. Again, if three strands of this 2/50's were put together, the twist for the six-fold would be as follows :—The three together would be 25's divided by $3 = 8.33$; the square root of this $= 2.9$ nearly, $\times 3.75 = 10.875$ or, say $10\frac{3}{4}$ turns per inch; therefore, a general rule is that the square root of the number formed by all the threads combined, multiplied by 3.75, will give turns per inch of twist. For hosiery yarns, multiply by 2.25; yarns for embroidery—four-folds—and crochet cotton, multiply by 2. We might give more calculations, but the above rule will provide for all possible combinations.

Doubling Winding.—Doubling is performed on three distinct machines, although they can only be considered as spinning modifications. They are the throstle doubling-frame, the twiner—a sort of mule—often made from a worn-out mule; and the third is the ring doubling-frame. To describe these in detail would require several pages, but we can give an ample explanation of the process by the particulars of the Doubling-Winding Machine (Figs. 36 and 37). We take this only as an illustration of the operation, as there are other and equally good machines in work, which give every satisfaction.

Machinery for winding several threads of yarn together, twisted or open, to form one doubled strand, enables doublers to place yarns upon the market as nearly as possible free from " singles "—that is, places of one strand only. This perfection is obtained by improvements in the " detector " or " stop motion," one of the most important adjuncts to this class of machines; in fact, owing to its

Fig. 36.—Doubling-Winding Frame.

WINDING AND REELING. 89

adoption, " singles " may reasonably be considered an impossibility. The illustration (Fig. 36) represents, in perspective, one of these machines. The stop motion is simple and not liable to get out of order. This is a primary consideration, without which, however well

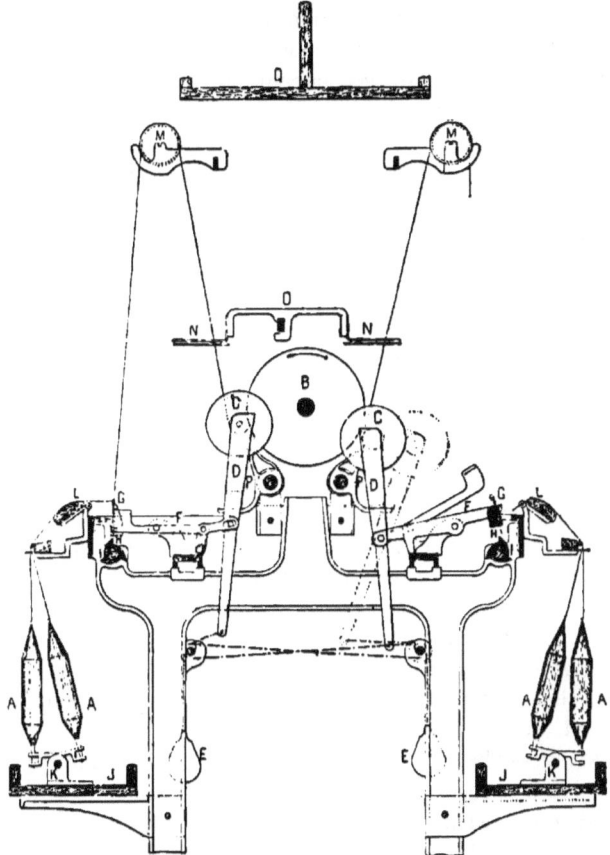

Fig. 37.—Doubling-Winding Frame.

the other parts of the machine might be constructed, yarns would be imperfect and of little value.

By reference to the transverse sectional illustration (Fig. 37), it will be seen that A, A are the cops, B, a drum, of which there is a series extending the length of the frame; each drum drives two bobbins, C, C. The traverse motion for actuating the guide wires is the same as is usual in other machines, motion being imparted by a mangle wheel fixed at the end of the frame, D, D, cradles holding the bobbins, C, C. The most important feature in this winding and doubling machine is the stop motion, which is made so sensitive and quick in action that, though the bobbin may be winding yarn at the rate of 5000 or more inches per minute, the broken thread is arrested 18 inches from the bobbin. E, E are weights attached by chains to the lower extremities of the bobbin holders, D, D; connected to these holders are the frames, F, F, containing a number of detector wires equal to the number of threads required to be wound. G is the top and H the bottom end of the wires, and I, I are cams or wipers with three projecting teeth, as seen in the drawing, revolving at a great speed. The cops, A, A, are fixed to the boxes, J, J, by the brackets, K, K. The cop threads pass upwards over a rail, L, L, which preserves the exact tension by means of a covering of loose woollen cloth or flannel. The threads then pass through the eyes of the detector wires, G, G, and thence over the rollers, M, M, through guiding wires, N, N, on the bracket, O, to the bobbins, C, C. Whilst the operation of winding is proceeding, the bottom end, H, of the detector wire is free from contact with the wipers, I, I, and when an end breaks down or becomes slack, the lower end of the wire, H, comes into contact with the wiper, I, and the bobbins, C, C, are at once stopped in their revolutions by the brakes, P, P. A box, Q, is for the purpose of holding the finished bobbins. Every action is simultaneous, reacting upon each other in every part without complication. The left hand of illustration shows the parts in working position, and the right hand, when a breakage has occurred.

Twisting.—The twisting machine is similar in general appearance and size to a throstle or ring frame, but the spindles at the front have two sets of wharves, one set a little above the spindle feet, and the other near to the lifting rail when it is at the bottom.

These spindles are driven by bands from the driving drum by the lower wharves, at 4000 revolutions per minute, but the higher wharves, larger in diameter, drive spindles situated behind the rollers by wharves of the same size and at the same level. On these spindles are placed the bobbins of doubled yarn from the winding machine, and they are caused to revolve at the same speed as the spindles, and are twisted at the rate of 4000 turns per minute. The inner part of the bobbin head is round in shape for a depth sufficient to allow the threads to pass easily, and the eyes above, through which they are drawn, are fixed at a proper height for the same purpose. Descending from the eye, the thread, half twisted, is drawn by the rollers through a trough of water, and the twisting is completed by the front spindles. As in the ordinary plan of twisting, the whole of the work is performed at the front of the rollers. The saving that results by increased production, by the decreased cost of labour, and by economy in waste, etc., is increased one-fourth, and the quality of the work is undeniable. After many tests, the strength of the twist is proved to be greatly superior to the old system of doubling.

In the ordinary machine, the distance from the roller to the flyer being small and the amount of twist large, thin places are often formed; but, in this machine, the longer distance from the creel to the rollers allows the threads to be more evenly spread or laid, and the fact of putting in twist at both ends makes it more regular. Counts from 150's down to heavy carpet yarns, with cops half-a-pound in weight, can be produced. Old frames can be altered with lifts from 2 to 8 inches. In the case of ring frames, the speed of the front rollers can be easily increased. A solid, wiry, and glossy thread is regarded as the best result or test of a good doubling process. It would be difficult to demonstrate the advantage secured by the adoption of any one of the present improved doubling machines as compared with another. It may safely be said that they all have some special feature rendering them worthy of consideration.

Gassing Yarns.—Fine cotton yarns used for lace sewing threads and hosiery are generally singed, by passing through gas jets, to clear them from loose fibres, by which means they acquire a rounder form, and by the decreased weight of the material, become finer in the counts to the extent of 5 per cent.

This process, like all others in cotton manufacturing, has been the subject of many patented improvements, too numerous to be mentioned here, but it may be said to consist, simply, of winding with a series of gas jets, through which the threads are passed several times, with a velocity in proportion to the quality of the material. As we have already mentioned (p. 89), in referring to winding machinery, there is a contrivance for arresting the bobbins, whenever a knot or

Fig. 38.—Gassing Frame.

lump occurs in the thread, so as to prevent the flame from burning the thread. The gas flame is, by this means, turned quickly on one side, until the defect is repaired. Several upright tubes are connected by joints with a small stop tap screwed into two main gas-pipes, which extend the length of the machine, and terminate in the larger supply pipe. A gassing-frame is shown in Fig. 38. The work requires only a moderate degree of attention, and the machinery in its operation is as near perfection as human ingenuity can devise.

Waste in Cotton Manufacture.—From the first process of mixing the raw cotton material, to its final production in finished yarns, a great amount of visible and invisible waste is made, but it is almost impossible to give the exact loss, though, in many cases, as much as 20 per cent. occurs. There are many causes for this serious loss, some preventible—such as too much twist in the spun yarns, winders' waste, sweepings not thoroughly picked or riddled, etc., etc. In many blowing rooms the loss could be easily avoided by the adoption of very simple measures. Carelessness and a want of thorough supervision may be considered the main factors for waste throughout many establishments. Whoever may be in charge ought to enforce care and economy in the use of all materials, and should detect as well as rectify any fault that may occur, for at no period in the history of cotton manufacturing has the trade had greater difficulties to contend with than at present, as the growth of competition at home and abroad, together with the impediments in the way of retaining or creating new markets for our productions, has reduced profits to so low a margin that the utmost caution is needed on the part of every responsible person in this industry.

CHAPTER X.

COTTON BLEACHING.

Use of Bleaching Powder.—We now proceed to the bleaching of cotton, which is a very different operation from that employed some decades ago, when chemists were endeavouring to discover a system better adapted for the process than exposure to the open air of a bleach ground. The introduction of bleaching powder paved the way for the present effective methods. This powder, which is now well known to be one of the most common and important chemical elements—chlorine, which in combination with sodium gives common salt (chloride of sodium)—is one of the most plentiful and widespread substances in nature. A great economy of time and labour was effected by its use, and, in one form or another, it is still a reliable agent in bleaching cotton material. Raw cotton imbibes a certain amount of grease and dirt, which are added to by the process it undergoes, up to the spinning of it into yarns and the final consummation into cloth. It is, therefore, of the utmost importance to remove these impurities, especially the grease, before the actual bleaching operations take place. As it is not worth while to recount the numerous experiments and so-called improvements which have claimed attention from time to time but vanished as quickly as they came, we will confine our attention to those methods which are not only satisfactory and economical, but are likely to retain a precedence, and, therefore, to make the process of bleaching as complete in description as possible, without any after-reference. We will state the best methods for the treatment of yarns and woven cloths.

Washing.—The mere dirt is easily removed by washing—this is the first step; but grease, oil, and other stains are not easily removed. The simplest way, in cases of this kind, is to boil the material in lime-water or any alkali, such as potash, soda, or ammonia. The soap which is thus formed can be easily separated, and the material is quite ready for bleaching. Of course, repeated washings for extreme cases of soiled or stained goods will be necessary, as well as, between each washing, a thorough rinsing in pure water. The water purification is of the greatest importance, and the more rinsings the materials undergo the greater the economy in time and labour in the further operations. The next process may be bringing the goods under the influence of chlorine as a gas or dissolved in water, or using the chlorine in combination with calcium, the metallic base of lime, lime being its oxide. By the use of the gas the operation is at once effective and rapid, but it entails the risk of burning the fibre, or what is known as "rotting" the goods. By the combination with calcium there is less risk, because it can be regulated at will, if the strength of the chloride lime solution and the temperature and time employed are properly adjusted. The colouring matter being the first point of resistance, when this has been sufficiently operated upon, the materials ought to be removed at once. The quality, quantity, and character of the goods must determine the strength of the bath. This requires experience and good sound judgment. The chloride of lime is generally absorbed by the material in the bath in something like a period of six hours ; thence passed into a very weak bath of sulphuric acid, which combines with the calcium, leaving the chlorine a free agent to act upon the colouring matter.

Steeping.—A steeping for a day or night removes the sulphate, and the decomposed colouring matter is boiled out by carbonate of soda. To obviate any trace of yellow tinge the material is generally passed through a weak acid bath, and the boiling with caustic or soda carbonate may require a day or, perhaps, only eight hours, according to circumstances and quantity. The whole process does not exceed two days. If the goods are to be dyed or printed, it would be prudent to allow more time rather than use stronger solutions, as the colours are, to a certain extent, impoverished in brilliancy by undue haste. These details are general. Different well-known bleaching establishments vary their processes according to the class of goods in hand. The details followed by one of the largest firms will be found not only useful but exhaustive.

Bleaching Kier.—The improved kier much used at the present time consists of a steam-tight vessel of a cylindrical shape, made of cast-iron, heated by steam at a low pressure (see Fig. 39). The goods are placed on a grating, and a vertical pipe, with a cap at its upper opening, causes a distribution of the lime liquor. An iron pipe, in connection with the kier below the grating, has two branches, each possessing a valve, one supplying water, the other steam. The

Fig. 39.—Bleaching Kier.

mouth of the kier is flanged, so that the cover can be bolted down and thus made steam-tight. One ton and a half of goods can be operated on, the process occupying from ten to fifteen hours. For this weight of material 2 cwts. of lime, previously dissolved in water, and used as milk of lime, is necessary. This is poured upon the goods as they enter the kier, and a quantity of water added to cover the material.

The high-pressure steam process is also used, and is said to be more energetic, and the goods have a more complete impregnation of the liquor; but constant attention is required, and, to ensure success in the management of the kiers, they can only be entrusted to the care of a very competent operative. No such extra care is needed in the low-pressure treatment.

Souring.—Souring consists in passing the goods through a weak solution of hydrochloric acid, and then through water, to remove lime and insoluble lime soap, together with any metallic oxides that may be attached to the fibres. A few hours' steeping allows the acid to thoroughly decompose all impurities. The soda ash process is performed in the same kier as described. A certain quantity of resin soap, along with the soda ash, is placed at the bottom of the kier, and is soon dissolved by the water, about 1·5 cwt. of soda ash and 80 lbs. of resin soap being allowed for 1·5 ton of cloth. By this boiling all fatty acids are converted into soluble soda soaps, which are easily removed by washing. The cloth, having passed these operations, is cleansed from nearly all impurities, but is wanting in the requisite whiteness. To destroy the least trace of colouring matter the pieces are passed through a very dilute solution of bleaching powder at specific gravity 1·0025, or $\frac{1}{2}°$ Twaddle, till completely saturated with the liquor; then passed through squeezing rollers to remove the excess of solution, and afterwards allowed to lie for some hours. This process is commonly known as "chemiking." The bleaching powder is a mixture of hypochlorite and chloride of calcium. The second passing of the goods is through a weak acid bath —sulphuric acid, gravity 1·075, or, as before alluded to, 1·5° Twaddle.

The Final Operations in Bleaching.—The final operations are washing and drying, this last being performed by passing the pieces, opened out to their full extent, over and under a series of hollow iron or copper cylinders, heated by steam, the greater portion of the water being previously removed by squeezers. A vertical drying machine is illustrated in Fig. 40. Cotton yarns are bleached by the same process as cloth—the skeins being first boiled in water, then in a weak solution of bleaching powder, then again in weak acid, and finally washed and dried.

The Twaddle Test in Bleaching.—Sulphuric acid is not equal in action as a lime scour to hydrochloric acid, which is very soluble, and is thoroughly removed from the goods by washing. The sulphuric acid forms with lime insoluble sulphate of calcium, which is very difficult to remove entirely. In all cases the acid should be of a uniform strength to give satisfactory results. The Twaddle test is no guide of the strength, as the lime dissolved by the acid raises the Twaddle, having reduced the strength of the acid.

The following test is best, and most easily carried out:—Take a solution of caustic soda, known as 77 per cent. Newcastle (this make of soda to be always used as a standard at a strength of 20° Twaddle); then a narrow and tall white glass bottle, with a mark at the top, may be filled with this solution. A five-ounce measure of freshly prepared souring liquor, at 2° Twaddle, is poured into a jar, and some of the soda solution carefully added to it, until a portion of cloth dyed with turmeric is turned brown when dipped in; this shows that the acid is neutral, and the quantity below the mark on the bottle determines how much has been required. In every test of the liquor, five ounces must always take the same quantity of the test solution—if less, it is too weak; if more, too strong. The bottle would be better graduated 1°, 2°, etc., and also for 2° of Twaddle.

The Madder or Turkey Bleach.—Many bleachers use caustic soda instead of soda ash: 120 pounds of caustic soda will suffice for five tons of cloth. The details which we have given are simple, and the different operations are the most perfect kind of bleach at present applied to cotton cloths. It is known as the madder bleach. The Turkey red bleach is identical, so far as the operations are concerned, with those already described, and so is the market bleach, though every bleacher differs more or less in a few details; but the principle upon which the whole process depends remains, so far, unchanged. Water, in some instances, may be hard, requiring more acid, alkali, and resin; the steam pressure will vary, and the strength of an alkali will cause a difference in quantity. On these points it is impossible to give general rules, the nature of the conditions being so variable. We can only afford a brief glance at a few new systems now engaging the attention of bleachers and manufacturers. It is remarkable how the same operations are re-discovered again and again, and old specifications of patents rehearsed. Caustic soda for cotton was patented forty years ago, and we could give scores of instances, in every branch of the textile industries, where "something new," oftener than not, turns out to be as "old as the hills."

Bleaching by Peroxide of Hydrogen.—The bleaching of cotton by hydrogen peroxide, or, in other words, a highly inflammable gas, with a powerful oxide, containing the greatest possible quantity of oxygen, is supposed to be the most efficient of all agents. We are told that its superiority cannot be explained so far, but it is advisable, prior to using this dangerous compound, to treat the cotton with a weak acid, in order to remove the metallic salts and oxides from the

Fig. 40.—Vertical Drying Machine.

raw tissue. We have already shown that this is done by the ordinary process of bleaching, and further details give the following profound summing up:—" Bleaching cotton with hydrogen peroxide, the latter acts upon the cellulose, carbonic acid being always formed, the fatty bodies, remaining upon the fibre, as oleates, palmitates, etc., of magnesia, with an after treatment of alkaline lye." Very little information for practical purposes is conveyed in these statements, which are vague, and almost unintelligible. The process may be suitable for wool.

Bleaching by Electricity.—Bleaching by electricity and with a new bleaching powder called "fluor acid" may be passed over, as no satisfactory developed results, economical and really practicable on a large scale, have been obtained.

Recipe for Bleaching Samples.—We may more pertinently end our remarks by giving a useful recipe for the bleaching of samples, such as a few hanks, short pattern warps, etc., that are required quickly and with brief notice. Boil the twist well in water, with two ounces of soda ash to the gallon, and mix one pound of chloride of lime, fresh, in two pints of water, crushing all the lumps; add 43 pints more of water, allowing time for lime to settle; pour off the clear liquor, and immerse the yarn for ten hours in a cool place; keep from contact with iron. Wring out, and wash in cold water; the yarn must not remain in the air too long. Then steep in a well-mixed solution of 26 drachms of double oil of vitriol to 45 pints of water. The yarn must remain in this acid solution ten hours, or during the night; wring out, wash off in cold water. In order to thoroughly remove the acid, work well in a good white soap bath, and add a little marine blue to give the yarn any desired tint. Finally, wash in warm water to clear out the soap. These proportions will do the least injury to the fibre strength. If time is to be shorter, a stronger solution can be used. For soft mule yarns, the solution may be one-third weaker, but for double yarns, the strength must be increased according to perfection required in whiteness.

CHAPTER XI.

SIZING OF COTTON.

Economics of Sizing.—A great deal has been said and written on the question whether sizing is in any way lucrative, but there is no doubt as to its forming a very important branch of the cotton industry; the necessities of our trade and the call for sized cloths are imperative. It will not be disputed that unsized cotton warps demand the very best quality of raw material, and, further, a very large amount of twist; this latter condition is limited in practical application, because the twist has everything to do with the fineness and softness of a fabric. Sizing by manual labour involves a great amount of time, is far from beneficial to the weaver on piece-work, and, independent of this consideration, there are many defects which cannot be removed by the most careful attention; *e.g.*, the warper is unable to obtain an equal tension when the warp is divided into many parts, causing some portions to be longer than others. These defects have always been a source of vexation and trouble to manufacturers, because in the finishing process they become only too prominent and past remedy.

Again, it is of the greatest importance that weavers shall have warps so prepared that the fibres will withstand all mechanical strain during the process of fabrication. The solution of this problem has been attended with more or less satisfactory results. In saying this, we mean that results have not come up to expectations, so far as durability and evenness of the warp threads are concerned. Great efforts have been made to invent a sizing material and a process sufficiently effective and economical for all purposes, and without costly mechanical substitutes, the profitable use of which is restricted to very large manufacturing concerns. In practice it is well known that the sizing of yarns in skeins, or by the pound, is insufficient without resorting to after sizing, if durable and pliant warps must be produced. The various methods at present in use have the common defect, unless great caution is used, that the yarns in drying will form a mass or stick together, and the remedy for a mishap of this nature removes a quantity of the size, so that what is gained on one side is lost on the other.

A few years ago a machine was introduced, which, in spite of the greater consistency of the size used, brushed the yarn threads loose and smooth after drying. The solution of the problem seemed to be reached, and the question of preparing yarns in skeins or pounds appeared to be settled. The subsequent reeling, however, brought out defects in the shape of excessive breakages, which required so much piecing up that the knots rendered the weaving into cloth very imperfect. The sizing material is the main point for consideration, as our export trade in cotton goods depends upon the quality of the ingredients entering into the composition of these fabrics, especially when we bear in mind that the length of time they may remain unpacked may generate mildew and other evils. The views we are presenting have been carefully worked out and corrected by the experience of practical men, who thoroughly understand the business in every detail.

The Sizing Process.—In constructing the plant for a successful carrying out of the sizing process, it is necessary to have the mixing-room on a ground floor, with a storage room above for flour, farina, china clay, etc.; the clay boiler should be placed over the mixing beck; the flour-steeping becks elevated a few feet from the ground, so that pumping may be avoided until the size is mixed and ready for use; the mixing beck large enough to have a division in the centre, so that two mixings can be made ready for use simultaneously, thus allowing the use of one division while the other, which has been used up, is being prepared, and so on alternately. The flour becks, being elevated, enable the raw mixture to be rapidly drawn into the mixing beck without pumping, and thus time and labour are saved, and no mistake is likely to arise in the mixing by neglect or over-charging with raw size, causing a great difference in the material operated upon. The flour becks ought to have a shoot immediately above them in the upper room, by which the becks would receive the necessary charges of flour or farina; the service or taper's becks behind the taper's back, or within easy reach of the

SIZING OF COTTON.

operative, ought to be divided in the same way as the mixing becks, so that qualities may be changed without keeping the slashing machine at a standstill for want of size.

It is highly necessary that condensed water should be used—that coming from the tape cylinders being in every way useful. It should be collected in a cistern, conveniently placed, with pipe arrangement connecting the various becks and clay boiler. This would require pipes of something like two inches in diameter, or a little under; two size pumps, 3-inch diameter, 9-inch stroke; one pump fixed to the mixing beck, to send the size to the taper's beck, and the other fixed to the latter beck, to send the size forward to the slashing frame; the pipe for carrying the size from the mixing to the service beck 2-inch diameter, and from this service beck to the boiling apparatus a-pipe of $1\frac{1}{2}$-inch diameter. All pipes conducting size must be of copper, and carefully arranged, with as few turns or angles as possible.

This plant would be found ample for 500 looms making shirtings, allowing the steeping becks to have a capacity of not less than 25 bags of material. In a beck fitted up with screw dashers or agitators no portion of the contents can possibly remain undistributed.

The clay boiler is generally of iron, oval in form, 5 feet long, 4 feet deep, 4 feet wide, and covered with an iron lid—one half a fixture and the other loose—made to open with hinges, a perfect fit, steam-tight when closed and weighted. This boiler is fitted with an oval pipe of copper, nearly 1 inch in diameter, joined to a brass tie, in which a brass nipple is inserted to pass through a hole in the boiler near the bottom; the pipe is pierced for jets of steam to play upon the space in the boiler, which is supplied with steam from a range one-half inch overhead, and passing downwards outside the boiler to connect with nipple in boiler pipe, a steam tap being placed at the top of the boiler to regulate the pressure; an exhaust trunk, about 6 inches by 4, rising from the fast half of boiler cover, is carried through the roof; over the orifice, to which this trunk is fitted, is a brass slide to open or close the exhaust outlet, and by this means, when the boiler is in use, a steam cushion is maintained over the boiling mass, doing the work more effectively and in less time than with a wide-open vent, which allows the steam to escape before the work is accomplished, the heating power being wasted, leaving only a very imperfect boiling. A shoot from the floor above, about 4 inches square, will be found sufficient to feed the boiler with clay. This shoot is large enough for all requirements, because there is less steam surface to protect, and it prevents large lumps of clay being delivered, causing breakdowns and other evils.

Recipes for Sizing.—The following recipes for cotton sizing will be found useful, giving satisfactory results, if carefully followed out:—

No. 1.—Farina, 120 lbs.; china clay, 589 lbs.; tallow, 25 lbs.; dulcin, 89 gallons; chloride of magnesium, 149 gallons at 56°; chloride of zinc, 6 gallons, at 92°; 2 pennyweights of blue.

No. 2.—Flour, 280 lbs.; china clay, 500 lbs.; tallow, 100 lbs.; chloride of magnesium, 20 gallons, at 56°; chloride of zinc, 2 gallons, at 92°; blue, 5 pennyweights.

No. 3.—Farina, 80 lbs.; china clay, 589 lbs.; tallow, 25 lbs.; dulcin, 31 lbs.; chloride of magnesium, 56 gallons; chloride of zinc, 28 gallons; blue, 2 pennyweights.

Use of Tallow or Oil in Sizing.—It may be observed that tallow is not absolutely necessary in sizing cotton yarns, as there are other agents which may be used with equal advantage. Tallow is merely used for maintaining the pliability of the yarn, for without it, or some softening compound, the different elements—starch, farina, corn-flour, rice-flour, china clay, barytes, etc.—would make the fibres too brittle, and the size would fall off during the friction in weaving into cloth. If too much, however, be added in sizing, the yarn, when stored in a damp place for any length of time, will become mouldy; and in the woven fabric also it is injurious. If, for instance, the cloth requires to be dyed in very delicate tints, the goods must be well purged by washing with soda, when the tallow becomes saponified and soluble. A very good substitute for tallow will be found in a preparation of castor oil, or any soluble oil, as oleine, etc. No doubt, for softening the yarns and for sizing generally, these soluble oils possess many advantages over tallow or other fatty matters, as a more uniform mixing can be made, ensuring a better impregnation, and one easier to remove in the after processes of bleaching and dyeing the woven goods.

Quantity of Size Required.—The quantity of size necessary will vary according to the class of goods ordered, and may be from 50 per cent., or under, up to 250 per cent., or over. Very accurate calculations must be made to determine the percentage ordered, so that neither more nor less may be put on the yarns. Let us suppose the order to be a fabric 72 ends per inch, 36 inches wide when finished; that would mean, at least, a little over 37 inches in the reed, 60 picks per inch of weft; warp counts—24's, and weft, 20's; length of piece 40 yards, and, allowing for milling up, the length would really be $43\frac{1}{2}$ yards of yarns; net weight of pieces to be 12 lbs. each. The total weight of warp and weft required to make one of the pieces, from particulars given, would be 11 lbs. 5 ounces, leaving a balance of 11 ounces for size—a little over 11 per cent., allowing for waste.

In every sizing department of a cotton mill, with few exceptions, tables of percentage for all classes of cotton goods are in use, but it may not be out of place to give the principle on which these tables are constructed. Five per cent. is allowed as a rule for waste in winding, warping, etc.; and for contraction in width of reeds and yards in length, 5 per cent. each. Of course, it is not possible to give a hard-and-fast line, because the counts, picks, and widths vary to such an extent; but where very coarse yarns are used, as in heavy domestics and kindred fabrics, there will be more waste contraction in reeds and milling up in lengths—therefore special allowances must be made, and these are soon noticed by a little careful supervision.

A general rule for obtaining weight of warp and weft is as follows: —For weight of warp twist, × number of ends per inch in reed by the width in inches and by the length of warp; divide this product by 840 yards in one hank, and this quotient by the counts, and the result will be weight of twist. For weight of weft, × width in reed by the picks per inch and by warp length; ÷ by 840 yards per hank, and this result by the weft counts; this weight and twist weight added together give total weight of material; let this be subtracted from the weight ordered, and the given amount of size is the difference. So far as allowances are concerned, when once they are thoroughly ascertained by actual experiment, they may be added to the width in reed and length of warp material before multiplying. It will thus be seen that there is no necessity for tables, as the calculations can be made almost as soon as hunting over a list of figures.

The percentage of size is obtained by subtracting weight of yarn from weight of sized yarn, and adding two ciphers to this difference, divided by weight unsized—the percentage is the quotient. Again, if the difference between the weight as calculated and the weight ordered has two ciphers attached, divided by the weight of twist, only the result will give percentage of size in all cases.

Slashing Operations.—The machinery used for tape-sizing or "slashing" is being continually improved, every improvement, perhaps, having special features; but for present purposes it will be sufficient if we take a typical one for description. No matter what form of slasher may be used, careful attention is necessary, as the warp may become brittle because of bad sizing, or from being over-dried on the cylinder of the machine. The usual place where the elasticity of the warp is seriously damaged is between the squeezing rollers of the slashing frame and the drying cylinders. The sized yarn, before it is dried, is very liable to be "ratched" on many makes of slashing machines, or through negligence, but, with good machinery, a careful attendant, and proper sizing ingredients, no mishaps need occur, and the system is far beyond ball warp sizing, both in point of economy and ease to the weaver.

The first thing that attracts attention to one of the most improved sizing machines (see Fig. 40) is the unusual height of a large cylinder; but this constitutes the key to the great departure from more common arrangements. The size box is of the ordinary construction, but fitted up with two strong seamless copper rollers, instead of the usual one of copper and one of wood. If required, a separate size-boiling compartment is provided, with a self-feeding apparatus for the size, and in every case the rollers work outside the box in broad brass bearings. The steam-pipes for boiling, instead of proceeding from a single pipe at one end and branching into two forks, are separate, receiving steam—one from the right hand and the other from the left of the machine; this means that the boiling will be equal over the whole width of the material, instead of being, as usual, stronger on one side.

The drying cylinders are of tinned steel, four and six feet in diameter; the centre of the smaller cylinder is a little lower than the line of the framework, but the larger one is raised two feet above it, so that its lower surface is further away from the floor than usual; but there is a further object for this change—that is, to make room for a fan between the two cylinders, a position found to be the most effective for its use; and the yarn, after it has passed over the small cylinder, instead of going immediately forwards from its lower surface parallel with the floor, is guided by a small roller, and kept in contact with a greater portion of the heated surface, and then descends to another roller below. The fan works within the

SIZING OF COTTON.

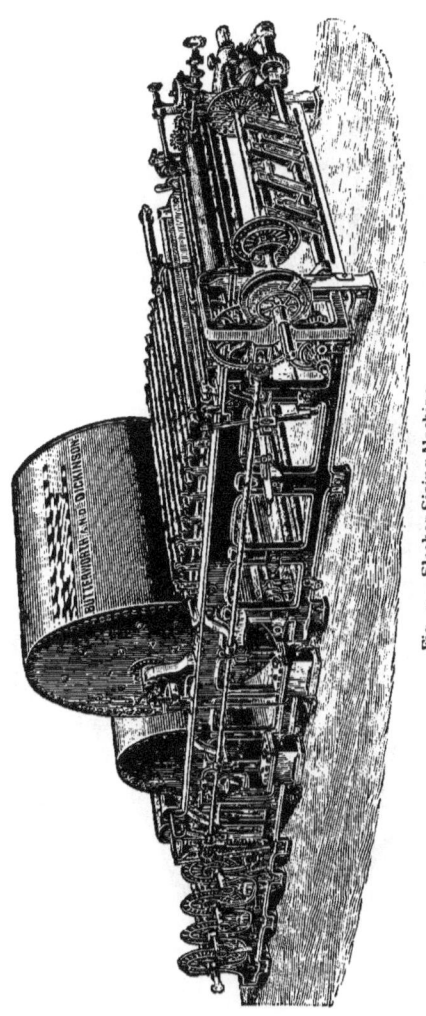

Fig. 41.—Slasher Sizing Machine.

roof-covering of yarn. Another fan at the front of the large cylinder allows of a return of the yarn under these fans, so as to expose

Fig. 42.—Slasher Sizing Machine Headstock.

the largest possible extent to their drying influence; this, of course, really means drying with sufficient speed at a low steam pressure. The headstock is illustrated in Fig. 42.

Without going into further details, it may be mentioned that an adjunct, in the shape of a small additional size-box, is situated above the lesser cylinder; this is for dhooties and other goods specially fabricated with borders. In this box the coloured threads are sized separately, and then join the other threads of the warp in drying. This prevents the mixing, or rather scattering, of colour, and, in fact, the machine is as complete for general requirements as it is possible at present to entertain.

Sizing is really the backbone of cotton manufacturing, so far as a profit and loss account is in question. We have nothing whatever to do with the right or the wrong of over-sizing, our business being to deal with details; but we may say that ever since merchants have stipulated that goods should have a certain weight, beyond the actual particulars given with their orders to the manufacturer of the goods, it could only be secured by some sort of a conditioning apparatus, so that this uniformity of weight could be obtained in order to obviate rejects or light weights, owing to the imperfections in spreading the size ingredients, thus reducing the possibility of pieces, positively better in quality, with more yarn, being returned, or sacrificed by an allowance, perhaps, of $4d.$ or $6d.$ per piece.

The method that had to be adopted when the stipulated weights were found wanting was to lay the pieces on a very damp floor, so that the imbibed moisture would produce the standard at which they could pass.

Damping Machine.—The improved damping machine will be found a great acquisition for the conditioning of cotton goods, and bringing the light pieces to the proper weight. Every manufacturer of cotton cloths knows too well that the fabric comes from the loom hot, dry, harsh, and stringy, having no appearance or feel suitable for sale. Perhaps, after 24 hours in the cellar, we could make something by way of a show; but here, as we have said, a machine with a tank and a series of tension rollers at once paved the way to a complete success.

The machine (Fig. 43) consists of a framework, in the midst of which is fixed a tank. Behind is a series of tension rollers; over and across the tank are two horizontal rollers, and at the front, a calender and delivery roller. The bottom of the tank contains the conditioning liquor (provided by the firm), into which a fluted roller dips, and above this is a cylindrical brush driven at a high speed by a small pulley. Above the brush are two folding boards, for regulating the amount of moisture discharged upon the piece; the wider the opening the greater is the quantity thrown upwards on the cloth, which is arranged behind the machine, passed round the tension rollers, and extended underneath the rollers upon the tank, thence to the calender, and finally run upon the batching roller. The damping brush is so arranged that it does not come in actual contact with the liquor, and drains itself when the machine is stopped. Perfectly even damping is obtained by the liquor being thrown up in a fine spray. Any weight can be put in—from 2 to 6 ounces

Fig. 43.—Damping Machine.

per piece—and 50 yards a minute passed through. This machine is qualified to serve a very large output in a weaving establishment.

The Mote Clearer.—This machine (shown in Fig. 44) removes all leaf and impurities that obtain in fabrics woven from low grades of cotton, leaving a clean, saleable cloth; both sides are thoroughly brushed at the same time, and all foreign matter, loose threads, etc., are swept away. Any desired pressure can be put upon the texture. A fan drives out the dust, which by means of a pipe is conveyed outside the building. The cloth treated by this mote-clearing machine always has the appearance of being made from superior cotton.

SIZING OF COTTON.

The Filling and Finishing Machine.—We have already stated that there is, and always will be, we suppose, a demand for heavy sizing in cotton cloths. The goods required for our export trade to India, China, etc., demand this condition of weight, and it is absurd to suppose that the customers in these markets were not long ago acquainted with the proportions of materials entering into their

Fig. 44.—Mote-Clearing Machine.

composition. They desire the cheapest clothing attainable, and we are prepared to supply the demand, with the conditions thoroughly well known, without any mystery whatever or any improper motive. The weight is put in by the sizing process, so as to carry the percentage demanded; thus one part of the cloth can be dealt with, but the weft goes in pure. Why ought not this portion to bear a share? The passage of the warp threads from the beam over the back rest, and through the lease or clasp rods; then the friction of healds and reed, and the passing by each alternate thread to form an open shed, demonstrates that this amount of friction causes a large loss of percentage in size, which may be seen on the floor under all the looms—this is actual waste—then we have the wear and tear of the healds, reeds, etc. Weft sizing has been resorted to, but the process is unreliable, the pieces being either too heavy, too light, or otherwise unsatisfactory.

The question for solution is, Cannot the goods be made, in the first instance, in a pure condition, and afterwards filled to the desired weight? If this could be done, an enormous advantage in many ways would accrue. With this view, a unique machine has been designed to enable manufacturers to fill up goods to the required standard of weight in their contracts. A brief description of the machine (Fig. 45) will show how effectually this most desirable object has been perfected. The cloth is passed through three straining rollers or rails into a size box, under an immersion roller, thence between pressure rollers, and to an expansion roller, which spreads the cloth out to its proper width without stretching; and a series of drying cylinders, along with a plaiting arrangement, completes the operation. This is simplicity combined with economy and efficiency.

We have, by the use of this machine, a saving of the healds and reeds, no mean item; warp and weft equally receive the proper proportion of size; dust in weaving is reduced; no necessity for steaming exists; the feel of the cloth is improved; a full body, along with any amount of weight, is ensured, and whatever feel is ordered can be obtained by modifying the amount of size, whilst the colour of the cloth can be brought out clear and clean. One person can attend to a delivery of 70 yards per minute, and the size used is a guarantee against mildew. Another advantage gained is economy in finishing. When it is necessary to send goods from the weaving department elsewhere to undergo treatment, not only is the cost increased, but delay, with consequent inconvenience, often results.

The Process of Dhootie Sizing.—We have given a few recipes for size compositions. The following is one very much in use for cloths having coloured edges:—Four sacks (or 960 lbs.) of flour, containing 13 per cent. of gluten, fermented and aged for three weeks, 1½ bag china clay, boiled for sixty hours, and added to this 100 lbs. of tallow and 70 lbs. of soft soap, the whole boiled for

SIZING OF COTTON.

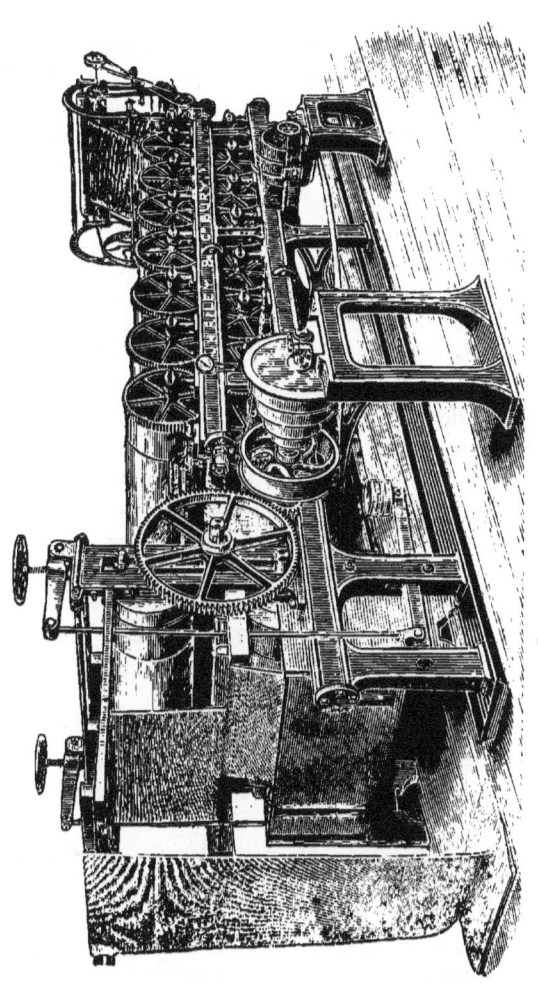

Fig. 45.—Filling and Finishing Machine.

twelve hours; then 9 buckets of antiseptic, 56° Twaddle, are put in. The flour paste and clay being mixed together, the size will be ready for use. Above all, good soft water must be used, the most suitable being the condensed water from the drying cylinders. The general allowance is about 5 quarts of water to 1 lb. of farina. When the calculation is made from the number of ends to make a piece of how many beams are required, they are then placed at the back of the slashing machine (sometimes called a taper); the yarn on the beam next to the machine unwinds from the top, the second from the bottom, the third from the top, the fourth from the bottom, and so on, in this order, according to the number of beams.

These yarns pass through the boiling size mixture, then between calender rollers, round the copper cylinder filled with steam, and are dried thoroughly above and below the fans, and, finally, wound on the loom beams. Care should be taken that the rods dividing the yarns are kept free from crossed threads and removed at stated periods, or when the beams are filled.

Rules for Slasher Sizing.—We have given full particulars of the machinery for slashing, with the process, and may conclude this method of sizing with a few necessary rules that will be found useful.

To find the length of a mark, × bell wheel, stud wheel, and circumference of tin roller together, ÷ by tin roller wheel, and the quotient gives inches in a mark.

To find number of teeth in tin roller wheel, × bell wheel, stud wheel, and tin roller circumference together, ÷ by the inches in a mark, and result = tin roller wheel.

To find stud wheel, × the length of mark by tin roller wheel, ÷ by circumference of tin roller, × by bell wheel.

To find bell wheel, × the length of mark by tin roller wheel, ÷ by circumference of tin roller, × by stud wheel.

To find circumference of tin roller, × the length of the mark in inches by tin roller wheel, ÷ by bell wheel, × by stud wheel. (Should it occur that the stud wheel and tin roller wheel have the same number of teeth, the length of a mark will be the product of the circumference of the tin roller and bell wheel.*)

To find bell wheel (this gives the signal), *to mark any given number of yards when tin roller wheel and stud wheel are alike*, ÷ the

---

* This calculation shows clearly that whatever number of yards is cut, or length, the change stud wheel must have the same number of teeth.

inches in the length of mark required by circumference of tin roller, and the quotient will be the bell wheel.

To change only one wheel, and this wheel *to denote number of yards* per cut, *or one yard* per tooth, supposing tin roller 14·4 inches in circumference, bell wheel 45 teeth, then $14·4 \div$ by $36 = ·4 \times 45 = 18·0$; this 18 becomes a permanent pinion on the end of the tin roller: the stud wheel to be the yards required. If 80 teeth, it would be equal to 80 yards; this × by $36 = 2880 \div 14·4 = 200$ revolutions of tin roller for 80 yards; then stud wheel × by bell wheel, $80 \times 45 = 3600 \div 200 = 18$ pinion for tin roller.

Ball Sizing.—Ball sizing is now considered rather antiquated. As in spinning and carding, so in this system of sizing, there are many opinions, some contending that the thread comes out stronger and rounder than by the slashing process, but the cost is the most important factor. The quantity of cotton twist, 32's for 500 looms making shirtings, may be taken at 14,500 lbs. The difference between the cost of ball warping and beam warping is nearly 2·621 per 55¾ lbs. Taking the lowest base—14,000 lbs.—this extra cost is nearly £2 15s. per week, and to this must be added the further cost of beaming. These figures, obtained from the most reliable sources, are sufficient to show the advantage of slashing. Again, whilst ball warping is almost stationary, beam warping is becoming universal.

The process of ball sizing may be briefly described and understood as follows :—The warp is taken into the box containing the size, through which the ropes are passed. At the bottom of the vat, two lines of rollers are fixed, between which the warp passes, and by which the air is pressed out of the yarn, so that the size may be even. The whole is kept hot and fluid by steam being blown into the size trough. The warps, after passing through the trough, are squeezed by large rollers, for the purpose of removing the excess of size which may be adhering to them, and are then passed over a delivery winch and folded into warp boxes. Each box is supported on wheels, and when full is pushed along to the drying machine, where a series of warps is passed from the boxes over pulleys to flatten them down, so that a larger surface may be exposed to the action of the drying cylinders. These cylinders are made of sheet iron, and heated by steam at something like 15 pounds pressure. There are guiding rails and pegs, by which the different warps are kept in their proper places and free from entanglement, whilst the delivering winch folds the dry warps again into the boxes. The dry warp from each box is then passed over a balling machine, and wrapped into balls if required.

Irregularities are much more common in this system than in slashing, and this arises from the fact that in the former all the threads are pressed together in a mass, whilst in the latter each thread is acted upon by the rollers separately. Much larger quantities of size are kept boiling in the ball-sizing cistern than in the size box of a slasher; therefore the size is more subject to change by evaporation, or by being diluted with condensed water from the steam. Again, the threads being all put together, it often happens that, the size not being of a proper fineness or the threads being pressed so close to each other, the size cannot thoroughly penetrate the core of the bundle. Cloth made from warps with this defect will always have a streaky appearance.

Prevention of Putrefaction in Size.—During summer weather, size is very apt to putrefy, especially in the absence of antiseptics in the mixing, and, by the "retting" of the size, warps become heated, either in the thread or in the woven fabric, causing spots and other defects. A capital preventative is creosote, of which about 50 drops will suffice for 12 gallons, and will keep size in its original condition for a long time. Warps that have been prepared without this precaution may be saturated with five drops of creosote to every three gallons of water, and the weft may be damped with solution. This recipe is cheap and effective.

Hank Sizing.—Hank sizing is a process in use for ball, section warping, and dry slashing. The yarns, both for warp and weft, are reeled into hanks and bleached or dyed. The dyeing is separate, and so is the sizing, in order to prevent the colours shading into each other, as they would certainly do if dyed and sized at the same time; hence it has been found desirable to have special machinery for this process, such as hank-dyeing, hank-wringing, and hank-sizing machines.

The colouring matters of many dye stuffs are very harsh, and cause the threads to become matted together, and the dipping of the threads into a liquor bath causes them to overlap each other. Again, when yarns have been unduly twisted or reeled in a dry condition, numerous curls occur; the cost of rewinding is increased,

SIZING OF COTTON. 117

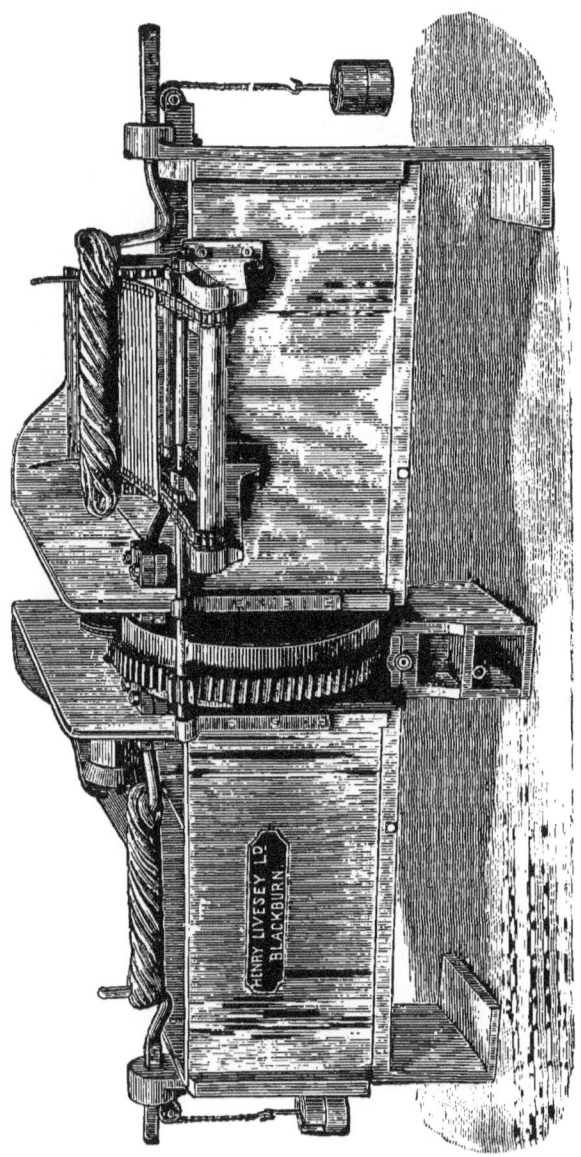

Fig. 46.—Double Hank Sizing Machine.

with less production and more waste. The old method of hand-shaking is far from satisfactory—the straightening is imperfect, with breakages of the threads. It is obvious that when curls exist in the hanks, the threads which are thus linked together in these curls become shorter than the others, and must, therefore, bear more strain in the various operations before winding, and this accounts for so many broken threads. Stretching and brushing has been found too expensive.

To obviate these defects and prepare hanks for thorough sizing, various machines have from time to time been introduced, but they have been defective in principle. A machine now in operation, and considered most effective, which we illustrate in Fig. 46, has the hanks placed upon saddles and passed under rollers, having eight sides in place of being circular. The shaft carrying these rollers steadily descends, gradually stretching the hanks without jerking them, and when this extension is sufficient, a break acts, and the machine is stopped; but, though the stretching ceases, the shaft continues to revolve, carrying the hanks round the saddles, so that they are perfectly straightened out, and in this condition are ready for sizing or winding. The process of hank-sizing is similar to ball-sizing, but is very much facilitated by lately improved machinery.

CHAPTER XII.

WARPING OF COTTON.

Object of Warping.—Warping may be considered a simple process, having merely for its object the arrangement of threads in parallel order to compose a warp for the loom. The invention of beam-warping machines has, to a great extent, disposed of manual labour, which in the old system required both strength and dexterity.

We may particularly mention a warp-balling machine, by a well-known maker, which has become an established favourite, dispensing with a great deal of trouble, time, and labour. It has a reed and curved creel, and the balling arrangement is composed of a steel flyer, carried by a series of four grooved rollers. In connection with the flyer is a small roller with a deep cut flat groove; this is the guide for the warp as it is passing through to the balling mechanism, which consists of a spindle suitably arranged, carrying the balling shank, and actuated by a cam giving the necessary oscillating motion.

A beautiful piece of machinery is the automatic differential tension motion, which ought to be seen to be duly appreciated. There is also a measuring and marking motion with indicator, so that it is easy to ascertain how many cuts or pieces have passed into the ball. The differential motion equalises, or rather provides for, the difference in the great circumference of the ball during the winding on. The warp proceeds from the bobbins placed in the creel, passing through a heck to the measuring rollers. The piece-marker is in front of these rollers, and marks the required length. The yarn then passes upon the tension arrangement and to the flyer, which balls it upon a spindle, and from this the ball is taken and is ready for sizing or dyeing.

The comparison between this invention and the ordinary mode of ball-warping is at once quite obvious. The advantages are manifold: less space; one uniform tension in balling; no strain upon any portion; the largest possible ball can be made with as great facility as a small one, the production being at the rate of 140 yards per minute, and in point of economy it is far and away beyond the old practice; as the machine is almost, if not wholly, automatic, no labour or skill is requisite, and the saving in cost between the two systems is sufficient to decide the manufacturer in favour of a machine so useful and desirable.

For beam or section warping, the same machine-maker has also a most excellent contrivance for making the sections of a warp of a uniform diameter, tension, and length, whatever may be the variation of yarns or number of ends.

The process of ball warping is, in many respects, one that requires considerable strength, dexterity, and, for coloured goods in fancy patterns, no small amount of calculation.

The Warping Mill.—The warping mill is a large skeleton drum or reel—in many cases 20 yards in circumference—having a central vertical shaft, from which arms extend to the outside reel, and the whole is moved or rotated by motion from the line shaft. The creel—or, as it is called in Scotland, the "bank"—is an upright frame forming the arc of a circle, with divisions from top to bottom wide enough to hold something like 500 bobbins in a horizontal position. This form of creel is in use where large patterns, such as bed tickings, etc., are required. The bobbins revolve freely upon either wooden or iron spindles, which are supported in notches of the partition framing, and can, therefore, be easily removed or placed as required. The threads from all these bobbins pass through the heck box, which consists of finely polished, hard-tempered, steel pins, with a smooth eyelet-hole at the upper part of each to receive and guide one thread. The heck is divided into two parts, either of which may be lifted by a small handle, the eyes being placed alternately. A lease is thus formed—that is, the warp is divided into two equal portions, so that the weaver may be able to keep the threads separate by the insertion of rods. This heck is placed nearer the mill than the creel, and as the mill revolves it slides up or down a fixed support, which has a smooth edge, and is either black-leaded or lubricated occasionally. The heck itself is suspended by a cord passing over a conductor to the centre shaft of the mill, where it is secured, and made, by this means, to act in uniformity with the direct or reverse revolutions of the reel. When it rises to the height of the upright support, it is so set that, when the creel reverses, it descends by means of a weight.

Warping Calculations.—We will give a few calculations showing how patterns are formed by the bobbins in the creel.

Multiply number of ends required in the breadth of the fabric by yards in length, divide by 800—this allows for the percentage of waste; then, suppose a pattern, 20 black, 4 red, 2 white, 6 green, 10 black, 14 red, 10 white = 66 threads in all. Now it will be seen that there are 30 black, 18 red, 12 white, 6 green. If the warp has 1000 threads 20 yards long = 1000 × 20 = 20,000 ÷ 800 = 25 hanks; and as 66 is the total of all the coloured threads in the pattern—

Black $30 \times 25 \div 66 = 11\frac{4}{11}$ hanks.
Red $18 \times 25 \div 66 = 6\frac{9}{11}$,,
White $12 \times 25 \div 66 = 4\frac{6}{11}$,,
Green $6 \times 25 \div 66 = 2\frac{3}{11}$,,

25 hanks total ÷ by counts, and we get the weight.

In warping large patterns, with solid stripes of one colour between, it is necessary to obtain an equal portion of this solid bar or stripe at each selvage: therefore, divide the extent of the pattern into the number of warp threads—for instance, 32 orange, 40 blue, 16 light blue, 40 blue, 32 orange, 100 blue = 260; number then of warp threads = 1,248 ÷ 260 = 4 and 208 over. Then, excluding the 100 of blue, we have 208 − 160 = 48; and if from this are taken selvage threads as 32, the difference leaves 16, so that 8 of blue at each side of the cloth width will give the proper proportion. The circumference of the mill is a divisor; the length of the warp the dividend; the remainder, if any, after the quotient, is multiplied by the number of staves in 1 yard of the mill or reel (these staves are the wooden uprights in the circumference of the skeleton drum). This calculation will be the round and staves to complete the length of warp:—thus, 3,860 threads, 186 yards, 6 cuts or pieces, circumference of mill 16 yards, 3 staves to a yard; then 186 ÷ 16 = 11 rounds of the mill and 30 staves to complete the length. These cuts are stained with a mark when the first round from top to bottom of the mill is done, commonly called first half-knot.

Where these staves are not divisible into the circumference, it would have to be reduced into inches for a divisor, and the length of the warp also into inches for a dividend, and the quotient afterwards divided by number of cuts. In warping odd-sided patterns, the bobbins would be placed in the creel with one-half the colour. Suppose pattern, 2 blue, 4 white, 8 blue, 2 white, 8 blue, 4 white, 2 red, 6 blue, 2 white, 6 brown, the creeling would be 1 blue, and to end off after the 6 brown, 1 blue, making the pattern exact in taking the lease. Any pattern, whether equal or one sided, may be turned, say, 4 white and 4 blue will make no difference from 4 white, 3 blue. If the number of warp threads is divided by the number of threads required in the half-beer for the beamer's ravel, the quotient will be total number of half-beers in warp; the beamer is, therefore, saved a loss of time in finding the count. The warp half-links are found by dividing the number of half-beers by the greatest number of half-beers in the creel or number of bobbins, say, bobbins 250, warp threads 3,860, quotient 15 half-links, and 110 over. These would have to be run in at the last when the 15 half-links had been got on; generally 3 staves are allowed for fenting to a warp. Very often a flat may occur—that is, two ends running one way in the lease. Many warpers are careless in this respect, which causes both drawers-in and weavers—especially in complex patterns—to be confused in the choice of the right colour, if broken out or otherwise displaced. Very little trouble would remove this difficulty. Suppose a pattern of any odd number, say, 27, 13, 17, 33, etc., then to remove a flat, or two threads running in the lease one way, make the pattern even by doubling—thus the $13 \times 2 = 26$. Every bobbin is equal to two threads on account of the repetitions required to make the breadth. If warping 12's in count, there may be 10 hanks on a bobbin, which is $1\frac{1}{2}$ inch in depth and 5 inches long between the flanges (12 bobbins to a bundle); total, 120 hanks in a 10-lb. bundle.

There is one particular feature in warping odd-sided patterns which is worthy of notice. When the repeat is placed on the lease-pegs of the mill it is taken off, or, before being put on, is turned inwards owards the mill. Plainly speaking, that portion of repeat yarn which, in an even-sided pattern, would be on the right hand is turned over so as to become on the left on the lease-pegs, both at the top and bottom of the mill. By means of this peculiar process, the repeat joinings fit in their proper order. Warpers call this class of patterns "skews," and do not care to have many of them to operate upon, because any neglect in the turning, either at top or bottom, would give a complete twist to the whole of the yarn, which would have to be cut and put into proper position by the beamer.

Sectional Warping.—We have given sufficient information to convey a practical knowledge of ball warping, which is a necessity to small manufacturers, because no large stock of material is requisite and any length can be obtained with facility. Invention may be compared to hunting. When a good idea is started, there will soon be a field in full cry upon the track. Improvements in warping-machines afford a conspicuous illustration. Like every other process of manufacturing, warping has passed through many phases of development before the present degree of perfection in the sectional machine operation was attained. Ball warping, at its best, has many defects, such as wrong lengths, entangled slack half-beers, and many other evils, owing to mistakes in the calculations, coupled with negligence. In sectional warping-machines, the warp is made in several sections in the form of a cylindrical bobbin, or, as it is called, a "cheese," having a hole in the centre.

In taking one of these improved machines as an illustration (Fig. 47), we may say that this patent is a fair type to describe. There are many novel features, the most prominent, in our opinion, being the method of regulating the diameter of the sections. It is evident that the main object is not only to have the length of every section alike to avoid waste, but also to keep them of equal diameter so as to preserve level cloth by equal tension of the threads. The measuring and regulating is, therefore, important. The yards in length measure are effected by a dial with bell, to which movement is given by a measuring roller; and at any time this can be checked by comparison with another dial on the opposite side of the machine from the one with the bell. This dial registers the number of layers or revolutions, and is worked from a shaft by means of a cam moving a wheel one tooth per turn. When the bobbins are full, the thread tension is least, and increases as they are unwound.

To check the effect of a varying tension on the diameter of the cheese, a weighted pressure bears against the under surface, and is so arranged that as the warping proceeds this pressure becomes less, and also compensates for any irregularity. The total pressure is given from a short lever with a weight upon it, and the diminished pressure by the action of a counter-balancing weight, varied automatically. This weight slides on an upright rod passing through it. The form is that of a cone, with the edge cut off or truncated. It is sustained by two bowls, one on each side, which rest upon the inner surface of two levers. These, rising from the same fulcrum below, recede from each other at their higher ends, presenting in appearance something like a fork with the prongs forced outwards. The lever nearest the front is kept against the lower part of the bowl of the counter-weight by the pressure of the initial weight, while the position of the other is regulated during the warping. If the whole pressure of the counter-weight bore against the front lever, the initial weight on the presser would be reduced in proportion; but, if sustained entirely by the back one, the whole of the primary weight would bear upon the cheese. At the front of the levers is a pair of compound levers, one centred on the presser shaft and solid, with a curved arm; the other centred at the top, and vertical with the lower centre shaft. The top lever has a screw and screw-block working in a slot, and by raising or lowering the block, the curved lever is made to travel at the varying speed required to build up the section. The elevation or depression is regulated by a small mill-edged pulley at the top of the screw. From the warping shaft, a horizontal screw is turned, which takes hold of the two levers, at equal distances from their centre, by a pair of links, thus keeping a central position during its travel. This apparatus is drawn to the front, at the beginning of every warp section, by turning the screw that is fastened to it, and it works through the machine at the front. As the warping proceeds, the screw, being automatically turned in the reverse direction, pushes the regulating lever backwards. The outer lever nearest the weight has, cast to the same hub, an arm, rather longer than itself, down the middle of which is a slot. A stud fixed in the screw-block of the vertical screw passes into this slot, so that, as the stud moves, it moves the arm and the lever attached to it, causing the latter to recede from the counter-weight. If the regulator moves too quickly for the warping, or rather the diameter of the cheese does not increase in the desired proportion, the hind lever may be caused to recede from the front one, so as to leave the whole counter-weight bearing on the latter, and easing the pressure of the cheese to

Fig. 47.—Sectional Warping Machine, with Singleton's Stop-motion Frame.

the greatest possible extent. This motion is very sensitive in its action.

Another novelty, and an excellent arrangement, is that of throwing off the driving strap by means of a projecting finger, which comes into action when a warp section is completed. The block on which the cheese is formed can easily be varied to serve different section widths, one of them taking the place of four ordinary solid blocks. To say that the machine is neat in appearance is not enough—it is worked out with consummate skill and care, yet, withal, so simple that a girl can easily tend it. It should be seen in working operation to be fully appreciated by coloured-goods makers who desire an economical, effectual, and accurate system of warping.

Calculations for Sectional Warping.—In the great demand for fancy goods, very complex or intricate patterns are designed. In sectional warping machines this class of pattern requires repeating several times to complete the width. Should a pattern be 5 inches in width, then it is obviously desirable to have a "cheese" suitable for this section, and to place the number of these sections side by side on an axle, until the width of the entire warp is obtained, and then to wind all off on to the loom beam. In this way every cheese, as far as width goes, has its proper complement of threads, according to pattern. If the bobbin creel is arranged for one section, it is sufficient for the whole warp, because the full width is merely a repetition.

Taking one type of these warping machines, the usual number of change wheels is dispensed with, and all calculations previously necessary in changing the details of the warp are avoided, the warper being able to adjust the machine for a new warp without any supervision whatever. The number of patterns for the proper width, with all the colours and their numbers, is given with each warp, and fewer or more cheeses and a larger or smaller number of patterns are taken as required. If a warp be 40 inches wide, 60 ends per inch, 120 ends in a pattern, then the calculation would be, with a 5-inch cheese, $60 \times 40 = 2400$ ends in warp breadth; $2400 \div 120$ ends pattern $= 20$ patterns divided by 5-inch cheese $= 4$, number of cheeses required.

The operation then is, when the creeling of the bobbins is done according to pattern, to run a few revolutions without yarn, and to notice that the ratchet wheel is impelled one tooth for every revolution of the section spindle. The empty section block or cheese having been placed on the spindle, great care must be taken that the flange, which holds the block in its place, is adjusted quite close to the edge of the block, and is securely fastened on the spindle by the screw in the flange boss. A few waste yarns tied round the block, or a strip of felt or cloth round the edge of each block, will prevent the warp threads falling down between the block and flange. The warp yarns are fastened in the hole of the block, and the machine turned one-half revolution, or until the set screw in the boss of the flange is at the bottom; then the lease of the threads or equal division is taken, and another half revolution made, which brings the screw of the flange to the top, and the lease just over the measuring roller.

In setting the measuring motion, the intermediate wheel is lifted out of gear and the numbered dial turned in the direction of the figures, until the finger points to the highest figure or number. This being accomplished, the screw which bolts the two sectors together is loosened, the presser falls against the section block, and the end of the toothed sector is brought forward until it is opposite the beginning of the index of the fixed sector. This is done by means of the handle attached to the worm shaft. The machine is then ready for the section block to be filled with length of yarn required for the warp. The first section determines size of wheels necessary on the regulator for all the other sections. Caution must be used in placing the numbered discs which record the revolutions of section spindle. When the length of yarn requisite is obtained, the number of revolutions made by the section block, as shown by the two discs, must be noted, along with the number indicated by the pointer on the sector index. This number recorded will be the number of teeth required for a change wheel on the ratchet wheel shaft. If the pointer stops between two numbers on the index, the highest number is taken. Suppose it is between 42 and 43, the latter number is taken.

As we have said, the record of revolutions regulates the number of teeth in the change wheel for the worm shaft, which is obtained by dividing the numbers shown on the two discs by 10; as an example, if the numbers were 375 and 18, these added equal 393, which, divided by 10, gives 39 teeth required. Before the second section is commenced, this change wheel is put on and geared by means of an intermediate or carrier wheel, and the toothed sector is run back as far as possible by the handle of the worm shaft, the numbers 25 and 1000 appearing on the respective discs outside the shields. This is done by drawing the carrier wheel out of gear,

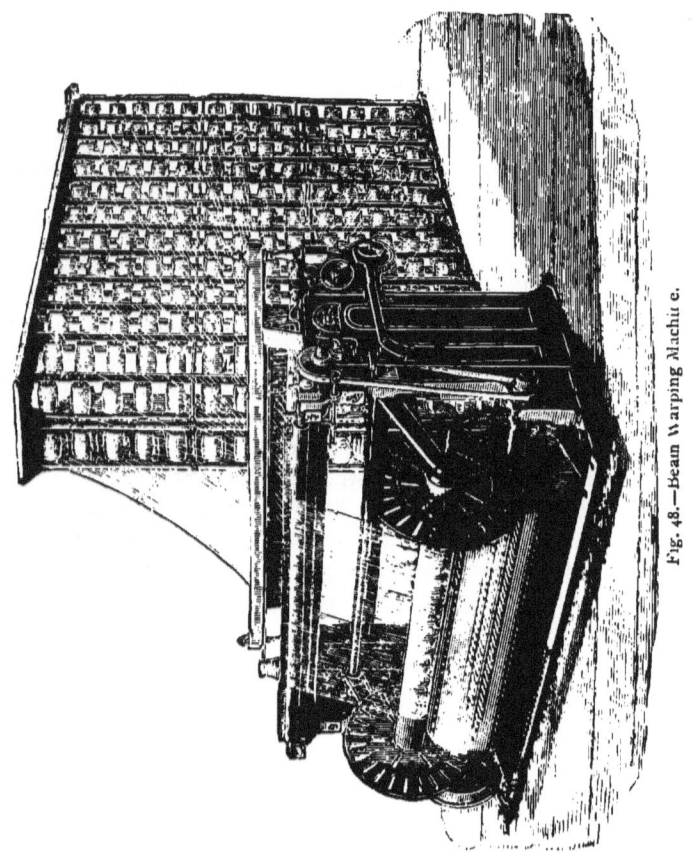

Fig. 48.—Beam Warping Machine.

turning the discs to the position required, and replacing the carrier wheel. The two sectors are then securely bolted together by the screw, and all other sections can be run off. No further alteration is required so long as the same sections are made, whatever may be the amount. Finally, the screw of the flange must always be in one position when the yarn is cut at the completion of a section.

Such is the operation of section warping, which is easily managed, after one or two trials, by a girl, and any material—linen, silk, worsted, etc., sized or unsized—can be warped with equal facility. There cannot be any twisted yarns—all have an equal tension, and experience proves that the production is increased, waste reduced, less expenditure incurred, and an improved quality of cloth obtained by the use of sectional warping machines. There are some of these machines with a reversing motion, by which the pattern can be reversed and run on the section block in an opposite direction. This is a very necessary operation in the case of large patterns.

Beam Warping.—This system of warping gives good results. The machine shown in the illustration (Fig. 48) has an automatic stop motion. The operation of warping on these machines is very simple. The necessary length is indicated by a dial method of calculating. Let us suppose a warp of 3000 threads is required; the hanks, sized and wound on bobbins, are placed in the creel; taking then 6 beams to complete the warp, 500 bobbins will be placed in the creel—*i.e.*, $3000 \div 6 = 500$. The dial being set for the length ordered, one-sixth of the warp is completed, the lease taken and the beam removed. Six beams are thus filled and removed to a frame, where they are placed in rotation—one slightly above the other—and weighted. This frame has a measuring roller, and a marker tipped with dye stuff, for the division of the warp into cuts. All these beam warps pass through an expanding reed, and are wound on to the beam for the loom. Many thousands of yards of warp material can be beamed in a day by this very economical process.

CHAPTER XIII.

COLOUR DYEING.

Colouring Stuffs.—Before entering upon weaving and finishing, we consider it necessary to give some particulars of the latest improvements in dyeing and printing—two very important branches of the cotton industry, involving, at times, considerable loss or profit. It would require a rather extensive volume to enumerate the various materials used in dyeing cottons. Most of the colours now such great favourites were unknown some few years ago, whilst many old ones have been rendered more permanent and less costly. So far as materials are concerned, we can only refer to those commonly in use.

The principal portion of our colouring stuffs are from the vegetable kingdom, such as barwood, orchilla, annatto, alkanet root, barberry root, Brazil-wood, camwood, logwood, Saunders wood, indigo, madder root, quercitron, safflower, weld, woad, and turmeric. The animal materials are cochineal, kermes, and lac. The metals are iron, copper, chromium, arsenic, etc.

There are also many artificial products, furnishing some of the most beautiful of all dyes—notably alizarine and its relatives, dispensing to a large extent with the use of madder root. The mordants and alterants are alums, salts of iron, and tin, together with tannin, in the form of sumach, catechu, gall nuts, divi-divi, etc., Some of these have a dual effect, and might be considered as colouring matters. By boiling or steeping, a deep-coloured liquid may be easily obtained, and cotton fibre or cloth immersed in the solution will absorb a certain quantity, but this is fugitive and can be washed out again; but to fix this colour—that is, to render it permanent—a mordant is required, and an alterant, if it is found necessary to change the tone or tint. This alterant will give any number of shades, and often produces colours impossible from any combination of dyes.

Bright Colours.—To produce good, bright colours, it is necessary that the material to be dyed should be freed from natural colouring matter and impurities. Cotton requires the aid of some substance to effect adhesion of colour matter. Animal fibres, as wool, etc., mostly retain the colour medium without a fixer, or, rather, a mordant. We may simply point out that the ordinary process of dyeing is merely immersing in a solution of colouring matter, gradually heated to a boiling point. If the matter is attracted to the goods and rendered insoluble in water, the colour will be retained, hence called "fast": this is only a manual operation, not needing any degree of responsibility. Several immersions, stirrings-up, wringings, and dryings, make up the sum of such a process.

The dyer—who is, or ought to be, something more than a labourer—has to deal with the science and practice of dyeing, his main object, if he is properly posted, being a thorough knowledge of the fixation of colouring matters, and the permanence of colours —that is, the power to resist air and light. The experienced chemist or dyer may satisfy himself in his laboratory with having produced positive effects on some two or three small samples, but his recipes would not only be useless, but, perhaps, too costly, if not ruinous, when applied to tons in weight of fibre or fabric.

There is a mighty difference in the two operations—the small and the large—which the hard-working practical dyer knows to his sorrow, by following these small experiments, which will not even bear the test of a proportionate value. That is to say, if a certain quantity of ingredients produce on one hank of yarn a desired result, what quantity would be required for 1000 hanks? Naturally, one would expect a proper equivalent result; but in almost every case there is bitter disappointment.

We give illustrations of two machines in use for colour-dyeing. Fig. 49 shows the dye jigs. Fig. 50 illustrates a dye wench with engine.

Alkaline Leys.—In boiling alkaline leys, so strong is the affinity of the water for the alkali that a small quantity of the latter is carried away in the steam, and this has such a deleterious effect on the colours that it is necessary for these boilings to be separated from any other part of the dye-house where coloured goods may lie exposed. Safflower and Prussian blues are damaged by alkaline steam mixing with the atmosphere, and other colours are more or less injured. The dyer must be acquainted with the chemical effects of heat, because, in making solutions, a certain temperature, varying for the different colouring matters, will be found to give good results.

We will suppose a case: The decoction of quercitron bark gives a purer and finer yellow, at a temperature of 90° Fahr., than when the liquid reaches a boiling point; but as the amount of colour so obtained is smaller than by boiling, quality is sacrificed to quantity.

Fig. 49.—Dye Jigs

Safflower red on cotton will stand a high temperature when the air is dry; but heat, with moisture, changes it very rapidly into a yellow brown. Cotton dyed with Prussian blue in a moist atmosphere and raised to a high temperature fades away in a few brief hours. Indigo blue on cotton is permanent when exposed to heat and moisture, but changes on silk. Goods placed in the dark and cold may be dried without change, and then subjected to a temperature of 200° Fahr., whereas they would not actually stand a heat of 95° if the precautions mentioned were not observed. Colouring agents which are volatile may, as a rule, be considered permanent when fixed on fabrics, and they resist heat action best; but those colours which do not sublime are prone to decomposition when subjected to air, heat, and moisture. Light plays havoc on the dyer's finished colours and goods. Reds dyed with Brazil wood and tin mordant, when exposed to light, turn a brown orange and fade away. Yellow turns brown,

Fig. 50.—Dye Wench, with Engine.

Prussian blue becomes reddish, and passes into a dirty grey. When grey goods are put into bleach, and kept in the dark, the whitening process goes on more slowly than when exposed to the light.

The principal point for any dyer to study, if he is not already well

up in the science, is to find substances which have an affinity for the cloth and colouring matter, as, by a proper combination, permanent colours are obtained; and it is not too much to say that the whole art of dyeing operations consists in a proper choice and application of mordants, because, between the colour agent and mordant, a new element is generated, differing, not only in properties, but in colour, from the original substances—so much so, indeed, that a very slight difference in the strength or quality of a mordant gives a new shade or colour. Logwood alone, by means of different mordants, will give every shade, from a creamy white to violet, from lavender to purple, from blue to lilac, and from slate to black. Simple processes will often give some excellent examples. Mordants alone have no affinity for the fabric, but are strongly inclined to join issue with colour matters, and are acted upon by some agent which produces the required affinity for the cloth; but it often happens that the agent which has induced a combination may destroy the affinity, and then it becomes necessary to neutralise the action of the agent, when it has done its share of the work.

Cop Dyeing.—It will be interesting to refer here to a machine which has met with a considerable amount of success—viz., the Cop Dyeing Machine, introduced by a well-known Lancashire manufacturer. The illustration here given (Fig. 51) furnishes a general view of the machine, which consists of a dye-vat, a large and small receiving chamber, shown to the left, and a treating chamber, to the right of the view. The vat is supplied with a coil of steam pipes arranged on the inner side, so as to be out of the way. There is an annular cavity round the exterior of the treating chamber, which is provided with a number of small holes, so that air or steam may be supplied thereto through a valve on the outside of the chamber. The large and small receiving chambers are suitably connected with the cop-treating chamber, and, in the former, a most efficient valve mechanism is provided. On the outside of the large receiving chamber there is also a double valve chest, one for steam and the other for air.

In the large receiving chamber a vacuum is formed—the liquid following. Thus the dye liquid enters the cop chamber in which the cops are placed, and the hinged lid is closed. It then passes through the cops and by a conveying pipe into the large reservoir, the operation being repeated as many times as required, until the desired

COLOUR DYEING. 135

shade is obtained. There is a suitable plate which exactly fits the treating chamber; this is removable, and is furnished with a large number of perforations. The cops are placed upon metal spindles, perforated closely throughout. It will be seen that the spindles fit perfectly into the holes in the above-named plate. Any number of

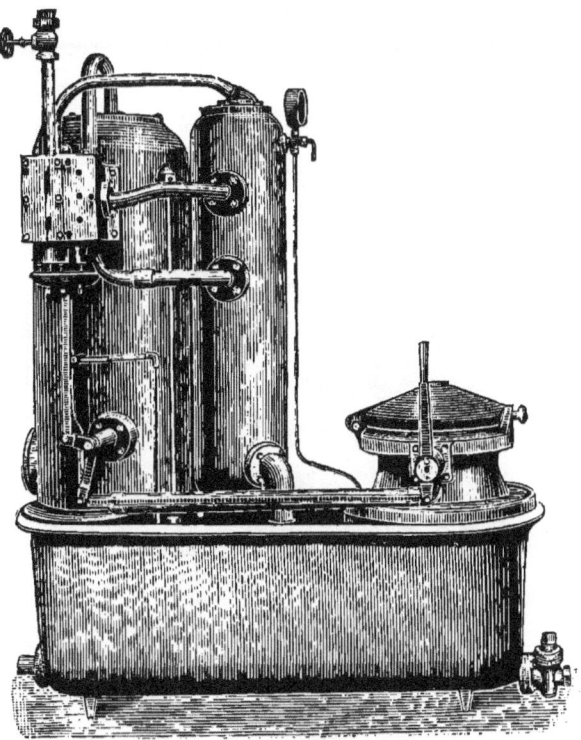

Fig. 51.—Cop Dyeing Machine.

these plates may be employed, so that, whilst one set of cops is being dyed, others may be in readiness to be placed within the treating chamber immediately upon the completion of the first set.

To the right hand of our illustration a lever is shown. The operator, having placed the cops in the treating chamber, turns the lever, when the dye liquid flows into the large receiving chamber,

136 COTTON MANUFACTURE.

and from thence back again into the bath, ready for another operation. It then enters the cop chamber, passes through the cops, entirely permeating them, and flows away through the spindles. It will be readily understood that a certain shade can be procured, time after time, for an indefinite period, with the most unerring accuracy, always providing that the dye liquid is prepared according to the specified formula for that shade.

Hank and Slubbing Dyeing.—Another very excellent machine

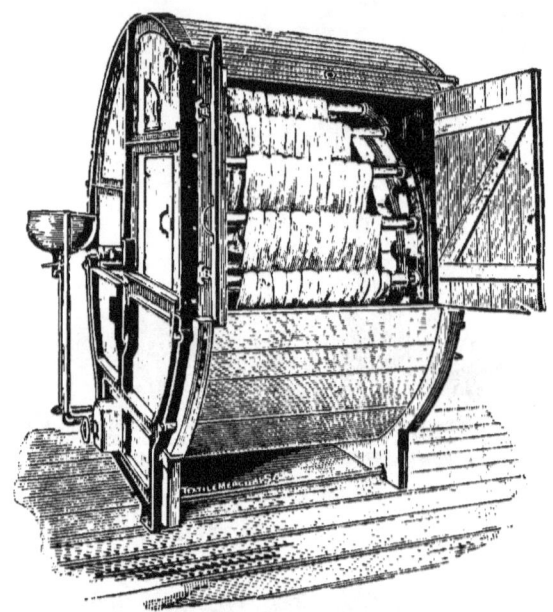

Fig. 52.—Hank and Slubbing Dyeing Machine.

(Fig. 52) is that for dyeing hanks and slubbing which was brought before the notice of the trade two or three years ago, by a well-known Yorkshire dyer. It is the invention of a practical dyer, who endeavoured, with a considerable amount of success, to overcome the objectionable features of some other machines for the same purpose. The apparatus is enclosed, and has a door at each side, although in the illustration we show it with the cover removed. There is an advantage in the enclosed, over the open vat, as when dyeing by hand in the latter, the temperature cannot be raised above 204°, without blowing and tangling the yarn.

By the enclosed bath, the yarn may be dyed under pressure, and such a high temperature may be gained that, in about five minutes, an effect can be produced equal to that obtained in fifteen minutes in the open vat, whilst the steam, not being able to escape, keeps the yarn at almost the same temperature in revolving in and out of the dye-bath. The yarn is placed on the sticks in the usual manner, 100 pounds being put into the machine in three minutes by a man and a boy. When this is accomplished, the boy can attend to two machines, and unskilled labour only is required. The machine can be emptied in the same space of time.

Whilst on the subject of time, we may say that 100 pounds of cotton yarn can be boiled out in about three minutes. The dyestuff is dissolved and poured into the pan on the side of the apparatus, and allowed to run into the vat while the machine is in operation. The yarn need neither be removed nor disturbed when dyestuff is being added, and the rotation of the machine diffuses it throughout the vat and facilitates the process of dyeing. The yarn is turned by an automatic trip, and as it is not exposed to the steam-pipes, cannot become steam-blown. The machine will take in from 54- to 90-inch reels.

We have especial pleasure in mentioning the two machines just described, as having been introduced by men actually engaged in the trade, who have them in use in their own works—a safe guide as to their reliability and usefulness.

CHAPTER XIV.

ACTION OF INDIGO IN DYEING.

The Alkaline Vat.—Aniline was formerly produced from indigo, until Mr. Perkins found out how to manufacture it from coal-tar. In dyeing blue with indigo the cold vat is used: it is made up with sulphate of iron and newly slaked quick lime. The hot vat is only used when the dye is to be extra fast; for instance, in goods that have to withstand bleaching. The hydrogenising of blue indigo changes it into white indigo. It is, therefore, necessary to determine the most favourable conditions under which an alkaline bath can be ormed.

White indigo is soluble, and remains in a state of suspension with ammonia, potash, and soda. The best materials for giving hydrogen are bran and madder—one of these for the warm vat, and sulphate of iron and copperas and lime for the cold vat. The vitriol vat is made by putting into the water green vitriol or the protoxide—coppered vitriol would oxidise the white indigo—and when the vitriol is thoroughly dissolved some newly slaked lime must be added and kept in motion for nearly half an hour; then powdered indigo is put in, motion being again kept up for a quarter of an hour, and the whole allowed to settle. After a period of four hours it must be stirred up, and this repeated every three or four hours, day and night —that is, for twenty-four hours. If this advice be followed, the vat will be in perfect order, and present a clear amber colour with copper-like sediments and scum covering the whole surface. A vat holding 200 gallons would require 26 lbs. of green vitriol; new slaked lime, 36 lbs.; powdered indigo, 13 lbs.; and, for a vat of less dimensions, these quantities in direct proportion.

We may now consider the chemical action which takes place. The lime acts upon the vitriol, precipitating from it protoxide of iron, which is really a combination of oxygen with the base. This protoxide decomposes the water in the bath, taking hold of the oxygen and freeing the hydrogen, which, combining with the blue indigo, changes it forthwith into white indigo, which remains in solution in the alkaline bath produced by the excess of lime used. A vat, when in the best order, will have a great quantity of scum—bunches of copper-like atoms, closely joined together, easily blown about by the breath. The bath ought to be of a clear transparent red-brown colour, streaked with blue. At this stage a piece of cloth dipped in will be stained a clear green, very bright and decided, which will slowly turn into blue by contact with the air. This is the perfect bath. Should it, however, appear blackish in colour, then it is not sufficiently alkaline, and lime must be added and stirred up for a quarter of an hour and covered over. It will be found all right in the course of a few hours.

Again, suppose the bath shows a greenish colour, it would indicate want of hydrogen, and the blue indigo remaining unchanged or only imperfectly changed, therefore, vitriol must be added, well stirred, and left to settle for a few hours. A cold vat may be renewed by successive additions of indigo, lime, and vitriol. This is the secret of getting an indigo vat into proper working condition; and, stripped of all technicalities, it is or ought to be clear to any person of ordinary comprehension. There is no difficulty in making this liquor, nor in working it. The yield is almost an absolute certainty, especially if due care be taken not to have an excess of lime; and, although this would not materially affect the fibres or fabrics dyed, it would waste the indigo, by the white indigo, in part, being changed into blue indigo.

Of course it will be requisite to pay attention to the proper proportions of the materials employed. The best quality of indigo will require more copperas and lime than the common brands, which contain useless matters introduced as adulterants. When a sufficient time has been allowed for all to settle, the deposits must be removed, as they are of no value. The liquid is drawn off by a pipe, one end longer than the other (in fact, a siphon), into another vat, which may be used as a new one.

The Hot Vat.—The hot vat may be formed thus : A large kettle, holding 150 gallons, is put over a fire, with sufficient water heated to 120° Fahr.; 50 lbs. of burnt ashes, or the equivalent in potash; 10 lbs.

of coarse bran, 9 lbs. of powdered indigo added; stirred every three hours; at the end of twenty-four hours it is fit for use. This vat requires some experience in the work. It shows its condition by the odour, because the bran undergoes a kind of butter fermentation, forming butter acid, known as butyric acid, with a peculiar rancid smell. If the bath is not actually yellow, there is a want of hydrogen, proving that the blue indigo is not completely changed into white indigo. The signs of perfection given for the cold vat are the same for the hot vat—clear, yellow-blue streaks, violet top, and a crumpled scum. The work can be done by this bath without intermission. The temperature ought always to be at or near 120° Fahr. In the evening the vat must be supplied with new indigo, potash, or soda, and a small portion of bran. The putting in of a fresh 10 lbs. of indigo does not require the first proportion of 50 lbs. of ash; 6 lbs. will do, with 4 lbs. of bran. If the vat appears of a black cast, or, say, dark blue, without streaks or scum, there is a want of hydrogen; then a dose of bran, a heating up, a few stirrings at three hours of an interval each, and the remedy is obtained. If green-looking, same remedy, only less bran. The bath may appear as clear water, and in this case it will be found that the indigo has fallen direct to the bottom of the vat. To obviate this, add potash warmed up to 110° Fahr., give three or four stirrings at intervals of three hours, a settling down of four or five hours, and it will then be found ready for use.

The cotton fabric, yarn, or fibre, is now dipped into the liquor, and becomes saturated with the solution. The goods are taken out of the vat, hung up exposed to the air when the oxygen in the atmosphere enters into combination with the white indigo in the fibre of the material, and restores its blue colour. This blue indigo becomes a permanent colour, and cannot be washed out. Other methods may be used in which madder is a principal ingredient with indigo; but in all cases the result is due to the chemical action we have described.

The indigo dyeing machine is illustrated in Fig. 53.

Chemical Reactions.—The action of indigo opens out a very wide field for inquiry, but here we can only give the primary principles underlying the curious chemical change which takes place with this dye stuff.

The dyeing of blue with indigo is to hydrogenise it, so as to transform it into white indigo, which is soluble in alkaline baths;

ACTION OF INDIGO IN DYEING.

then by oxidising, when put upon textiles to be dyed, it is restored to its original colour. It is, therefore, requisite to discover the most favourable conditions under which an alkaline bath can be prepared, and also the reagents which furnish the amount of

Fig. 53.—Indigo Dyeing Machine.

hydrogen in a constant and regular succession, as well as the reagents necessary to supply the oxygen. It must be borne in mind that white indigo is soluble, and remains in a state of suspension in alkaline solutions, such as soda, ammonia, and potash, but in the solutions of earthy alkalies, as baryta, lime, magnesia, and strontia, these are not so easily soluble ; alkaline carbonates are soluble. The best materials for a supply of hydrogen are madder and bran, or one of them for the warm vat, and sulphate of iron, or copperas and lime for the cold. The oxygen can be obtained by the air when the dyeing is done. The alkali is supplied by treacle, or paste, which evolves more than a sufficiency when fermented. Decomposed urine, ammonia, soda, potash, or their carbonates, are also useful agents for this purpose. The fermentation of putrefying matter evolves ammonia; this is required to supply the liquid in the vat with alkali, so as to give the bath the required power for dissolving white indigo ; butyric fermentation is further required for producing hydrogen, in order to transform blue indigo into soluble white indigo, the oxygen of the air being sufficient to take away from the white indigo the hydrogen it has absorbed, so as to restore it to the state of blue indigo in combining with the hydrogen and producing water.

As to the chemical reactions—the indigo formula is $C_{16} H_5 NO_2$; white indigo $C_{32} H_{12} N_2 O_4$, which is really the same as $2(C_{16} H_5 NO_2) \times H_2$; in simple language, white indigo is equal to 2 of blue indigo multiplied by hydrogen. If this last formula is taken and considered as that of white indigo fixed upon the material to be dyed, bringing it out of the vat and again entering it, then we have $2 C_{16} H_5 NO_2 \times O_2 = 2 C_{16} H_5 NO_2 \times HO$, an equation giving blue indigo and water. In considering these deductions, it will be sufficient in actual practice to have an alkaline bath, in the midst of which hydrogen will be produced, to be able to transform blue indigo into white indigo, and, consequently, to render it fit to be employed usefully as a dye.

Difficulties of Indigo Dyeing.—It is a moot question whether alizarine blues can furnish the good qualities of a perfect indigo dye at a competitive price. In all the varied processes of dyeing we are of opinion that the successful working of an indigo vat is one that requires the most careful treatment and experience ; in all other dyes, the defects are obvious. The indigo vat may look in good condition, but, when the material is taken out after immersion, it falls short of what has been expected, without any apparent reason. It is often the case that fermentation sets in rapidly during the night, and, before it can be used, the material (so to speak) destroys itself, and is, therefore, of no value. We have been careful to give the best possible instructions to prevent this failure. Overworking produces a flat vat, little lime is taken, the indigo is not properly dissolved, because the fermentation ceases to act, and additional bran, madder, etc., is then required. Many dyers renew every day, others once a week, but, of course, very much depends upon the amount of material passed through.

Recovery of Spent Indigo.—A very important and economical point in indigo dyeing is its recovery from spent vats. When the lime and copperas are exhausted as far as possible or practicable, there will remain a quantity of indigo, both in solution and with an insoluble residue, at the bottom of the vats, which is too valuable to be thrown away. The soluble portion may be allowed to settle well, and the clear liquor pumped into an empty vat; it can then be employed in the setting of a new vat instead of the same quantity of water.

The insoluble matter at the bottom is treated in many ways. The following process will be found most satisfactory, as almost every particle of indigo is recovered, and especially is this so where there is no room for settling pits; but in any case and under any circumstances it is the best so far known. The vat bottoms are heated with caustic soda and orpiment. The boiling vats are provided with plug holes at various distances from the bottom. The stock-reducing liquor is made by taking in the proportion of 1 lb. of orpiment, 1 gallon of strong caustic soda, and $\frac{1}{2}$ gallon of water. The orpiment is finely ground, and the whole boiled by steam, in an iron vessel, until solution is effected. It is usual to prepare 30 gallons of this reducing matter at one operation. From 3 to 4 quarts of this liquor are added to the contents of the boiling vat, which is then well raked up, and thoroughly boiled. The steam being turned off, the contents are allowed to settle, and drawn off by the plug holes in the bottom to a cistern, and from thence pumped into spouts leading into receivers placed about 12 feet above the dye-house level. These spouts are made broad and shallow, so that the liquid may have all the advantage of contact with the air. The indigo, in solution, is speedily oxidised, becomes insoluble, and settles to the bottom of the receivers. The clear and hot liquor from which indigo has been deposited is used instead of water in the boiling of a fresh lot of bottoms, or re-boiling a lot already treated.

If the indigo vats have been well set and well dipped out, it will be found that something like six or seven boilings with orpiment liquor will remove all the indigo from the bottoms, but, in obstinate cases, perhaps, this number of boilings will require to be doubled. This hot process is more rapid in action, and extracts from the bottoms more indigo than any other form of treatment.

To ascertain quickly whether a given specimen of vat bottoms contains much or little indigo, let the vat bottoms be raked up as much as possible, and then take a quart sample, mix this with 250 grains of orpiment and 2 ounces of caustic soda, boil in an iron pan, let the liquor cool, and test by dipping small pieces of calico, etc., in the clear part, and the depth of shade which is dyed in a given time will indicate sufficiently well to an experienced eye the probable amount of real indigo present. This test at once determines the requisite number of boilings.

CHAPTER XV.

CALICO PRINTING.

Dyeing and Printing.—The dyeing of cotton is associated with printing, and may be looked upon as a first operation. Printing is the production of a dyed fabric, with an impression of a pattern or figure of one or more colours. In block printing the colour or mordant is applied by hand blocks, with the pattern cut in relief on the wood. The block charged with the colour is transferred to the cloth; but of the many inventions introduced in the production of dyed and printed fabrics that of cylinder printing has become the most important. A separate copper roller is mainly in use for each colour, and the various parts of the pattern are engraved on the surface of these rollers. The rollers, when in the machine, dip into colour troughs, and this colour is by rotation applied to the cloth. When the surface of the roller is saturated with colour, before it comes in contact with the fabric, a fixed and thin metal plate, called the "doctor," scrapes all the superfluous colour from its surface, leaving only the engraved portion charged with the requisite amount. When the cloth leaves the printing machine it is subjected to treatment until the colours are perfectly dry.

Calico Printing Machine.—In order that this may be more fully understood, we give an illustration (Fig. 54) showing two rooms of a calico printing works. In the bottom room is shown a representation of an eight-colour duplex calico printing machine, the particular machine from which the view was taken being the largest one of its class extant, as it is capable of printing from rollers 15 to 48 inches in circumference and 60 inches wide. The engraving does not, of course, do justice to a machine of such magnitude; but as our object is to show the process rather than to describe any particular machine, it will be sufficient for our purpose.

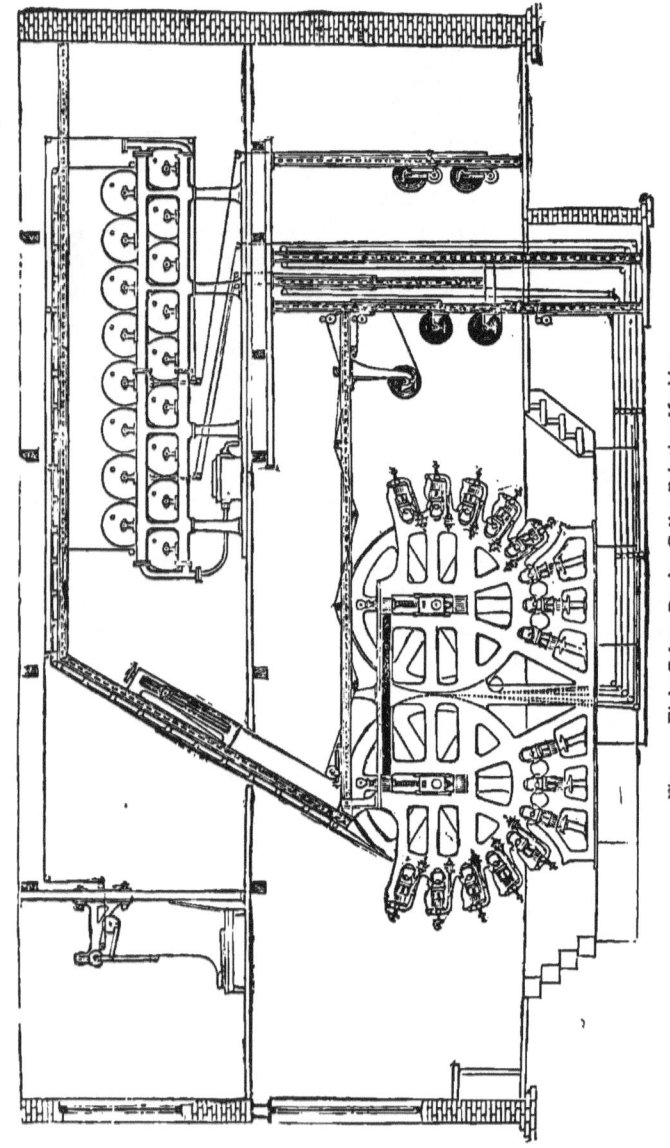

Fig. 54.—Eight-Colour Duplex Calico Printing Machine.

We have alluded to the method employed in printing, and it only remains to state that after the fabric has left the printing machine it is carried to the drying cylinders and steam chests on the second floor, as shown in Fig. 54.

Printing on Turkey Reds.—Some very fine effects are obtained by printing on Turkey reds. These prints being fast to all the ordinary influences of light and air, soap, dilute acids, and washing, their quality may be depended upon by purchasers, because they are the production of real indigo and alizarine red. The general process consists in hot watering the Turkey-red cloth thoroughly at a temperature of 160° Fahr., then drying and preparing in a solution of glucose at 16° Twaddle. The pieces are then padded in the printing machine with a good roller, dried as usual, and wound on for printing; this part of the operation in printing is in blue, dark or light, or blue and white. After printing they are properly dried, care being used that over-drying does not take place. They are aged in an aniline-black machine (Fig. 55) in an abundance of steam from 60 to 80 seconds. Afterwards the cloth is hung for a night in a cool room, well washed in cold water, squeezed, passed through sulphuric acid at from 6° to 8° Twaddle, and again washed in cold water until all traces of the acid are removed; the soaping is done at a 10 minutes' boil, squeezed dry, and finished. The fabric is passed through sulphuric acid (vitriol) to erase the alizarate of soda underlying the blue and white portions of the pattern. The caustic soda in blue and white colours, on steaming, decomposes the alizarine red lake, forming aluminate and alizarate of soda, which, though soluble, cannot be washed out without considerable difficulty; at all events, by the action of cold water only traces will remain. On running the cloth through vitriol the aluminate and alizarate salts are decomposed, and every atom is then washed out with ease.

In patterns containing no white this vitriol bath may be omitted, if the goods are treated with hot water followed by soaping; and it will be found that rather brighter colours are obtained, especially in blues, even if the red be not entirely discharged, nor the soda aluminate and alizarate thoroughly washed away. The fact is that the alizarate remaining in the cloth, being itself a blue to a certain degree, does not in any way injure the shade, but free alizarine remaining unwashed out in the acid treatment would materially alter the shade of blue, because alizarine is yellow, but subsequent alkaline

Fig. 55.—Aniline Ageing Machine.

soaping of great strength would no doubt counteract it; still the strong soaping would again not be advisable on account of the red and blue.

Padding and Topping the Fabric.—Pale and medium indigo colours are best suited for padding on white cloth, but perhaps the best effects are obtained with very pale shades of blue. The cloth is padded first in the ordinary way with glucose; the colour being made sufficiently thin and padded in the machine with the roller, is then taken through the same process which we have just described for blue and white. The quantity of indigo used is small, because it is only on one side of the cloth, but dip-dyed indigo is on both sides, therefore a large quantity in such a process would be necessary. Sky shades can be more cheaply produced in printing than by dyeing. On the blue padded cloths the ordinary discharge indigo effects by the chemic and by other processes can be resorted to for the indigo discharge colour styles. The most commonly used toppings for pure indigo yarns are logwood and violet, but other methods have been introduced with more or less success. One process for cheapening indigo yarn dye is to treat the material with copperas, and to afterwards dye in alizarine and a little sumach. This gives a deep though dull blue.

A good method of detecting this is to place it in a test tube, add some strong spirits of salts, and warm; then pour into another tube, dilute with water, and add caustic soda in excess. If alizarine has been used either as a topping or bottom, the dirty orange-coloured liquid will turn reddish-violet. Benzoazurine topping is detected by the action of strong nitric acid, which destroys the indigo and dissolves only the topping to a deep green liquid. In a solution of caustic soda the topping is dissolved out to a reddish liquor, leaving the indigo untouched. Aniline black topping is often used for very deep blues, and is easily distinguished by touching a thread of the yarn with nitric acid, the indigo being bleached. Many other blues can be made to combine with indigo; for example, alizarine blue.

There are hundreds of methods in use for the production of "indigo blues"; and although we have several "fast" artificial blues in use, giving every satisfaction to the consumers, yet many buyers have a disposition to insist on having pure indigo, though they are far from willing to pay the indigo cost. As we have just said, there are good cheap fast blues, as may be seen in the many printed fabrics with white spots on a blue ground, both for home and export trade, the principle of which is "alizarine blue," sometimes termed "anthracene blue," not usually employed for yarns or plain piece goods, but for prints. Deep blues of real indigo shades, and also green shades of sky blue, are produced with it both in "blotch" colours and "objects." It is a very beautiful colour, and extremely fast, and cannot be distinguished from indigo without testing. It may be said to be preferred to indigo, because in some respects it is faster. Strong nitric acid will leave the cloth orange, but on adding a solution of caustic soda it will become green, and in washing with plenty of water it is restored to blue.

The old-fashioned Prussian blue is generally employed for cheap yarns, where dark shades are required, also for cheap prints—single blue and white—and in combination with dyed alizarine in forming the present popular style of "dark blue and red" prints in imitation of the printed indigo and alizarine reds, but Prussian blue, even the dark shades, will not oppose soap, though they are fast to light, air, and acids. On treating with strong soapsuds, or with an alkali, such as soda, the blue disappears, leaving behind a rusty sort of red; the colour can, however, be again restored by treatment with an acid. A drop of strong caustic soda is a sufficient test. Logwood blue on yarn was at one time extensively used; boiling with hydrochloric acid will turn it red. Indigo dyeing must, however, be considered the most permanent of the various branches of the trade, and is at the same time the most satisfactory. It is not really so remunerative as many other dyes, except where there is a great output. The large influx of labour and capital into the business of indigo dyeing has perhaps been a great factor in reducing the profits, but it is in the production of moderate light shades of indigo that the greatest competition prevails with the least profits. The dyeing of the very darkest shades on cloth or yarns requires considerable skill, experience, and attention, and is made a speciality by some firms who can always command a fair price for good shades of pure indigo.

Alizarine Red Extract.—Alizarine red extract for calico printing is now to be found in nearly all classes of prints in which red appears, excepting in very loose work. It cannot be combined with all colours, but in conjunction with most of those now in calico prints. The increase of this extract is greater than that of alizarine for dyeing, because it is cheaper, shorter, and simpler in practice particularly in the production of red and pink. There is, however, no class of prints that varies so much in quality; the method of printing, colour mixing, length of time in steam, all make a considerable difference for good or bad in the appearance of the print. Sometimes the shade of red is dull brick-red, sometimes brownish, or too yellow in cast, or too blue or crimson; but, on the other hand, good workmanship gives the alizarine red extract as a bright, deep scarlet, nearly resembling a Turkey red.

Colours which are used in combination with extract alizarine colours in printed goods are, for the most part, fast, or moderately so, and are never made use of in such work as furniture prints, cretonne, etc.; many of the colours used in extract red fade in the light, such as bright sky blue and bright violet. In combination with alizarine red extract, there are to be found other alizarine extract colours, as pink, salmon, maroon, chocolate, brown, lilac, or purple, logwood black, aniline, and many other artificial colours.

There is at times a most unpleasant smell in some prints and in dyed alizarine work, which is mainly due to bad oil employed in the manufacture of the soap used for soaping the pieces. A very small percentage of rancid green olive-oil present in the soap imparts this disagreeable odour, which adheres to the fibre with the greatest tenacity. This smell is the chief reason why artificial indigo printing on calico is not a commercial success; but a very slight odour may be corrected by the addition to the finish of some essential oil. In this, however, as in so many things, prevention is the best cure.

Extract Red.—Extract red should be bright, deep scarlet in shade, and possess the peculiar bloom which is so distinctive of good manipulation. The colour should be printed in such a manner as not to penetrate to the back of the cloth, showing almost as much colour on both sides; such colour as is on the back should be red, and not yellow or amber in shade. This last shows faulty work—too much acid colour—so that, although the shade of red on the face is no yellower than it ought to be, yet, on washing the prints, the red will "bleed" into the other parts of the pattern. Good extract colours will but slightly rub off on being pressed with clean damp linen or cotton. Recent improvements have brought extract printing to such a perfection that dyed alizarine work in prints is nearly a thing of the past.

Variety of Reds.—A great variety of reds is in use for cotton, and they may, for all purposes, be classed as follows:—Ordinary aniline series, fast series, wood reds, including barwood, peachwood, etc., the Congo and benzo-purpurin series. The application of Congo and other reds diminished the consumption of such anilines as ponceau, crocceau, but not such as eocine and saffranine. A very great number of aniline scarlets are sold, differing more or less in their reactions or conduct with reagents.

As many varieties are simply called "aniline scarlet," which thus differ, the washing test ought to be used with a strong solution of soap upon the yarn or fent. Of course, it is not likely that such colours as magenta, eosine, or saffranine will be mistaken for a scarlet by reason of a difference in the shade. Whatever skill the dyer may employ in his work, or whatever expedients he has recourse to, the best results cannot be obtained unless the best materials are used, and we should say the most important of all is the red liquor. Nothing is so liable to upset every calculation if the quality is indifferent, and none, moreover, is sold in so many different qualities, good and bad. The ordinary red liquor does not always yield a red which will give a clean discharge—a fault often laid on organic impurities present in the mordant.

A good red liquor for dyeing Turkey red and ordinary alizarine red yarns with alizarine will shorten the dyeing process; and the work produced is so nearly equal to real Turkey red yarn, that it is difficult to distinguish the one from the other. A very beautiful bright scarlet is obtained as follows:—Prepare the Turkey red cloth, or black and red print, in a solution of acetate of tin, at 9° Twaddle for a heavy topping, dry and dye up, warm in aniline scarlet (ponceau or croccine), to depth of shade required; this will give a brilliant scarlet, which will be found remarkably fast.

Indigo Printing.—By a recent process goods can be printed with a solution of indigo, and fixed by the combined action of air and water. The solution is prepared in the following manner:—1 lb. of indigo is ground in water, to which is added 3 lbs. of caustic soda, but the soda must be kept stirred until completely dissolved, $1\frac{3}{4}$ lb. zinc powder is then added by slow degrees. When the zinc is dissolved more caustic soda is put in, until it is found that the yellow solution produced will remain a yellow when its surface is exposed to the oxygen (air) for at least five minutes. A mixture of gum water and soda solution is added, until the mixture becomes a gummy paste. With this solution, unprepared cloth is printed upon, and led over rollers in such a way as to cause exposure to the air, which partially oxidises the indigo, and thus more or less fixes it. The cloth is carried in front of a perforated water pipe, so that the jets are thrown against it, then it is led up and down through water tanks and between the tanks, still exposed to the air, for less than a minute. The water jets may be omitted if this process is properly conducted.

After the indigo has been thoroughly oxidised in this manner, the cloth is led through an acid bath, by which the superfluous material is removed, and it is then treated with soap and water, and washed and dried in the ordinary way. If it is found desirable to *colour both surfaces*, the cloth is run through the ordinary indigo vat, or through the vat containing the solution given above, but weakened to the desired *tint* by soda gum solution, and various new effects may with ease be produced by the use of such a vat in combination with this process. Thus the cloth, after being printed, exposed to the cold air and cold water, is run through the vat, and again treated as before, the result being that the portion *printed* upon receives a second coating and becomes darker, whilst the *unprinted* portions of the *face* and the whole of the *back* are dyed a lighter shade.

In this process, as well as in the simple printing process, two or more printing rollers may be employed, printing with solutions of different strengths.

Oxidising Print-Work Chambers.—Considering the revival of the print-work industry, we believe our remarks on the latest processes will be appreciated. By allowing a current of warm air along with the steam to enter the chambers of cotton print works, they would not only become fixing rooms for the mordants on superior goods, but also regular steaming apparatus for inferior fabrics. This really deserves some degree of attention. So-called *oxidations* do not fulfil all conditions.

Let us suppose, practically, that the first object is to fix the mordant by application of heat and moisture, or, in other words, to liberate it from a portion of its acetic acid, so that it will remain on the cotton fabric in the form of a *basic salt*, insoluble in water. Only with certain shades, or with such mordants as are wholly or partly composed of acetic or protomuriate-hydrochlorate of iron, is there a simultaneous oxidation. It is, however, a mistake to suppose in both cases that ventilation is altogether wanting or insufficient in oxidising chambers; they become filled with a strong acetic atmosphere soon after the goods are hung up, which at once prevents a further liberation of acetic acid from the *mordant* taking place; and in a more rapid degree the supply of *oxygen* is used up in rooms where the printed cottons remain from one to three days, and yet more rapid is this the case in chambers where the goods remain probably from half an hour or more; indeed, in this way, it becomes impossible to obtain with certainty certain strong shades, even if the printing colours are taken doubly concentrated. These shades demand, and must have, to meet with any success, a corresponding amount of *oxygen*, whereas, the addition of nitrate of copper is limited. This is no theory—it is well known in practice; and, therefore, it becomes evident that the passing in of a warm current of air would not only help to remove a defective ventilation, but also ensure a complete oxidation. The fixing process would be more rapid, and the putting up of a simple blower is all the expense necessary.

The present temperature of an ordinary oxidising chamber is generally, say, from $104°$ to $112°$, whilst the two thermometers of the hygrometer will show a difference of $7°$ or $9°$. Now, if a temperature of $212°$ could be maintained by means of a warm air current, this evil would vanish. No doubt, for the mere fixing simply of mordants, this high temperature would be of no real value; but practical men are very much impressed with the fact that some such mode of procedure would fix steam colours on the fabrics with a very great saving in fuel, and, by simple means, less costly than the cumbrous expensive steam chambers now in use. By the ordinary system of oxidation, a full catechu brown is almost impossible, and real oxidation is positively necessary in mordants consisting of acetate or muriate of protoxide of iron.

Injury to Cloth.—A very concentrated solution of caustic soda in making the printing colour will no doubt tender the cloth; but experience shows this only can take place when the *glucose* prepared cloth has become damp before printing, so that the caustic colour penetrates the cloth, and so, acting strongly on the cotton fibres, contracts and hardens the fibres.

Arsenic in Prints.—A large proportion of prints contain small quantities of arsenic—so small, in fact, that there is not the slightest cause for alarm. Many of the anilines, such as ceruline blue, aniline greens, etc., and many of the vegetable colours, are fixed on calico by printing the colour with a salt of alumina and a solution of white arsenic in glycerine, or in a borax solution. The reaction that takes place on steaming the goods is a double compound of arsenic, alumina, and colouring matter, or, briefly, double arseniate of alumina and dye. This compound constitutes the insoluble lake which colours the fibres of the cloth with a more or less insoluble and fast colour. This at once removes any danger in the wearing of the material. As a matter of fact, arsenic colours contain a considerable excess of alumina, and this is a *preventative* against the possible presence of uncombined arsenic. In extract alizarine colours, the soluble arseniate of alumina is sometimes added to brighten the colours, but, on steaming, the insoluble compound is obtained.

When used properly there is no harm in the use of this drug, and the cry which, no doubt, has been the cause of a decline in some classes of prints has originated from experiments on a small scale and conducted on false premises.

Colours.—Auramine is a colouring matter which gives a very pure shade of yellow, whether dyed on yarn or printed on calico; for yarn chrome yellow is, however, cheaper and more readily produced. Auramine has all the advantages of aniline yellows, without their deficiencies, as it is moderately fast to soap and light. If dyed by means of alumina, a good green shade is obtained, but it is loose; when fixed with sumach and tin, or tannic acid and tartar-fustic, a most beautiful shade of pure maize yellow is obtained. The more tannic acid employed—up to the limit of 12 ozs. good tannic acid to 3 ozs. auramine — the faster will be the yellow obtained; but it must be observed that the colour is not so bright as when less tannic is used, because the brown dulls the yellow. The same must be understood with equal force as to the use of auramine in printing, where a fine colour can be obtained by fixing the auramine with 12 ozs. per gallon of pure tannic acid, and from 3 to 4 ozs. auramine; this will stand strong soaping, but not so well as *berry yellow*.

Auramine is, at present, largely adopted in many styles, as it is much less difficult and a more regular colour to work than the berry yellow, and it will work well with many aniline colours. A yellow shade of green is got from it, and aniline crystal green.

To detect it on the fibre, the following tests are reliable:—Caustic soda turns the colour white very rapidly, dilute hydrochloric, same result. It can be readily ascertained whether the colour is fixed with tannic acid or alumina by boiling a piece of the cloth in a dilute solution of chloride of iron. Blackness will show tannic acid.

A very fine shade of yellow, possessed of extraordinary *fastness*, in fact, the fastest artificial colour as yet discovered, is chrysamine, used very much in dyeing, rarely in printing. It is not materially affected by acids, soap, or alkalies, and even caustic soda, light, air rubbing, or chemic have little effect, except that alkalies turn it to an orange shade, whilst acids will restore it to a pure yellow, with a slight tendency towards light green. The present prevailing features in some print dress goods of pale yellows and buffs, as well as in cotton hosiery and laces, are produced by chrysamine, or mixtures of it, with benzo-purpurine, and other *azo* colours. In dyeing, the shade is obtained at one bath and without a mordant operation, etc., necessary in other dyes, and which are so injurious to the fibre of fine cotton lace goods. The reason the use of this dye is so much restricted is that a deep shade of yellow cannot be obtained so far. It is, however, found useful for buffs, and for every delicate shade of pale yellow, salmon, etc.; it is not readily soluble in cold water, but dissolves freely in hot, and is still more soluble when in boiling water with a small atom of caustic soda. There is little doubt that this is the yellow of the future, when science unfolds nature's mystery.

Resists.—When deep bronze grounds are required, it would be well, in order not to tender the goods whilst being dried in the hot flue, to transform a part of the manganese used for the process (well known to printers) into pyrolignite of manganese, by adding a solution of pyrolignite of lead and making use of the clear liquid. If the finished pieces are steamed blues become brighter, but this brightness must be preserved, not by *subsequent washing before steaming.* Let this remark be noted particularly. Drying over copper cylinders by steam will also render all blues brighter than air drying. In the dyeing of dark blues, at the end of the first day's work, put into the vat a few pounds of lime and copperas, and, after the second day, a few pounds of indigo, lime, and copperas. By this process, for the *red* and *white* discharge—after the pieces have received a medium blue in the vat, they are steeped in chromate of potash, with proportions as follow—vat in 6 to 8 dips; steep in bichromate of potash about 4 troy ozs. to a quart, and dry on rollers in the shade; for the white discharge—water, nearly 1 gallon, white starch 3 lbs.; boil, and add, when lukewarm, tartaric acid, 2 lbs., then, oxalic acid, 19 troy ozs., dissolved in 1 quart of water. For the *red* discharge—acetate of alumina, 4 gallons, white starch, 16 lbs.; boil; allow one half to grow cold, and add 7 lbs. of oxalic acid, then the other half of the hot mixture to complete the solution of the oxalic acid.

It will be found for this printing mixture that actual practice will not require more than 8 lbs. of white starch, if of good quality. A reliable process is to use the following instead of printing on the red :— Dissolve oxalic acid in hot white starch paste, and, when cold, add *bichloride of tin;* print, and do not dry too strongly. *Dung* with *chalk* and silicate of soda, or with chalk alone; then wash and dry up in garacine.

Orange Discharge on Dark Blue Vats.—Give a deep vat blue, and print on—water $3\frac{1}{2}$ pints, white starch, 12 troy ozs., acetate of lead, 4 lbs., and precipitated manganese, 6 lbs.

Preparation of the Precipitated Manganese.—Water, 5 pints, chloride of lime, at 14° B., 3 pints. Add the following mixture :— Water, 2 pints, chloride of manganese, at 80° B., $1\frac{1}{2}$ pint, muriatic acid, 1 troy oz. This precipitate must have a deep brown colour. It is washed with filtered water three or four times, the pieces are then passed into muriatic sours, and into weak copperas water, thence into lime water; steep in bichromate, and rinse in boiling chromate of lime.

Preparation of Acetate of Alumina.—Alum, 2 lbs., pyrolignite of lead, 2 lbs., water, 3 pints. Deep puce is obtained by printing on a red with acetate of alumina. Print on the red and white discharge with the perrotine, or with a two-colour cylinder machine. Do not dry too strongly. Hang out in hot, but not moist, air. This is strictly a condition of success. Next morning, dung as follows :— Charge a beck, having rollers, with neutral arseniate of potash, 6 lbs., chalk, 24 lbs., water, 215 gallons. Pass the pieces slowly through at a simmer, to keep the chalk in suspension. When taken from this beck the pieces must be strongly squeezed between rollers that are covered with cloth; the beck must be fed, after 7 or 8 pieces are passed through, with $1\frac{1}{2}$ troy oz. of arseniate of potash, and a little chalk per piece. After cleansing, dye in garacine bark and woods, and pass through boiling bran. The oxalate of alumina formed in preparing the *red* is very hygrometric; the pieces must, therefore, be thoroughly dry, before passing into the beck with rollers, otherwise the reds would be poor and spotty.

CHAPTER XVI.

PREPARATORY PROCESS OF WEAVING.

Drawing-In.—This is a preparatory process to weaving, and for plain fabrics is very simple, merely requiring a degree of dexterity in placing the warp threads in succession through the eyes of the healds on each shaft or stave, from the front to the back, or the reverse, according to custom; some drawers-in commence from the back. In fancy drafts, however, great care is necessary in placing the colours and patterns in their proper order and according to the directions given by the designer. Now, however complicated the draft may be, or the multiplicity of colours, the figures given with other details for reeding are sufficient to prevent mistakes, if the drawer-in is at all competent. We give an example of a plan generally used for a guide.

DRAFT PLAN.

8		14				34	38	40	385
7	9	13	15			33			385
6	10	12	16			32			385
5	11		17			31	37		385
4			18	24		30	36		385
3			19	23	25	29			385
2		.	20	22	26	28			385
1			21		27		35	39	385
									3080

The figures on the margin of the horizontal lines, meaning heald shafts, indicate that the warp ends are drawn through the heald eye, on each shaft, in the succession shown; the figures on the right hand give the number of healds required on each shaft for the breadth of the fabric. *Reed* 48—4, 32 inches wide, 8 shafts, white selvages, 3,080 ends. *Pattern* 12 red, 4 white, 8 blue.

Calculation for Number of Healds.—The calculation for number of healds is as follows:—48—4 means 24 dents per inch, 4 in a dent, equal to 96 warp threads per inch, multiplied by breadth —32 inches—gives, including allowance for selvages, 3,080 threads. The draft plan shows 5 figures on each shaft, and 40 threads show one draft repeat or gait over; then we have 3,080 ÷ 40 = 77 repeats, multiplied by the 5 on each shaft, gives 385 healds for every shaft. In reeding or slaying, there will be 3 dents of red, 1 dent of white, and 2 dents of blue. The colours must occupy their own dents and not mingle with another colour in another dent—commonly called *split denting*, and should this occur, the reeding would have to be done over again. In many cases, the healds might be finer than the number given—that is, containing more healds than are required. Suppose 390 on each shaft, then these extra 5 would have to be left empty at every repeat of the draft, or in any way that would prevent a *strive* between healds and reed.

It very often happens that the number of ends of warp and breadth in inches are given to find the reed. In the example just given, 3,080 ÷ 32 inches = 96 ends per inch, but we have only 32 reeds at liberty—that is, 32 dents per inch—then in place of 4 in a dent, we should have 3 in a dent—32 × 3 = 96. No doubt extreme cases will now and again crop up, which must be dealt with according to circumstances. A good rule for finding the given reed is to divide the number of warp threads by the number in the draft, and divide by the breadth required. This quotient, multiplied by dents in draft, will give dents per inch—thus 3,080 ÷ 40 (extent of draft) = 77 × 10 (number of dents in draft, 40 ends at 4 in a dent) = 770 ÷ 32 inches breadth = 24 dents per inch, with an allowance for selvages.

It is scarcely necessary to multiply examples of the numerous drafts, because, as we have already shown, the numbers given with every draft plan at once points out the progressive order of the warp threads as taken from the lease by the reacher-in, and given to the drawer-in. Designers or pattern weavers may not, at all times, give full details to drawers-in, although they ought to do so, and, in every case, simplify the drafts as much as possible, not only for the drawer-in, but also for the convenience of the weaver, to whom, in a very complicated draft, a shuttle smash is a question of serious consideration.

It is little short of gross negligence to find, when the warp is inducted into the loom and every preparation made for weaving the fabric, wrong drafts, etc., so that all has to be unloosed and taken back again to the drawer-in; this is simply wasting time and creating expense, which cannot be endured in these days of keen competition and little, if any, profit. The reduction of drafts may be carried to excess—so far, indeed, that sheds cannot be properly formed in the loom for the shuttle to pass through; and it is here where the difference is at once discovered between theory and practice. An extra shaft or two would clear away the difficulty, although an alteration might be required in the draft, and some shafts may have to bear more warp threads than others, perhaps, twice as many. It becomes necessary, therefore, to prevent, by all possible means, this overcrowding; and, failing to be able to do this, the best way out of the difficulty is to arrange the shafts in such a manner that those having the least number shall be placed at the back, so that the shed may be clearly formed, as the smallest angle is in front or nearest the slay. Of course, judgment must be used, because, if the warp threads are fine and very few in number, or of silk, by being placed directly at the back, they might slacken and not take up in the weaving. In such a case, the shaft or shafts would be better placed near the back, between others having more threads on; but experience can alone determine this, which is a rather vexed question under any circumstances.

Reeds, etc.—Although it is becoming every day more generally understood that the dents per inch is by far the best standard to adopt, yet there are different calculations still in use throughout the manufacturing districts of England, Ireland, Scotland, and Wales. It would be impossible to go into full details for every system, if, indeed, any practical benefit would result from so doing; we will, therefore, merely notice a few by way of comparison.

The Stockport cotton count is reckoned by the dents per inch, two in a dent; therefore, a 50-reed in this count is divided by 2, so that 25 dents per inch, 2 in a dent, contain 50 warp threads, and so on with any other counts in this system. The reeds for fustians, velveteens, cords, and heavy beaverteens are mostly 32, 34, or 36 beers on 24·25 inches in breadth, each *beer* of 19 dents; taking any one of these reeds, we merely multiply the number given by the 19 dents, thus 34 beers × 19 dents = 646 dents, two in each dent = 1,292 warp ends on 24·25 inches breadth, and, if required on 30 or 36 inches, the proportion would be as 24·25 gives 1,292, what will 36 inches give? = 1,918 threads. In these reeds, 27 inches is often the breadth for velveteens, 30 inches for beaverteens, and, in some cases, only 19 inches for certain fabrics, as trousers goods. The English *beer* and Scotch and Irish *porter* are alike; 20 dents is commonly the measure of the Lancashire beer, and thus 20 dents, 2 in a dent, is equal to 40 threads. The Scotch porter, a 25 Scotch porter = 500, because 25 × 40 threads, each porter = 1000 ÷ by 2 ends per dent gives 500 dents on a 37 inch scale of breadth. Now, if 40 × 37 inches gives 1,480 ends, then this Scotch count is equal to a Stockport 80-reed, which has 40 dents per inch × 37 inches = 1,480. Rule:—Multiply dents in Stockport counts by 37 for Scotch, or divide Scotch by 37 for Stockport. In Bradford, 36 inches is the basis of calculation; a 40 set would be the number of sets of 40 threads each on 36 inches. $\dfrac{40 \times 40}{36}$ = nearly 44·5 threads per inch.

There is also a system of 50 threads to the porter, calculated by multiplying the set by the width and adding a cypher to the product. The whole is divided by 9, which will give the total of warp threads for any given width. In Leeds, Huddersfield, Batley, and other woollen districts, the portie is reckoned at 38 threads, and so many of these porties on 9 inches determine the set or gears. The rule is to multiply the given set by the threads in a portie, divide by 9 inches, and multiply result by number of heald shafts required. This will give quantity of healds per inch required on each shaft.

In Scotland the woollen reeds are determined by the number of dents or splits on $1\frac{7}{8}$ inch; a 20-reed, 2 in a dent, would equal 40 threads on $1\frac{7}{8}$ inch in Scotch counts. Silk reeds are calculated by number of dents on 36 inches; a 50-reed per inch would be called 50 × 36 = 1,800 reed. These calculations are very necessary, because many of the districts insist upon their own systems being followed. It is high time that some one universal mode of reckoning should be adopted. In America the 1-inch standard is common; on the Continent dents per centimetre = to ·39371 inch.

To sum up, an equivalent in any system for a given count in

another system may be found by multiplying the number of dents per inch in given system and dividing by number of dents per inch in the required system. Reed tables might easily be constructed in this way for ready reference.

Twisting or Looming.—This is merely a mechanical operation, requiring little or no skill. A long fringe, perhaps 20 inches of warp yarn, is left in the heald when a warp is finished, and to each separate end of this fringe or *thrumbs* every separate end in the fresh warp, as taken from the lease, is twirled together, and made to adhere by whiting or chalk ground down, but in linen, woollen, and worsted the threads are generally tied together.

Placing Looms.—To place looms in a proper position, a line should be drawn on the floor as follows:—Fasten a cord with a weight suspended to the end of the line shaft that is to drive the looms, make a mark where the weight touches the floor; the other end of the line shaft is gauged in a similar manner; a cord well chalked is stretched tight and fixed at each end, passing directly over the marks previously made; the centre is taken up a few inches and, if let fall, a straight line will be marked on the floor from end to end. If the gearing is accurate, all the shafts will be parallel with each other and also with the chalked line, and this line gives the position of the looms parallel to the line of shafting. Another line at right angles to the first is required. Place three marks upon the first line, at equal distances from each other, and if the distance between the centre mark and the two on each side be, say, eight feet, a cord of 12 or 14 feet, with a piece of chalk at one end, will be sufficient to describe circles cutting each other on the first line, giving the points for the second line to be at right angles. This is really erecting a perpendicular upon a horizontal base. The second line gives the measurements for placing the looms in rows.

If this plan is carefully carried out, the driving belts will have more power, with less friction, wear, and tear. The distances that the looms may require to be set from the second line will depend upon the space to be allowed for a passage between the ends of the looms. If the driving pulleys are made to suit the ends, all will be in a straight line, but if the driving pulleys are all at the same distance from the ends of the looms, then the looms must be placed to suit the belting, one being a little out from the other. If this is not attended to, the belting will not have room to work, and the evil will be that, when one loom is stopped, the belts will run upon each other.

The looms being all placed in a proper position, next, get them level, with a straight edge or spirit level, and securely fastened down to the flooring. All bolts must be carefully examined and tightened up if loose; pickers put on; shuttles fitted to the boxes; reed put in; loom made to pick at the proper time and with sufficient force; tappets set to tread neither too late nor too soon; the driving belt put on; loom well lubricated in every stepping, and allowed to run for a few hours.

CHAPTER XVII.

THE POWER LOOM.

Varieties of Power Looms.—The power loom, as at present in use, has reached a very high state of perfection, and there is no lack of really excellent machinery in the various makes. One hardly knows which to admire most, the beautiful action of a loom, or the faultless work it turns out. There are looms for every speciality, from the weaving of handkerchiefs, neckties, ribbons, dress goods, to canvas cloths for our ships. In all the varied makes of looms many are all that can be desired, but it would be useless to disguise the fact that there are others capable of great improvement.

A loom, however, may be made in strict accordance with the rules of science, of the very best materials and workmanship, with all the modern improvements that experience can suggest; but put into the hands of unskilled workmen in the weaving department it will have all its best qualities destroyed. We know from practical experience, amongst hundreds of looms of every possible construction, how a good loom, like a good watch, may be ruined by careless treatment or ignorant workpeople.

The Hand Loom is nearly obsolete, except in a very few districts and for some special classes of fabrics or pattern weaving, for which it is better adapted than the power loom. Changes can be made in weft colours to an unlimited extent, without loss of time and economically; in fact, the hand loom is generally used for experimental weaving. It is so well known that a description of its parts is unnecessary, and it has received but slight improvements during the hundreds of years it has been in operation.

Kay, of Bury, introduced the fly shuttle, which gave a greater production in checked fabrics. Nevertheless, there are to this day hand weavers who still use the primitive system of changing their weft by hand, notably the Spitalfields silk and Dewsbury carpet weavers. Any one observing the action of these weavers in fabricating their cloths would at once see, by their dexterity in changing colours without stopping the motion, that the fly shuttle would only be an encumbrance, at all events, on their class of fabrics. By a very quick movement they can change all the shuttles and pull the weft shoot at the same time to its extreme tension, and by this

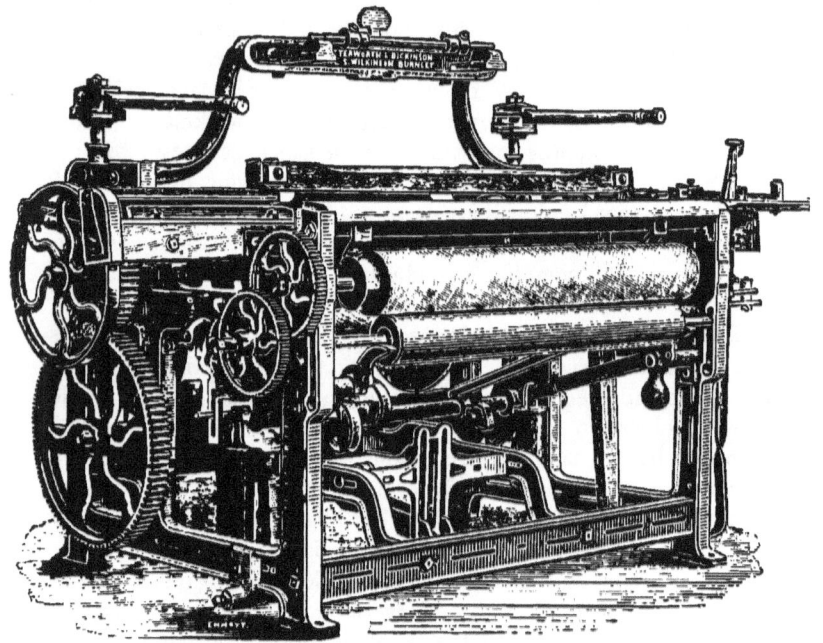

Fig. 56.—Calico Loom.

means the selvages are clean and almost equal to the edge of a knife; the looping is scarcely visible, and in silk goods this is one of the principal features in connection with a well-made cloth. A modern calico loom is shown in Fig. 56.

Dividend for Picks.—On new looms it is necessary to determine the weft-pick dividend, and not to take for granted what the makers may give. This is a most important consideration, and ought to be calculated forthwith.

Suppose, as in the ordinary calculation in use, the circumference of the taking-up roller to be 15 inches, beam wheel 75 teeth, carrier wheel 120 × 15, rack wheel 50 teeth, give a dividend of 507. With an allowance for shrinkage, the calculation would be:—Multiply rack wheel 50, carrier wheel 120, and beam wheel 75, all together = 45,000; and divide by pinion wheel 15 × 60 (number of ¼ inches in circumference of taking-up roller) = 900 divisor, and the result = 500. Add 1½ per cent. (say 7) for shrinkage, and the result, 507 = the dividend.

To find the picks given by any change wheel, divide this dividend by number of teeth in the wheel.

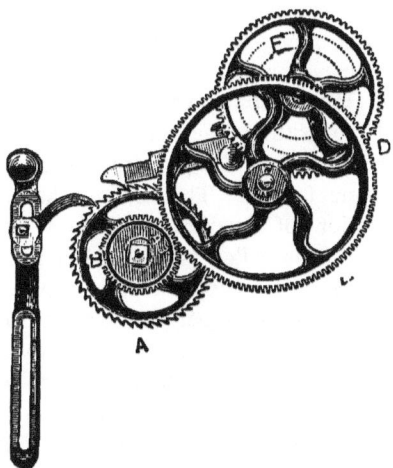

Fig. 57.—Taking-up Motion.

The train of wheels is not always the same in every class of loom, but the dividend can always be obtained by multiplying all the driven wheels together, and dividing by the drivers and circumference of the roller.

Taking-up Motion.—The object of a taking-up motion in any power loom is simply to take up the woven fabric. The illustration (Fig. 57) very plainly shows the train of wheels attached to the cloth beam. The vertical lever on the left, with catch, is known among weavers as the "monkey tail"; it actuates the wheel, A. The wheel, C, is a carrier, merely transmitting power to the cloth roller, D. This arrangement represents a constant number or dividend of the motion.

The action of a taking-up motion (see Fig. 56) requires setting correctly when the crank is at the front centre towards the healds; the taking-up catch should drop into one tooth only on the ratchet wheel; and when the crank is at the back centre, the holding catch should be so set that it will drop over the ratchet-wheel tooth about $\frac{1}{8}$ inch, and the stud which works the taking-up catch should move one tooth at a time. The finger which is in connection with the holding catch must be set so that it will permit the latter to rest upon the bottom of the tooth on the ratchet wheel.

On no consideration ought any of the taking-up gearing to be too deep—that is, the teeth too much embedded into each other, so that the wheels can scarcely revolve without a great amount of friction, sometimes stripping the teeth away, and in any case being a fruitful source of uneven cloth.

Prevention of Unevenness in Cloth.—In addition to the regular winding of the cloth on the roller as it is woven, there must be the prevention of thick and thin places, particularly in muslins, gauzes, and very fine fabrics. This is, and has been, a problem engaging the attention of inventors for many years past. The present form of motion is called positive; but in reality it is only comparative in practice, as manufacturers know to their cost. A mathematical dividend for picks per inch can be obtained, but the teeth of gearing wheels cannot be divided into fractions. Then there is the slip, allowances for milling up by very coarse wefts, etc.

The point requiring solution is by what means to allow the woven cloth to be let back a determined amount each time the weft gives out, or when a quantity of the cloth has to be unpicked for defects, as thick and thin places will occur when the weaving is again commenced; for, however perfect a weft-stop fork-motion or brake may be, the shuttle cannot be stopped at once on its journey, and the take-up is still in action.

In letting-off motions, the requisite points to be considered are:— the cloth and warp must be in a proper position for a rejoining of the two materials, without any special care on the part of the weaver; no adjustment should be necessary from the beginning to the end of a warp; in many beams one equal tension. The length of the warp threads being thus equalised, all slack yarn would be avoided, and the weaving strain would be properly divided, and the breakage reduced to a minimum.

A very great objection to positive motions, as at present applied, is that certain conditions of temperature affect the stretching power of the yarns. A very simple contrivance has been suggested to remedy this. A lever is hinged on a stud, and bolted to this lever is a casting running the length of the warp beam between the flanges, and bolted to a similar lever at the other side of the loom. The under surface of the casting is curved and covered with leather; at the end of the lever a weight is suspended and connected to the other side of the loom, where a similar weight is hung to the other lever. The *grip* or holding power of the friction in this arrangement diminishes as the yarn-beam diameter becomes less, and the power of the yarn, as applied to turning the beam, diminishes at the same ratio, so that when once set, by hanging the weights on the weight bar, no further attention would be required. The principles of weaving are limited in their adaptation by the qualities of the yarn and cloth texture.

In heavy fabrics power is necessary for production, and the object is to economise the power; but, in light textures, the power expended is comparatively small, and the saving, therefore, to be effected is chiefly in preventing the breakage of warp yarns, so as to maintain the utmost speed at which the machinery can be made to run.

The Slay.—There is no portion of a loom more important than the *slay*. The swords ought to be so set that they will be nearer to the loom front at the bottom, by at least $2\frac{1}{2}$ inches, than at the top. When the crank arm is upright—that is, at the top centre—the shuttle board or race ought to be 1 inch below the front plate of the loom. If a fast-reed motion, carefully make the stop rod or protector blades fit the frogs in a dovetail shape, so that they may not fly over when the loom bangs off. The fingers of the stop rod must be set so that the blades of the stop rod will always go over the frogs when the loom is at work; $\frac{1}{4}$-inch play from the *swell* in the box is sufficient when the shuttle is at rest. The spring that holds the stop rod down must, on no consideration, be too tight—a very common fault. Let the fingers be securely fastened to the stop rod, and the steps the rod works in be well cleaned and oiled. "Prevention is always better than cure." Shuttle traps or yarn smashes would be less frequent if the protector were carefully looked after and kept in good order.

In a loose-reed loom the fingers under the slay are for the purpose of making the reed fast when it touches the cloth or beats up the pick. They are generally fixed to go under the brackets attached to the breastplate of the loom—$\frac{3}{8}$ of an inch is allowed—they are set as fine as possible, so that they will not go over the bracket in place of under. The nearer they can be got in this way the more steady the reed when striking the cloth.

One particular point must be carefully noticed—that is, to allow sufficient scope between the spring that holds the reed when the shuttle is working and the brackets on the breastplate. This permits the shuttle when caught to throw out the reed, and prevents a *trap*. There is a rod holding the strip which causes the reed to fly out; it is held by carriers and brackets to the slay swords, and these ought not to be neglected, but kept well lubricated. If the slay has not a sufficient amount of bevel the carriers of the swing rail may be moved nearer the loom front, and if there is too much bevel *vice versâ*; but if the slay is too near the front of the loom it will be difficult to get a good shed for the shuttle to pass through.

The Picking Motion.—This is the most defective piece of mechanism in a power loom; there is a wear and tear in connection with it that is anything but desirable. In the majority of looms the picking motion transmits the force almost in a parallel line, causing an undue amount of harshness and wear that is injurious to every other portion of the loom, the uneven vibration loosening bolts and causing a general disorganisation. Now, practically, the force required to drive a shuttle from side to side of a loom, smooth and easy in action, should be as nearly as possible at right angles to the axis of the fulcrum on which it moves in the cone pick; this would be the upright shaft. The direction of this can be ascertained from the point of contact between the cone and tappet, at right angles to a line drawn tangent to the circumference of the cone at the point of contact. The movement ought to be graduated, practically, to come away slowly, until the slack of the picker leathers is drawn up and the shuttle about to move; the shuttle would not be pitched with a jerk, nor worn so much as it is; the pickers and leathers would last longer, and, more than all put together, there would be *less throwing* of the *shuttles out* of the *loom*; and another evil that might be prevented is the knocking-off of the loom. This is a cause of unnecessary labour and expense. If the picking movement is short and sharp in its action, or harsh, then this banging or knocking-off may be expected, and a loom is worn out before its time by violent concussions. The increase in the size of the shuttles is an unmitigated evil. The extra weight requires extra strength in the picking motion, so that, really, the loom would need to be strengthened specially for the sake of the shuttle, quite as much as for the fabric produced.

The picking motion being so defective, we have to apply the best remedies we can to counteract the evil. One very simple appliance is the check strap. With this addition properly adjusted, the shuttle can be passed backwards and forwards with comparative ease, but without it the shuttle requires to be tight to prevent a rebound, and a jerking motion ensues; but if a little flowing or fluff from the warp and weft gathers unobserved in the front of the box, the shuttle, instead of being thrown across the loom, is sent straight at the head of some unfortunate near at hand. The check strap is not generally understood by those employed in power-loom weaving. Perhaps it is too simple to be noticed; but we know, from long experience, that, by its judicious use, the throwing out of a shuttle is of rare occurrence, and a saving is effected all round of pickers, straps, shuttles, etc.

The Check Strap.—To prevent the shuttle rebounding, cops breaking, and shuttle wearing, let the strap be at least 1 inch broad, and the length of the slay within 6 inches. It can be easily kept in its position through staples screwed into the slay below the shuttle board, and by fastening a stout piece of leather in the middle of the check strap to work between the staples (which may be set 4 inches apart) it acts as a travis, and tabs working on the spindles are connected with the strap—each about 5 inches long and $1\frac{1}{2}$ inch broad, a hole being punched in one end to work on the spindle freely, and a slot, the breadth of the check strap, cut in the other end, to fasten to the check strap with a piece of wire as a buckle, so that the strap may be let out or shortened as required. Guards ought to be placed, 5 inches from the ends of the box, to prevent the tabs from coming too near the spindle studs, or, in place of guards, a piece of leather, 5 inches long and $1\frac{1}{2}$ inch broad, with a hole through each end, will be equally as good, one end put on the spindle in front of the tab, the other fastened to the spindle plate. All the tabs, guards, and check straps must work free and easy. If applied as we have described, the check strap will be found a "save-all."

The Shuttle.—The injuries received by operatives in consequence of shuttles flying out of the looms are often of a serious character. The main cause of these failures being found to be something radically wrong in some part of the machinery, the shuttle should be prevented from coming out of the loom; but often, when some one is seriously injured, a trifling disarrangement is sought for and remedied, whilst the real cause is left untouched, and the shuttle flies out again as soon as the next slight obstruction occurs.

The *pick* of the loom is principally the cause of this *evil*. A short, sharp, harsh-working pick requires little, if any, obstruction, to throw a shuttle out of its proper course. A very great fault lies in constructing the shuttle box back of wood, and, in cheap makes of looms, this wood is very inferior. It is soon worn by the action of the shuttle, and torn up at the entrance to the box by the point of the shuttle coming in contact with it when the weaver throws it in. Some loom makers face this part, within the shuttle box, and the back of it, with iron. When the wood of the box-back begins to shrink, or the iron gets loose, as very often happens, these iron plates become a greater obstruction than the wood, even when torn up. The most sensible arrangement is to leave the box without a plate on the sole, and to make the back of the box of cast iron. Wear and tear would be obviated, and this light casting would never be worn out.

The point of the shuttle will sometimes rise off the shuttle board as it is thrown into the box, and, when thrown out again, will go off in a straight line to the position it was in when the force was applied to it; and, instead of entering into the shed, will pass over it, and be brought to a standstill only by coming in contact with the first thing it meets, by which it is, perhaps, broken or rendered useless. To prevent this, place a small slip of wood on the back of the box, just above the shuttle, not as a binder, because a little room must be allowed for full play, but close enough to prevent it rising. Nearly all new looms have this appendage; it is a cure, but, when it gets worn, the evil again occurs. The best plan is to make the box in the shape of a dovetail, and to fit the shuttle to it.

An obstruction may, and will, take place in the shed by a broken thread becoming entangled with others, but, if the tip of the shuttle be placed a little under the centre, it will have a tendency to get under this obstruction and knock off the loom, without flying out; but being, as it is, placed above, or in the centre, it goes over the obstruction, and is thrown completely out of the loom. The tip of a shuttle should always be as near to the race board of the slay as it is possible to work with, without dipping under the yarn.

The erecting of shuttle guards is of necessity a very cumbersome affair, and such arrangements cause neglect of remedies, and evils keep increasing until the loom becomes thoroughly disorganised. The spindle stud and the spindle hole frequently get worn, so that the spindle has too much play, and the shuttle is projected in a line which is not parallel with the entrance of the box on the opposite side. A worn stud should be replaced at once, as it is a fruitful source of trouble. We are now speaking of the ordinary calico loom-overpick for shirtings, printers, and cambrics. A broken picker, a tight shuttle box, a picking band too long or too short, two odd shuttles, the reed loose in the slay, a worn raceboard—all, or any one, will cause a flight of the shuttle out of the loom.

In looms having either the drop or circular box motion, the difficulty increases. The box cannot always be in line with the raceboard of the slay, because of the colour changes; then, if they work above or below the line, when in position, the shuttles are thrown out of their proper direction, and the picker that projects the shuttles is irregularly worn by a movement of this kind, so that one evil produces another. (A drop-box loom is shown in Fig. 58.) The shaft of healds holding the yarn may not be tied up level, and, if the shedding happens to be late, the shuttle will rise to the top shed and be thrown anywhere. The shed must be clear, the healds not linked together by the shafts, and the reed free from standing too far out. The slay bottom giving way and rising in the middle or tending outward is a source of trouble not often thought about. When looms are empty they should be carefully examined, and all defects remedied before warps are put in.

Treadles.—The treadles ought to be level when the loom crank is at the extreme point towards the front of the loom, but, of course, this will very much depend upon the strength of the required fabric. Setting the treadles level when the reed is in contact with the cloth will make a difference of several teeth between a light cloth and a heavy one.

The best way for all strong fabrics, such as sateens, drills, etc., is to have the treadles level with the crank, two teeth from the

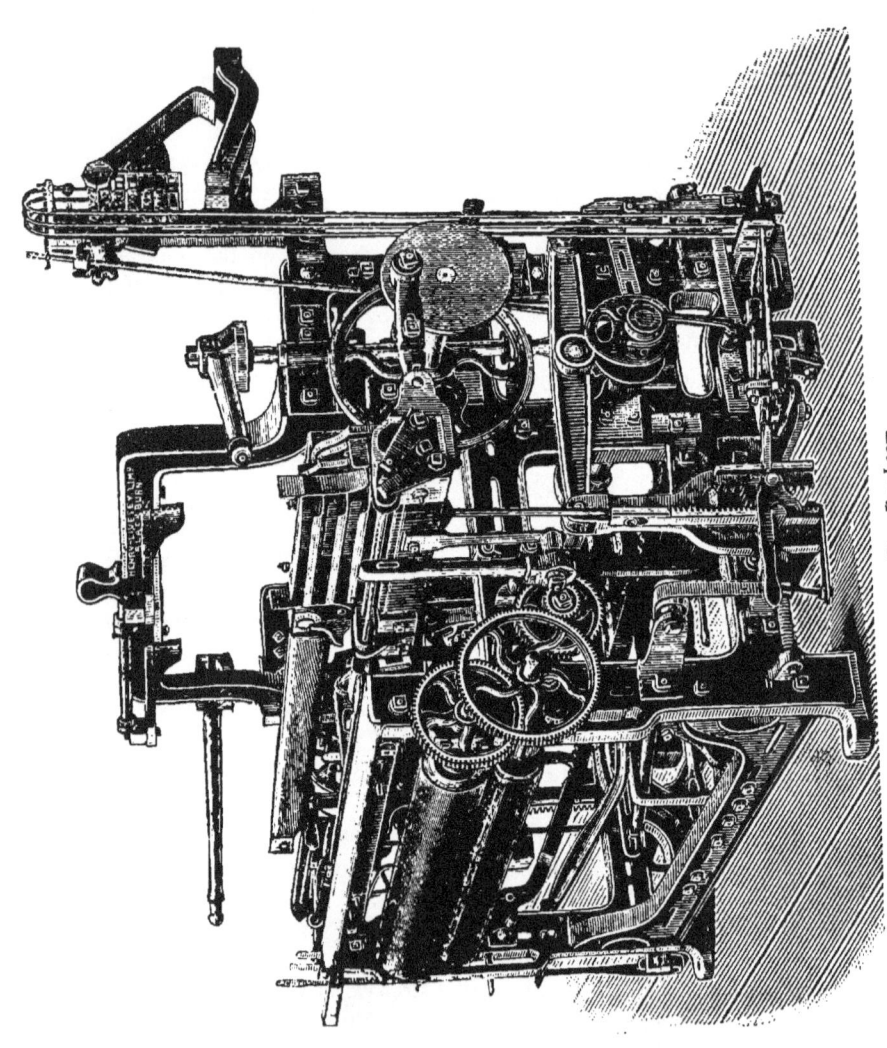

Fig. 58.—Drop-Box Loom.

extreme point towards the front of the loom. We are not dealing in these remarks with the scientific aspect of the question, but with the practical requirements necessary for weaving various fabrics of different weights, and with other peculiar features, which require an exercise of resource on the part of the loom manager. In plain weaving, the cams are secured to the shaft by set screws or keys, so that they may be moved for setting in the right time. The greatest care is necessary in this adjustment. It is very often done in any way. The ease upon the yarns, especially if tender, the steady motion of the healds, without straining and jerking, depend upon the true setting of the cams. If the treadle rollers are worn they are worse than useless, and ought to be dispensed with at once. Waste and dirt accumulate in these rollers so much, owing to the dropping of size and fluff from the warp, that they become choked up and cannot revolve; there is no portion of the machinery of a loom that requires more cleaning nor one that gets less, consequently healds are worn out before their proper time, more breakages of yarns and imperfectly formed sheds must and will take place, and this for want of a little time spent in clearing away the incumbrance. Diminished earnings through this gross negligence result, and the fault is attributed to bad warps, healds, etc.

It is an undoubted fact that the very best constructed loom will, in the hands of a negligent operative, become useless before it has turned off sufficient cloth to cover its first cost, the want of sufficient lubrication and cleansing being the cause. In many places a man is appointed to oil the looms, but it is often done in a slip-shod manner, some portions never receiving a drop of oil for months. We say this advisedly, because we know it to be only too true from personal experience. Set the cams so as to come in immediate *contact* with the rollers on the *centre*. The rollers should be a little forward, thus giving a smooth, gradual beginning of the tread; the shed is opened without *plunging*, and the *dwell* is allowed full time to get the shed duly opened for the delivery of the shuttle.

Time for the Pick.—Opinions differ as to when the shuttle should enter the shed or the box, many having an idea that its entrance into the box should be when the loom crank is at the back centre, others contending that the shuttle should be through the shed before the crank has passed the back centre. These are mere fancy notions. If all about the loom is in corresponding order, the pick should begin from the *bottom centre*, or when the arm of the crank points downwards; but there is often a confusion, or want of harmony, between the time of picking and treading, sufficient to damage a loom more in one year than if, in proper time and order, it had been running ten years.

The Scroll Pick.—This is a picking motion used on a great number of looms for the production of fancy cloths. There are slots in it to set the tappet or boss, to which the scroll is fastened in the centre of the slots, so that it can be made to pick sooner or later. The point which gives the pick must be set as near the end of the bowl on the picking shaft as possible, to give a gentle, easy pick; the bowl on the picking shaft must be set in the centre of the slot, so that it may be raised higher or lower, according to circumstances; all the nuts and studs must be securely fastened, along with the key on the boss; the wheel on the crank shaft is pulled off, and the scroll is set in position to give the pick about one inch before the crank gets to the bottom centre; the wheel is then put on and fastened. In this form of pick, on no pretext ought the picking pegs to be allowed to go too far back; they work best two inches from the end of the slay, and should it be found that the pick is too weak, let the picking peg be drawn a little towards the inside of the loom, when it will pick stronger, with less wear and tear.

The Tappets.—In calico or for plain weaves, the one side of a tappet is from $\frac{3}{4}$ to $\frac{3}{8}$ of an inch larger than the other; the larger side is for the back shafts because they are farthest from the reed, and the least side for the front shafts. Although there are four shafts for the warp threads, yet, owing to the draft being skip-shaft—that is, drawn in 1, 3, 2, 4—they are coupled, and work as two shafts; the front shed generally opens when the pick is from the right-hand side in a left-hand loom, and from the left in a right-hand loom. If a twill is required, the first and third treadles must be level with the tappet, and, in this case, the second and fourth are sure to be right.

This setting means a weft face, the back shaft to rise first, and the others will follow in succession.

We give illustrations of four tappets (Figs. 59 to 62). The first is for plain weaving; the second a four-leaf, two up and two down; the third a six-leaf, two down and four up in loose sections; and the fourth an eight-leaf, five down and three up in loose sections. In the drafts given below each tappet the black squares indicate where the weft passes over the warp.

The Weft Motion.—There must be a clear course through the grid in the slay, and the prongs of the weft fork must not touch in any part, either at the sides, top, or bottom, or, if they do, the loom will not stop when the *weft* is *broken*, and will *stop* when the *weft* is not *broken*. The fork ought to project through the grid so far that

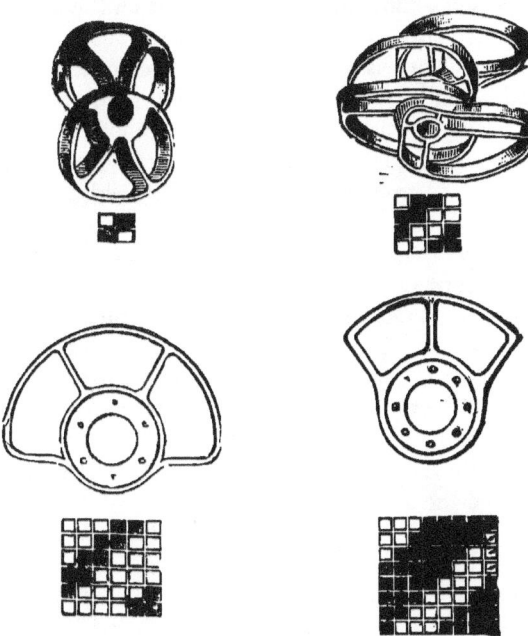

Figs. 59, 60, 61, and 62.—Tappets.

the weft will lift the fork so high that its other end will just *clear* the *hammer* ¼ inch when the loom is working. To set this motion in proper time, when the crank arm of the loom is at the front centre, set the bowl on the weft tappet close to the weft lever, which must also be set so as to be neither too late nor too soon; the slay must not be allowed to swivel backwards and forwards, nor the pick too soon from the weft fork side; if too weak or too late, a pick from the opposite side will stop the loom without broken weft, and the same will occur if the shuttle rebounds; and if the setting on the rod is too weak, so that it springs against the fork holder, the loom will not stop at the proper time; the weft fork requires to be very finely balanced, and the slightest defect will render it useless.

The Break.—On very speedy looms the break ought to be in a position to allow the loom to stop with the shuttle on the weft fork side; this may be regulated by the weight and lever. There is one important point in connection with a break arrangement which is much neglected—allowing the belt to run on the fast pulley when the loom is not working. A good plan is to cause the loom to stop when the loom crank is a little over the back centre. The value of this advice would be found in weaving light mulls or cambrics. The loom, in setting on again, would have a proper start; *cracks* and *cloudy* places in the fabric would be prevented; the loom would not *bang*, and *traps*, especially in loose reeds, would be effectually disposed of.

Waste of Weft.—A great amount of waste and loss of time often occur if the cops in the shuttle are smashed or ravel off; if the shuttle is delivered too solid on the left-hand side, or picks too hard and quick from the right-hand side; if the *swell* in the shuttle box be worn too far back, as it causes the shuttle to light too solid on the left hand; or if the picking-stick has to be forced back by the blow of the shuttle. These defects must be remedied at once. The mischief is often found in the shuttle spring requiring raising, or the cops may be too soft and the shuttle peg too small.

Unlevel Cloth.—The chains or ropes on the yarn beam must not be permitted to bind the flanges, and it is possible that the beam flanges are not true or the beam pivots may be loose, and the cloth roller pivots in the same condition. A slack belt will cause thick and thin places in gauze cloths, and setting the loom on with the crank at the top centre is another cause. Let the chain or beam rope hang as far from the fulcrum of the weight lever as it is possible to place it. Too much drag on the weft is an evil, and so is one shuttle with a long peg and the other with a short one. A careful examination will prevent many troublesome defects, and these remarks will be found of great benefit to those loom managers who, however skilful, may be short of experience.

Reed Motion.—When a shuttle is prevented from entering the shuttle box through some defect, the *banging off* or sudden stoppage of the loom is often a serious injury to the *swords* of the slay, and many other portions of the machinery. Weavers generally have a bad habit of knocking off the setting-on rod, not caring or heeding where the shuttle may be. The expense for repairs, owing to downright carelessness, is, in many instances, beyond all that is really necessary, causing the profits on the manufactured goods to be very microscopical, and there is also a loss in production and, therefore, in earnings; whilst repeated concussions of the loom, through the shuttle not lifting the *protecting finger* over the *frog*, shake and loosen every bolt and derange the sensitive going part, so that the loom comes to a standstill, and some casting being found fractured or smashed has to be replaced, and hours of idleness ensue, which can never be regained. In a loose-reed loom many of these disadvantages are avoided, the reed being liberated should the shuttle stay in the shed of the warp. Loose-reed motions were only applicable for light fabrics not requiring heavy wefting, but improvements have produced a loose-reed motion applicable for heavy fabrics.

The keen competition in the production of cloth, to which manufacturers are now subjected, forces upon them the adoption of the very best appliances, and much, no doubt, may be done in preparing the materials for cloth fabrication; but the loom is the main factor in giving results.

Improved Picking Motion.—Underpick looms with Woodcroft's section tappets have an unusual amount of wear and tear; the power for driving is excessive. The providing of leather for the picking-straps amounts to a large item, and picking-motions have been devised to dispense with the cost of picking-leathers altogether. We briefly describe one of these improvements in delivering a shuttle without the upright picking-shaft, picking-leathers, scrolls, and half-moons, usually employed. An auxiliary shaft is driven from the crank-shaft of the loom, and a picking-shaft is secured to it, a double lever being pivoted upon a stud projecting from a frame, and its upper extremity is in the path of the tappet, so that, at each revolution of the auxiliary shaft, the tappet strikes the end of this double lever and raises it, thus depressing the lower end, which is fitted with an adjustable piece set to operate on the picking-stick. A bracket, to which the picking-stick is secured, is pivoted on a stud projecting from a sleeve, carried on another stud forming part of a bracket bolted to the framing of the loom. The picking-stick bracket has what may be called a tail exactly beneath the adjusting piece on the lower end of the double lever; it has also a *snug*; one end of the spring is attached to the picking-stick, the other end being fixed to the framing. This arrangement keeps the picking-stick, at all times, in a normal position.

The operation of this improvement is as follows:—When the loom is in motion, the auxiliary shaft, with its tappet, on each revolution, strikes and lifts the upper part of the double lever, throwing the lower end sharply downwards, striking the tail part of the picking-stick bracket, causing the picking-stick to impel the shuttle through the warp shed, the shuttle's return being effected in a similar manner from the other side of the loom; and when the tappet has passed the lever, the spring brings the picking-stick to its normal position again. This improvement is applicable to all underpick looms, whether producing light or heavy goods. On ordinary calico looms the auxiliary shaft and its gearing can be dispensed with, and the tappet fixed on the lower treading shaft. Quick running, little driving power, and the action being communicated directly to the picking-sticks, are advantages worthy of consideration. The motion of the shuttle is certainly a very important object in weaving.

Let-off Motions and Warp Tension.—A certain degree of strain must be borne by a warp in the loom, but, at all times, the strain should be at its minimum, and disposed equally throughout the stretch. The quality of the fabric depends upon a proper tension. Raw, uneven cloth may occur through damp weight ropes on the yarn beam; the beam collar may be rusty; weights touching the floor, etc. The drag on the yarn beam of a loom is usually maintained by coiling a rope two or three times round the beam end, between the flange and pivot, the inner end of the rope, next the flange, being secured to the rail of the loom, and the outer end to a weighted lever hanging beneath and parallel with the beam.

This arrangement has many modifications. The drag being applied at both ends of the warp beam, very often the lever has its pivot or fulcrum beneath the centre of the beam. The full beam of yarn requires the greatest amount of drag, which, however, keeps decreasing—or ought to do—as the circumference of the beam decreases by the weaver shifting the lever weights; this is more "honoured in the breach than in the observance," and, in consequence, the drag may remain the same from the commencement to the end of a warp.

Many attempts to make this motion automatic have been tried for years past, but without any positive success; what might possibly have been suitable for light goods would be of no use for heavy cloths. One of the best results in this direction was introduced a few years ago. The lever used was a short one, with the fulcrum outside, the weighted end pointing inwards, and the resistance being placed between the two and near the pivot or fulcrum. Instead of the position being kept nearly parallel with the beam, the weighted end formed an inclined plane. With a full beam of yarn, the weight hangs at the end of the lever, and is kept in this position by a light chain fastened to and governed by a guide or feeler, some six inches long, arranged to rest on the surface of the yarn, on the beam, below the threads as they unwind. Now, the diameter of the beam becoming less as the yarn is unwound and made into cloth, the feeler has its position altered accordingly, and allows the chain to slip that governs the weight; this *slips* down the inclined plane of the lever, till, at the bottom of the beam, the lowest point is reached with a *minimum* drag; in place of ropes, chains are used for durability, but the method of coiling is altered to give a *double grip*. The chain is folded in two, the loop in the middle being placed opposite the beam centre, between the flange and pivot, and the attachment to the lever is made by means of a hook; then each side of the chain is taken over the top of the beam end and given a complete turn round it, and the two ends hanging down behind are fastened to the loom rail. Only one end of the beam is dragged.

It is claimed that this method of letting-off requires only one-sixth part of the weight usually necessary. A regular unwinding of the warp and uniform tension on the threads must be a desirable result. Too great tension wastes power, and the yarn is unduly strained; healds are worn out sooner, and the cloth runs up in the finishing more than is requisite. The latest and most effective improvement as a let-off motion is operated as follows:—When the warp opens for the passage of the shuttle, and the slay moves towards the back centre, the tension on the warp tends to depress the upper end of a lever and pull it forward; the lower end is pulled backwards, and through suitable gearing positively lets off just as much warp as is needed. After the weft is put in and is being beaten up, a connection with the sword of the slay pulls the bottom end of the lever and releases a tension spring; up to 100 pricks can be inserted, and, by a system of compound gearing, the claim is that the heaviest of fabrics can be positively dealt with.

We have little doubt that light and medium goods can be worked successfully with the improvement, but velveteens, cords, etc., will still require a negative motion.

The Trap Preventer.—It is very common, in fast-reed looms, for the shuttle to be caught in the shed between the reed and the woven cloth,' and to make a smash of yarn called a "trap," and, probably, the shuttle may also be broken. The stop fingers, on each side the loom, underneath the slay, are really intended to prevent traps, by coming into contact with the "frogs" opposite to them, whenever an imperfect pick is made; but, through want of attention, wear, and proper adjustment, they often fail in their action.

The trap preventer is constructed as follows:—On the rod beneath the slay board there is fastened, at the outer extremity beyond the ordinary stop-rod finger, another finger, slightly curved, projecting in a similar manner to the former, but more in advance, and having the same up-and-down motion, as the swell in the box is pushed back and released by the shuttle. This finger, as the slay oscillates, passes over a triangular piece of metal, fastened to the loom frame, when the shuttle is passing to and fro; but if, from any cause, the course of the shuttle should be arrested, the finger goes under the triangular metal, which acts as a bolt, fastening the finger down, or rather locking it, so that the loom is stopped at once, and the pressure on the shuttle box swells can be reduced, and positive security against "traps" is assured.

Electric Stop-Motion when Warp Yarns Break.—We can only briefly allude to this invention. It would seem that however valuable a device may be (and this is one), yet, if it is found to interfere with production, the operatives are certain to throw it out of action, under the belief they can do better without it. This stop motion has a series of india-rubber tubes along the upper edges of the heald shafts, and a thin metallic tape is cemented within them. On the loom top a bracket is fixed to hold a battery, one pole of which is connected with the healds, and the other by a wire with the magnet, the circuit being completed by the collapsing of the india-rubber tubing whenever an imperfect shed is formed by the yarns in the healds.

Without going into further details, loom weavers block it as they do shuttle guards, being more willing to run all risks than to have a loom stopped otherwise than by their own practice. In calico weaving, the *production* is the question, *quality* being often a secondary consideration for the cut-looker to determine; and a fine of 3*d.* may save the loss of 6*d.* At all events, these improvements do not seem to be appreciated. We well remember when the weft-fork stop-motion was introduced, and, for years after, it was not uncommon for weavers to take them off the looms and throw them away, because overlookers, in many establishments, did not thoroughly understand how to put them into proper order, the result being that looms were constantly stopped or worked away without weft; the disgusted weavers therefore considered weft forks an unmitigated evil.

Cover.—The main object in the manufacture of ordinary cotton fabrics—such as calicoes, drills, shirtings, checks, etc.—is to obtain "cover" on the cloth; that is, the spreading of the warp threads during the process of weaving, so that a very smooth face, or, as it is termed, a "skin," may be produced, with a minimum of warp yarns. "Cover" gives a more filled up appearance, and prevents *reed raking* on coarse cloths. It is accomplished, in many instances, by sinking the heddles below the warp line and elevating the back rest of the loom, or by weaving with the two portions of the healds or shed at an unequal tension, the lower half of the shed being tight and the upper half loose. The shed is to be open during two-thirds of the crank's revolution, the sheds passing each other, so that the shed for the next pick of weft is partly open before the first pick gets beaten up in the cloth. This was termed by hand-loom weavers "treading on the pick." The treading tappets require to be specially adapted for this object. The treadles generally move in a circle *instead* of in a straight line at their point of contact with the tappets, the consequence of this action being that when the sheds are opening and closing an undue stiffness or tightness takes place, and when the shed is fully open the treadles are not up to the point of contact. The treadles spring up and draw the upper part of the shed *tight* when it ought to be *loose*, and, despite all efforts, the cloth will be *reed marked.*

Many simple matters materially interfere with warp spreading—the position of the lease rods behind the healds ought to be determined; change the position nearer and further from the healds, and mark the effect on the cloth. This will soon be visible, and the rods can then be secured as a fixture. The raising of the back rest is a very poor, if not a really mischievous, method of obtaining "cover," because all the strain is thrown on that portion of the warp threads between the healds and lease rods, which must be injurious to the yarn and a cause of excessive breakage. The healds must have an eccentric motion, very fast in the centre of the stroke, and gradually slowing down until it merges into the full pause; but in this there is one difficulty to contend with—viz., the full weight of the weft has to be borne by the warp yarns during the time the pick is beaten into cloth; but a little experience will be sufficient in adjusting the opening of the shed in this beating up, so that shed and pick may both be in one time, or the shed a very *little later* than the *pick*. This obviates most of the strain, except on the selvages, which is caused by shrinkage, cloth always being narrower than it is in the reed. When the slay comes forward to beat the full weight, the tread is on the warp yarn, so that those threads composing and near the selvage will be strained in proportion as the cloth contracts. The selvage threads ought to be made strong enough or sufficiently elastic to withstand a portion of the contraction.

Cloth Contraction.—If the fibres consisting of warp and weft were inelastic and incompressible, there might be a possibility of formulating a set of rules governing the percentage of contraction in different fabrics, but practical and experienced manufacturers know how futile it is to lay down any fixed basis. In weaving specialities, with repeat orders, "year in, year out," it is possible to make a formula, giving a very close approximation of the shrinkage.

By way of showing that the various tables of percentages given in textile treatises are comparatively useless when applied practically, we may take an ordinary cotton thread, supposed to contain the full number of twists per inch, got by the square root of its count, multiplied by 3·75 or 3·25; then we have to take note of its oozy quality and what its probable diameter will be at every intersection, when subjected to the strain in weaving. This might be determined with some degree of accuracy, but we are met with the unknown in the shape of size, which may be anything ranging from 25 to 200 per cent.; and thus the diameter of the warp threads is increased, and the more size composition the more rigid and unbending of the yarns in intersecting, especially in plain weaves. The weft must then give way, and the bulk of the shrinkage is in the weft picks. The resistance to bending of round materials is *inversely* as the square of the distance between their places of support, and *directly*, as the square of their diameters. This is a mathematical theorem advanced, as applicable to textile fibres, on the supposition that they are dense, unyielding bodies, when we know the contrary to be the fact. A warp and weft all of wire might follow the rule, but wool, worsted, silk, and cotton cannot be gauged by any such means.

One evil in connection with the manufacture of plain goods requiring a smooth face and good soft feel to the hand of the buyers is having too much twist in the materials. This does not increase the strength of the yarn, as many are led to suppose, but is really a source of weakness, as a hard, twisted, wiry thread in calico goods cannot be made to produce a soft, velvety cloth equal to that which has a minimum of strength with the least possible amount of twist. With this class of yarns in any open reed, two in a dent, and a good "cover" put on, by the method already explained, the fabric would appear as if woven in a far finer reed. Goods made in a 40-reed would certainly look equal to a cloth made in a 60-reed if wanting "cover," and of hard, twisted yarns.

Shedding tappets, if of imperfect construction, will affect the quality of the cloth in weaving, as well as cause continual obstruction to the passage of the shuttle by broken warp threads. The proper size and dimensions of tappets must depend upon length of stroke and distance from the tappet shaft to the treadles. The length of stroke is of importance, because the smaller the shed and the less friction of warp threads. The *size* of the *shuttle* and *shed* should be almost equal to each other, for if the shuttle can just pass freely through the reed that is sufficient, and more than this is power uselessly expended.

The Pause.—The length of a pause to be given a tappet, or, plainly, the time during which the shed should remain fully open, and at what point of the crank's revolution it should take place, require some consideration, and this seems to be a matter on which nearly all loom makers differ, where no difference ought to exist, only for the necessary short and long pause. Those made for the short pause only require the shed to be open during the time the shuttle is passing through it. A difference of opinion exists as to what portion of the crank's revolution this will occupy. For all practical purposes it will take as near as possible one-third—that is, when the reed is in contact with the cloth, the crank will pass through one-third of the circle during the time the shuttle is going through the shed.

The long-pause tappet gives "cover," and may be used whether "cover" is wanted or not. The weft can be got on much easier when a great quantity of picks per inch are required, and with less tension on the warp threads, and a reduced strain on the bearings of the loom. This, in itself, is a great economy and a matter of vital importance with regard to wear and tear. We know, from years of

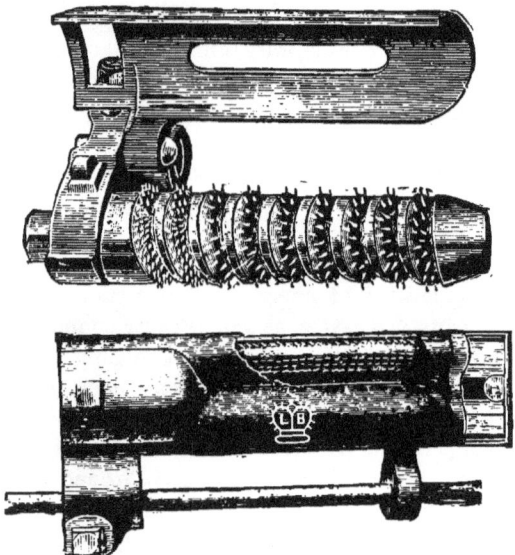

Figs. 63 and 64.—Loom Temples.

personal experience, the advantages derived from this form of tappet motion. In treading, the healds last longer, the reed is not so resilient or used as a hammer.

Temples.—The nature and uses of temples are generally understood by all who have even the remotest connection with the process of weaving, it is unnecessary, therefore, to enter fully into the subject, but we will content ourselves by giving a couple of illustrations (Figs. 63, 64), the former being known as the "ring," or "segment," and the latter as the "crown" double-roller side temple. There are, of course, many makes of these appliances, which perfectly answer their purpose—viz., keeping the cloth at its full extension during weaving. We give the accompanying illustrations merely to show fair specimens of perfect temples.

The Eccentricity of the Healds. — If the movement is gradual, or made to come slowly, as the shed gets open, the yarn is saved from a sudden plunge. The tappet, for this purpose, is constructed with a pause equal to two-thirds of the crank's revolution. It is necessary that all swinging and jerking movements in healds should be prevented; therefore, the tappets ought to be in continuous contact with the treadle rollers during the entire revolution of the crank.

Were we dealing with the scientific aspect of this problem, we could easily adduce proof by a mathematical figure and demonstration; but it is not requisite, as every practical loom tuner can vouch for the accuracy of our assertion. We have been discussing tappets for plain weaves, but the principle is the same for twills of every description, whether warp or weft effects, the only distinction being in the treadle rollers; but, taking the diameter of these rollers in ordinary use, the following information may be found useful:— Suppose a circle divided into 24 equal divisions, then, if we take the number of treads or rounds necessary to complete a pattern, and multiply by 4, or, more clearly, say 6 treads × 4 = 24, the round of the circle, the half of four will be found sufficient for the long pause to keep down a heald shaft during one pick, and, in a short pause, one part is enough, four for the second pick, and so on, four divisions representing one revolution of the crank.

Loom makers increase the length of the pause in proportion to the width of the loom, many of them giving a deeper shed on a broad loom, while using the same size of a shuttle as for a narrow loom; but neither of these practices is necessary. It may be said that the shuttle has a longer distance to travel, and requires more time in which to do it; but the slower speed and the greater eccentricity given to the movement of the slay, by the arrangement of the connecting rods with it and the crank, completely meet this difficulty in very wide looms.

Shedding Motions and Dobbies.—The section and every other form of tappets are limited in the round of picks and number of shafts, so that extensive patterns require something between these shedding motions and the Jacquard machine pegged barrels, called witches, indexes; and many other devices have been used with varying results. The pegged-barrel contrivances served the purpose required for a time; but the evils in connection with this class of shedding appliances were the necessity of having barrels of different

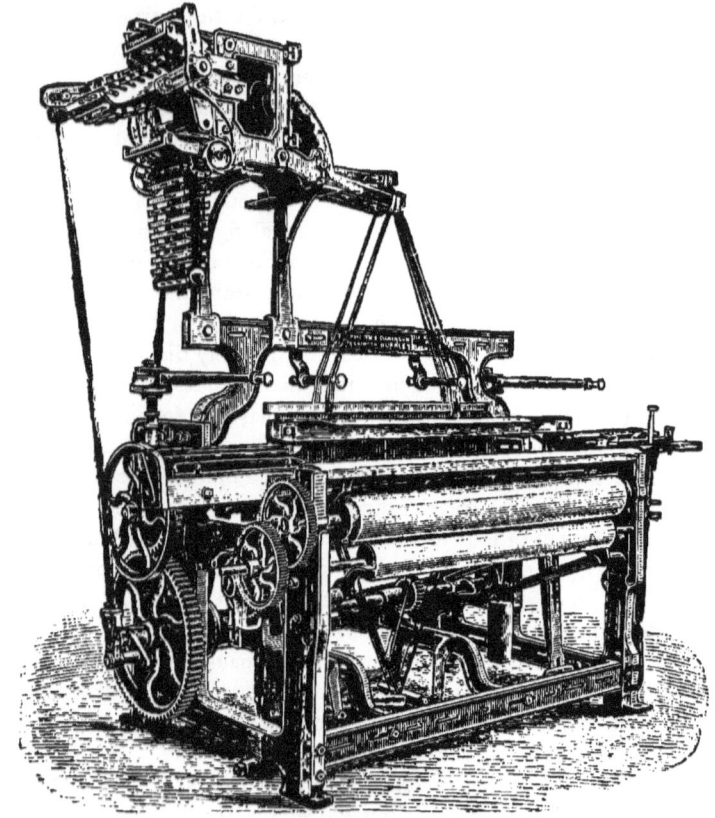

Fig. 65.—Calico Loom with Dobby.

sizes, or of enlarged circumference as the round of picks to form a pattern increased.

To obviate this difficulty, endless modifications were produced. It would occupy too much space to enumerate these or even to describe them at any length, though to intelligent and ingenious readers, with mechanical knowledge, they would certainly be suggestive. The development of all these shedding motions has given way to the dobby, which may be called a Jacquard machine variation; but of course it is needless to say that it has a much less range of work, being designed for shaft harness, and it is not so complex nor so costly as the Jacquard, and is most extensively used in nearly all textile fabrics of every material. A calico loom, with dobby attached, is shown in Fig. 65.

There are many types of dobbies, for all of which some particular merit is claimed, but there is a defect, or rather weakness, in their construction—the necessity in most of them of drawing down the heald shafts when lifted by the use of springs, which are unreliable. The healds ought to be operated by positive tappets; and if a small spur wheel were fixed on the crank shaft of the loom and geared into a larger one carrying a crank pin, the connecting rod could be attached to this pin, the upper end of the rod then being fixed to one of the horizontal levers mounted upon the rocking shafts.

The further arrangements would be as follows :—The levers being connected in the middle by a pin, upon which they move freely, the vertical lever extremities could be slotted, and carry the knife bars, these having the usual lateral movement in each direction. The knives actuate the horizontal hooks, which are notched at each end for the purpose of engaging the knives, all the hooks resting upon the ends of a horizontal lever, which is pivoted at two points and held down in the centre by a spring, and beneath the levers are placed the pattern chain barrels, moved directly from the crank of the looms. The horizontal hooks are pivoted from the under side of a sliding frame, carrying either a cam or tappet, having a horizontal movement backwards and forwards. This cam being cast upon the top of the sliding frame receives a bowl, carried upon a stud, on a projection on the under side of the heald levers. A corresponding series of levers is at the bottom of the loom, having their fulcrums in the centre, the same as the heald levers, and at one end the heald levers have a direct communication with the bottom ones by cords. A positive movement being thus secured throughout the two barrels gives the advantage of working borders on fabrics, as they can be put in or out of action when necessary. This description gives the working operation of the generality of dobbies.

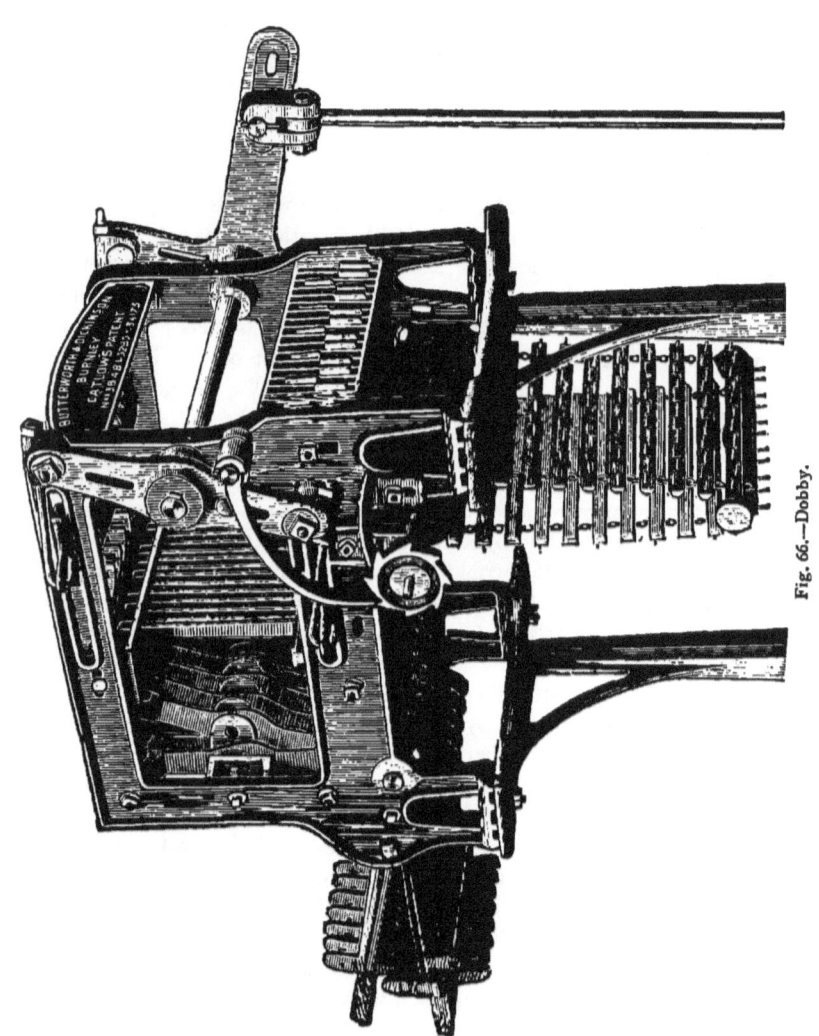

Fig. 66.—Dobby.

We give in Fig. 66 a neat illustration of an excellent type of this class of machines, known as the Catlow dobby. The single lift is worked from the crank shaft, allowing the healds to be brought to a level at each pick of the shuttle. The double-lift dobby, with one barrel, has an easing bar, which lifts the heavy-ended levers slightly, just before the pattern pegs begin to act upon them, thus very materially reducing the strain and liability of breakage. The double-lift dobby with two barrels enables broader pegs to be used, obviating any mistake in the weaving of a pattern, as is often the case, by the lever slipping off narrow pegs, or through pegs breaking. The two barrels are geared together by a toothed wheel, both running in one direction, making the working of the lags easy and certain.

Distinction in Fabrics.—The difference in cotton fabric construction (irrespective of ornamentation) is in the counts of the yarn and fineness, or rather measure, of the reed. In selecting warp yarns they should be well adapted for the goods required. Fine counts for crowded weft picks and in fine yarns should be of the best quality. A greater error cannot be committed than the purchase of indifferent materials because less in cost—it is not economy, but a direct loss, as those who have had any experience know too well.

To adapt yarns to the different sets of reeds requires more than ordinary skill, and experience must be brought to bear so as to make fabrics of different degrees of density proportional to each other. The reed, or, as it is called in the woollen industry, the "slay," is the measure, and the relative counts of yarn, the proportion to determine, sometimes known as slaying or setting, and in Scotland as "caaming."

The rule evolved by old textile writers, nearly a century ago, seems to be generally accepted by writers of the present day. Murphy, in his "Art of Weaving," gives tables of logarithms for reeds and yarns, but, as we have already remarked, the basis of such calculation cannot be otherwise than erroneous, to a certain extent, in consequence of considering all yarns, whether of animal or vegetable fibres, as solid cylinders: as a theoretical guide the rule may suffice, but, in actual practice, it is not altogether reliable. The rule, as given by early and modern writers on the subject, is that "Threads are cylinders, whose bases are the measure of the reed, to produce relative closeness of texture, as the squares of their diameters. As the squaring of any given set of reed is to the grist of yarn adapted for that reed, so will the square of any other set of reed be to the grist of yarn for making the *same fabric*."

We here give another rule :—Having fixed on the counts of yarn for the required fabric in the given set of reed, divide that number by the set of the reed, and to the dividend add the quotient, and the sum will be the counts of yarn for the next half set. Divide again the number thus found by the same set of reed and add the quotient as before, and the sum will be the required number, or counts, of yarn for the full set. In the ascending scale addition is used, but where the series is of a descending nature, from a higher to a lower, the quotient must be subtracted from the dividend.

This rule may possibly be found useful, and, as it is based upon the Scotch system of reeds, the following information will give full particulars :—Scotch reeds for cotton fabrics are made to a scale of 37 inches, which is considered a yard, and the reeds, as well as the cloth, are counted by the hundred splits or dents in that measure; 1,600 is the number of splits on 37 inches, and the difference between one hundred and another is called a set ; 1,200 reed would be equal to a Stockport 64—that is, 64 threads per inch, 32 dents per inch, on $37\frac{1}{2}$ inches $= 1,200$ dents Scotch. If, then, by following the latter rule, a 1,200 is made with 72's twist, the operation will be $72 \div 12 = 6 + 72 = 78$ for a $12\frac{1}{2}$; $78 \div 12 = 6\cdot5 + 78 = 84\cdot5$ for a 1,300, and $84\cdot5 \div 13 = 6\cdot5 + 84\cdot5 = 91$ for $13\frac{1}{2}$, etc. The descending scale $72 \div 12 = 6$; $72 - 6 = 66$ for $11\frac{1}{2}$, and $66 \div 12 = 5\cdot5$; $66 - 5\cdot5 = 60\cdot5$ for 1,100, and so on, down to $40\cdot5$ twist for a 900 reed. The table for cotton in general use is one round of the reel $= 1\frac{1}{2}$ yard $= 1$ thread; 80 threads $= 1$ skein $= 120$ yards; 7 skeins $= 1$ hank $= 840$ yards; 18 hanks $= 1$ spyndle $= 15,120$ yards. The spyndle is the highest denomination of yarn made up of separate skeins to form the hank, and 20 of these make a head of yarn in bundling it for the market; this practice is, however, fast giving place to the English system.

The Jacquard Machine.—When extensive designs for ornamental fabrics are required, and they are found beyond the scope of heald shafts or the capacity of a dobby, the Jacquard machine and harness are necessary.

The Jacquard machine (shown in Fig. 67) may be considered an enlarged dobby, though more intricate in its construction; and the harness differs from heald shafts to this extent only, that what is called a full harness operates every warp thread singly if needed. In heald shafts one may contain several hundred warp threads, which are operated in the bulk.

To make this clearer of comprehension :—A possible design might have 100 threads for a repeat; this would mean that no two threads could be found alike in the design, 100 heald shafts, for which no loom of modern construction could find room, especially power looms. In such a case as this, if a warp contained 3000 threads, 50 threads per inch, we should have 60 inches in the reed, and each repeat of the design containing 100 threads; the figure, or whatever the device

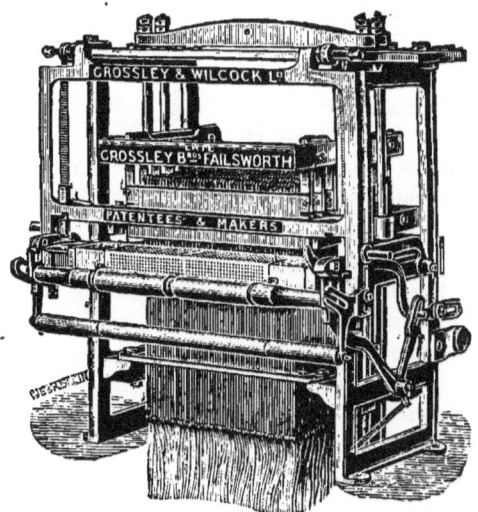

Fig. 67.—Jacquard Machine.

might be, would occupy exactly two inches—that is, 50 × 2 = 100 warp threads. This is the form of calculation which, as a rule, governs the general details of fancy designs for Jacquard machines, which are of various capacities, up to 600's, 800's, or two 600's—in fact, a number of machines working together in unison.

To go into details of how the harness is built for various fabrics would entail an amount of space sufficient for a large volume. We must therefore limit ourselves to a short description of the construction of a good type of the Jacquard machine. We have used these Jacquards in our occupation as weavers with every degree of satisfaction and confidence, knowing that our fabric was perfect in its ornamentation, and fit to bear the most critical inspection. This, beyond all dispute, is what manufacturers require in such an age of competition as the present. No rejects, no faults, and repeat orders must be the primary consideration.

The illustration (Fig. 67) shows the Jacquard alluded to. It is a 400 double-lift machine and is adapted for the production of almost any description of textile fabrics. It is arranged that the comber board may remain in its normal position—neither lifting nor falling—and this at once shows to practical experts the great reduction in wear and tear by friction. This one feature stamps the value of such a machine. The equal tension on the harness prevents the Jacquard hooks from twisting. In plush goods—a popular fabric—the full value of dispensing with the rising and falling of the comber board is most apparent in its effectiveness against friction.

Practical Operations in Weaving.—To sum up:—The practical operations in weaving cotton goods as far as the loom is concerned, allow no unnecessary space between the healds and the cloth; the shuttle must be as light as is consistent with the class of fabric to be woven, and the shed barely sufficient in size to allow the shuttle to pass through; the shuttle line must be as near the heddles as possible; the stroke of the lay,—that is, its proper length—about three times the breadth of the shuttle; the lay eccentric without angularity, and timed to the motion of the shuttle; the shed, in its eccentricity, must be in accord with the eccentricity of the lay; the shed ought to rise when the lay strikes the cloth; the shuttle action must commence when the lay is turning from the cloth.

The breakage of yarns is not always caused by excessive speed of the loom, and the only strain worth notice in this case results from the less time required for forming the shed. The strain on the yarn in shedding is as the square of the distance passed over by the warp divided by the time. If the shed is properly made, and of the smallest for the shuttle, the additional strain on the yarn from speed will be so small as to be scarcely worth notice when compared with the absolute strength of the yarn. The greatest source of yarn-breaking in any power-loom is over-shedding. The obliquity of the warp line to the stroke of the lay is very harsh upon even the strongest warp yarns, because the yarn is deprived of the elasticity it ought to receive from the stroke of the lay in the *turn* of the shed.

Time must be given for the shuttle to go through the sheds, though there is a limit; but, to gain the highest speed for the shuttle, the motions of lay and shed are made as eccentric as possible, so that time for the shuttle can be economised.

Many loom-makers obtain speed with short connecting rods; this increases the eccentric motion of the lay, and as there arises a degree of weakness or a pause of the crank in passing the centre of motion, the more so when they are shorter than in ordinary practice, it is necessary to accelerate the crank's motion by a small wheel inside the driving pulley, and fixed on the crank shaft. The crank shaft is on a lower plane than the studs of the lay on which the connecting rods are hinged, and the degree of obliquity thus given determines the amount of eccentricity imparted to the lay in connection with that which takes place from the revolution of the crank. The eccentricity from the shaft in working is as the length of the connecting rod to the diameter of the circle described by the crank.

The power-loom has now been brought to such a state of perfection, that the duty of the attendant is limited to keeping a loom going with the least possible stoppage. Skill is not so much required as quickness and dexterity of motion. The overlooker or tuner is, properly speaking, the *weaver*, and the production and quality of the goods depend very much upon his skill and management of the looms, providing that the looms are adapted for the work. It need hardly be said that a good practical knowledge of the business is indispensable in effecting these results.

CHAPTER XVIII.

FINISHING OF COTTON FABRICS.

Cloth Finishing.—To the eye of an experienced manufacturer, nothing is so pleasing in the many processes through which cloth passes in its production as the sight of goods well finished, soft, "lofty" to the feel, lustrous in appearance, clear and brilliant in colouring, with the designs pleasing, and in every way suitable for purchasers. The finishing of cotton goods has now arrived at a state very near perfection, principally through improved machinery, a few examples of which will be illustrated in this chapter.

Coloured cloths, such as ginghams, shirtings, and checks, must, in the first instance, be cleared of all loose threads, as far as is practicable, by shearing; the goods are then put through soap and water, with fuller's earth, which brightens up the colours, then sized in a light solution of starch, having a very little soluble oil—about 1 part to 75 parts of starch; this softens the fabric, and gives a good clothy feel. The goods now require stretching out to the required width. This is a very difficult operation in many cases, because the warp threads shrink the most and allow the weft threads to lie closer together. If a piece of cloth is subjected to any degree of tension whatever, it will become longer and narrower. Some pieces will finish longer or shorter than the average allowance, from slack or tight weaving. Any fabric in cotton that is singed, scoured, and hot-pressed, must, in the nature of the fibres, undergo some stretching and, in order to be sheared, the cloth must be drawn tightly through the machine; the greatest amount of shrinkage takes place in the drying machines. When it is imperative that the width be retained, the stretching machine (see Fig. 68) will widen it as it is passing through; and in the *best makes* of this class of machinery the amount of widening is placed under the control of the finisher.

FINISHING OF COTTON FABRICS. 197

Fig. 68.—Pin-Tentering and Stretching Machine.

Tickings.—These, whether sateen, fancy, or plain, are sheared. The sateen is simply sheared and calendered with hot steam rollers, the steaming being in front of the calender. Fancy and plain are sheared and calendered with a cold roller, and some few have a light size of starch. To get the greatest quantity of size into a cloth, the squeezing rollers in the size box on front of the drying machine are covered with from ten to twelve thicknesses of heavy unbleached sheeting which holds the size, and the goods are better impregnated with a quantity of size than would be possible with bare rollers.

The evenness and beauty of a finished cotton fabric depend very much upon the care in making the stiffening mixture. The following recipe is in general use:—100 lbs. of starch (corn or potato), an equal amount of either china clay or barytes, and 2 or 3 lbs. of tallow or octopus clay, as desired, for softening; the whole boiled for from fifteen to twenty-five minutes before putting into the size box.

The most effective preparation known is the following:—Place water in the boiling pan; after the starch is well mixed in, then heat is applied until the temperature attains from $130°$ to $140°$ Fahr.; perhaps the practice of boiling until the liquid runs clean off the stirring paddle is more suitable. When this starch is made, soap or oleine is added, with wax, and after this, glue size, mixing all well up by continuous stirring, and, lastly, adding china clay or chlorides of zinc and magnesium. The mixing must be constantly agitated whilst boiling. If iron vessels are used, the greatest possible care must be taken to prevent rust, or iron stains will most certainly spoil the cloth. The longer a starch mixing is boiled, the thinner it becomes.

The soap softeners are used only with starch, flour, etc., but not with any of the metallic salts. A little soap softening put into the above mixture gives the calender finish a gloss and closeness of texture with less pressure; oleine can be used for the finish of the most delicate tints, without tarnishing the colours, but it must be perfectly neutral, that is a clear, transparent solution. A good soap softening is made from cocoanut oil, retaining the glycerine. These soap softeners are used for cloths having a "kid" finish, that is soft, pliable, bright, full, and without greasiness. After leaving the drying cans, the goods are lightly calendered; calicoes are sized and run on friction glazing calenders, with hot steam rollers, the levers being heavily weighted.

Sateens are very highly glazed to appear as if the fabric were made of silk. These goods are filled with as much of the mixture as they can bear, but softness of feel is necessary. The mangling is done through hot water, and on the stiffening mangle, drying, damping, and calendering with wood and metal bowls, the metal bowl being highly polished and used as hot as is possible or convenient. The goods are afterwards folded and pressed between glazed cards.

Printed or *dyed goods* in diamine reds, purple, or other colouring matter belonging to this class of dyes require a starch with a small portion of soda crystals for a stiff finish, and soluble oil with soda

Fig. 69.—Rigging Machine.

for a soft finish. Colours will often run, through no fault in the finishing, because they are loose; the only remedy is to dry off speedily. The chlorides of zinc or of magnesium ought to be avoided in mixings for calender finishes, as they destroy the *gloss*, and, perhaps, *rot* the cloth.

Linen Finish.—In this process, the goods are run over a sprinkling apparatus, and put through a heavily weighted hot calender—a five-bowl calender is preferred; sateens or glazed goods are improved by gas singeing. Goods to imitate wool, such as suitings and flannelettes, have a short nap raised on both sides of the fabric by carding. The card rollers revolve at a high speed, and, as the cloth passes over them, a fussy appearance takes place, which can be increased by more runs, or by changing the rollers to a quicker speed; but, of course, the quality of the cloth must be considered in this operation: one run through may be sufficient for some makes, but others of good strong wefted materials, requiring the pile on both sides, may have two or three runs if necessary.

So that the nappers may be always in proper order, they should be kept well ground; if they are in any way blunt, take two of the

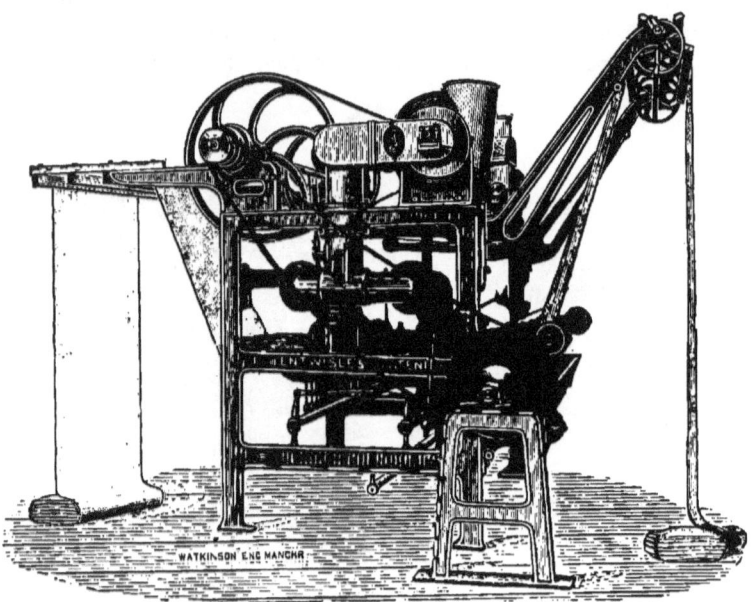

Fig. 70.—Gas Singeing Machine.

workers off at a time, place them in a frame made for the purpose, and run them *against* each other in opposite directions.

Moire or Silk Finish.—It is effected by giving an oscillating motion to the cloth by passing it through a calender, one of the two bowls being engraved with fine lines, the other, unengraved. The calender in ordinary use is a three-bowl machine, with top and bottom cotton bowls and one middle steel bowl, engraved with fine circular grooves heated by steam and arranged to run with friction on the bottom bowl. The waving motion is actuated by suitable mechanism.

Rigging.—Many fabrics require folding down the centre of their length; this is called rigging. If this is done by hand, it is usual to spread the cloth over a long table, placing the selvages

Fig. 71.—Four-Bowl Calender.

carefully upon each other. Machinery, however, obviates this slow operation by the cloth passing up a triangular table, which terminates in a knife-like edge, over which the cloth folds. It then passes through a plaiter or folder, as may be convenient. An example of a rigging machine is illustrated in Fig. 69.

Embossing.—Calenders for this purpose are generally made with two bowls—one with a raised design, the other with the same design depressed below the surface of the roller. As a rule, the cloth is heavily starched. The edges of the design on the roller must not be sharp, but rounded off slightly, or the cloth would be cut, owing to the great pressure required. Very great care is necessary to fit the two rollers properly in the machine. Diagonal slash lines are often engraved for what is strictly speaking the "real *silk*" *finish*.

Singeing by Gas.—The machine for this purpose, shown in Fig. 70, has the burners so arranged that the application of compressed air takes place at the moment of the combustion of the gas; and, by varying the pressure of the air, all the different degrees of heat can be obtained, so that heavy, light, or very thin fabrics can be singed with equal facility. The upper part of the burner is open along its whole length, taps and other means being provided for regulating the line of flame. The position of this line of flame can also be varied according to the effect required. Thus the flame applied directly under the roller singes the cloth thoroughly, bringing out the *grain* so much required and sought for in many classes of fabrics; in other cloths, where the dressing is very slight, as in shearing, the line of flame is placed tangentially to the roller, and merely takes off the projecting fibres.

The feed of cloth can be stopped instantly, with the line of flame reversed, so that fringed goods, such as cotton shawls, can be singed without injury. There is no smoke nor soot to interfere with the most delicate fabrics or colours. By this preventative, after bleaching, dyeing, or printing, no change takes place in the purity of the goods.

The most important advantage in connection with this useful machine is the great economy of gas with the greatest intensity of flame, so much required in the singeing of heavy goods. This intensity of flame is increased by augmenting the pressure of the air, without varying the *quality* of the *gas*, which only becomes more perfectly consumed.

Calender Finish.—Cloth reaches the finisher more or less twisted, and is first opened out by a machine called a scutcher. The two-bowl calender is used for simple finishes, such as lace curtains, etc.; but the most useful are made with four and five bowls. The heat is very often obtained by gas in place of steam,

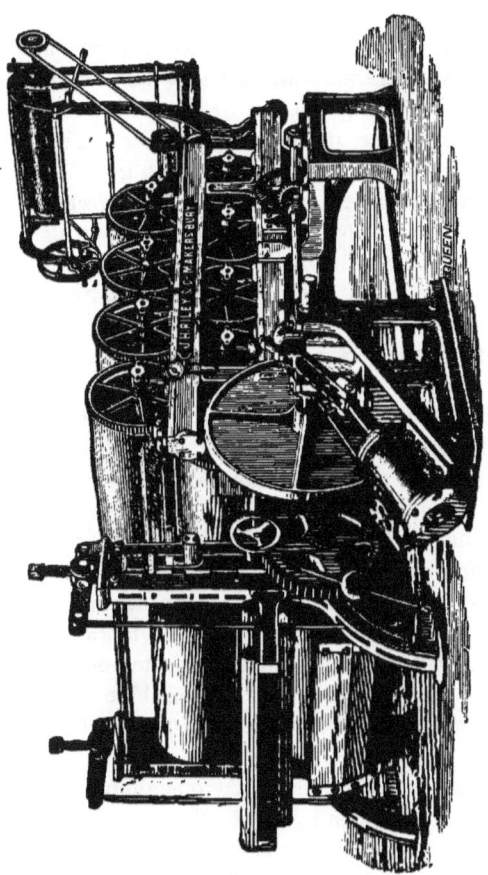

Fig. 72.—Combined Starch Mangle and Drying Machine.

because a higher temperature may be required. A calender finish is lustrous and glossy, but may vary from a very simple lustre to the highest degree of gloss possible. Goods for export, such as are intended for China, are finished bright and stiff—they are mangled and stiffened, dried and calendered, first, between cotton bowls and between metal and cotton bowls, and in some instances they are first filled with the mixing, which we have alluded to above, and when dried are stiffened a second time with starch on the face only.

The four-bowl calender (Fig. 71) may be called an all-round affair, as it is arranged for *glazing*, *swissing*, and *chasing*. Two of these bowls are fine large cotton bowls, having wrought iron centres, with 6-inch journals; one is a perfectly clean cast iron bowl, and one a bright chilled bowl, fitted with a gas heating and injector arrangement, causing a clear flame to be carried the full length of the bowl. The combustion is drawn out by the injector, and, by the use of compound levers with wrought iron pins, a dead set calender is furnished. The four bowls are in operation at once when the swissing is done. When glazing, the top bowl can be wound up from the hot bowl, so that friction or glazing can be done with three bowls, as well as the chasing. This excellent mechanical contrivance is under complete control, and can be stopped at once when the ends of the cloth are to be caught.

Cotton Curtains and Laces.—These goods are put first through the singeing process, the goods by the ordinary process being wrung or pressed and hung up in a hot room to dry. An improved machine now in use (Fig. 72) wraps the fabric round in a kind of coil between two copper cylinders, the outer one being perforated with holes. The revolutions of the machine are 1000 per minute, so that the centrifugal force thus obtained quickly drives off the damp and every particle of moisture.

The finishing of lace is a most important operation, requiring the utmost care and experience; and, in fact, all gauze cloths must be treated with due caution. In laces the meshes must be kept fully extended and in shape, and by a proper amount of stiffening a collapse is prevented. The bleached or dyed goods are passed through a hot mixture of gum and starch, the revolving cylinders squeezing out the surplus stiffening fluid; then taken from the cylinders to the stretching room, which is wide enough to allow two frames being placed at a sufficient distance to be worked side

Fig. 73. –Beetling Machine.

by side, and the heat is often above 80° Fahr., and never less. By means of the side of the frame receding, the lace is gradually extended its full width, the utmost care being taken not to disturb the *mesh*, either in length or width. On this point depends the quality and saleable value of the article.

The amount of stiffening and weight for single, double, and triple stiffness, also the colour, clearness, crispness, and elasticity must be considered. To ensure freedom from small blotches of stiffening and to prevent impurities clinging to the meshes, the pieces are lightly and carefully rubbed with flannels to equalise the stiffening, and then gently beaten by switches and rods as they are distending; this promotes rapid drying, with a clean face and elasticity. The goods are fanned by strong currents of air, and then carefully rolled up.

A piece of cotton lace or net in an unfinished state, weighing, say, 15 pounds will increase in proportion to the finish required. A "Paris" finish will make it 60 pounds weight, and the edges of a piece, with this particular finish, will cut equal to a saw. All nets for foundations as articles of female attire are thus weighted and stiffened. The value of the starch used for stiffening laces is tested in many ways:—for purity of colour; forming a good paste; freedom from lumps, and, in boiling water, swelling into a thick jelly; together with its powers of absorbing water. All these requisites constitute a first-class starch for lace-finishing purposes.

Beetling.—Perhaps this process has received the most attention of all those in use for the finishing of either cotton or linen goods, the most noteworthy of the machines which might be mentioned being a 14-foot roller beetle, with forty beech fallers of large size. In this excellent machine (shown in Fig. 73), a wiper shaft is fitted with forty three-lift tappets, of the greatest possible strength, in section, put on the shaft singly. The fallers rise 11 inches, and the cloth beams of this improved beetle are provided with a self-stripping arrangement, as well as with the regular cam motion. The machine is driven by a 5-inch shaft, on which are placed a pair of 30-inch friction cones and a bevel pinion, gearing into a larger bevel on the end of the wiper shaft. These wheels should be seen to be appreciated.

Turkey Reds.—These cloths are generally sold at so much per pound, and the finisher endeavours in finishing to make the 16 ounces of fabric weigh more than a pound, but the colour suffers in the attempt, and the ordinary process of weighting fails, as starching or stiffening impoverishes the shade to a great extent. In one part of the dyeing process, sulphate of baryta is driven into the fabric, but the ordinary method is the use of chlorides of calcium and magnesium. Epsom and other soluble salts give weight, firmness, and stiffness, far beyond the plastering of china clay or flour, but, in washing the goods, the salts dissolve invisibly, leaving a very poor specimen of the fibre to look at. Gloy is a starch preparation, gelatinised by a caustic alkali, then neutralised and mixed with a quantity of chloride of magnesium, but the great objection is that it forms a deliquescent chloride in the cloth, and more so if used as a sizing material on warps prior to weaving. If the chlorides of magnesium come into contact with soap, they will decompose each other, and when the chloride is evaporated, it liberates hydrochloric acid, and, in the process of hot calendering this salt, becoming decomposed, has a strong tendency to damage the cloth.

Shearing.—The main object with cotton flannels and other popular fabrics, is the napping and shearing to give them a good appearance to buyers, and we will give a few directions for the setting and keeping in order of a shearing machine. It is a matter of importance that this should be carefully attended to, as, in too many instances, want of judgment and neglect in this matter cause the total destruction of the goods. The larger the cylinder, the fewer the revolutions required, and there are more cutting edges; they will also keep longer in order, and the blades are not so apt to get hot with the friction, and a better finish can be obtained. The temper of the blades differs considerably. If too hard, they are brittle, and if too soft, a constant welting and grinding are required.

In commencing with a shearing machine, have a spirit level and see that it stands true, with the blade set so that its edge will be ·25 inch above the ¦cylinder centre. To get this centre, take out the cylinder and push the frame back to get the journals of the cylinder vertical—a straight edge, with one¯end to the centre of the journals, the other end on the edge of the blade. If then a spirit level is attached, the true centre will be found.

The cylinder being put in, the blades are so set that a thin piece of paper will draw evenly along the blade; reversing the cylinder belt and covering up all the exposed parts, the grinding may commence. The finest and best emery powder, made into a thin paste with a good clear oil, not too thick, must be used. This is spread on a strap fastened to a piece of wood a few inches wide and a foot long. The paste is thus transferred all along the cylinder as evenly as possible, pouring out thin oil frequently to wash the emery down on the blades. The grinding is continued until a thin knife edge is got on the ledger and the cutters of the cylinder are sharp. There should be ·125 inch bevel on the ledger blade. The cylinder is then taken out, and a hone applied to the ledger blade in the length, using a 25-degree slant towards the rest, and a very thin bevel will be produced on the edge of the blade.

The grinding being finished, the back of the cylinder cutters is run over with a whetstone or fine file to take off the feather edge, if there is any, and all must then be wiped clean. The blade is set $\frac{1}{16}$ inch below the rest; the cylinder is put in, and a trial made to cut wet paper; if not correct, the ledger blade is raised all along by the top set screws. The ledger may be honed once a day, but, with care, grinding need not be resorted to more than once in four weeks.

Raising.—The application of power to dispense with hand carding or napping is in the form of a cylinder set with cards or teazles. The latest development in "raising" (as this process is termed) is by means of card rollers, having a motion on their own axis, backwards from that of the cylinder itself, so that, while the cylinder's motion brings the card teeth directly into the fabric, the motion of the card roller on its own axis almost at once takes them out again, bringing the nap with them. This softens the action of the cards, and the very lightest of textile fabrics can be as easily dealt with as the strongest make, and a level raising surface is got.

On some improved machines, a double action is introduced, by which the weft, as in flannelettes, is raised from both sides at once, and the card rollers are moved by an oscillating lever in imitation of hand-raising, giving a good silky nap. There is, as in hand-raising, a varying speed of stroke; and in the slower movement a longer nap is pulled out. In its passage through the machine, the cloth is carried between tension rods and over guide rollers to tension rollers. These rollers can be raised or lowered so as to bring the fabric more or less directly into contact with the cards, and the cloth is prevented from slipping back by the drawing rollers being covered with plush, etc. The raising of the nap on cloth is now of great importance, the success of the Moser machine having

FINISHING OF COTTON FABRICS. 209

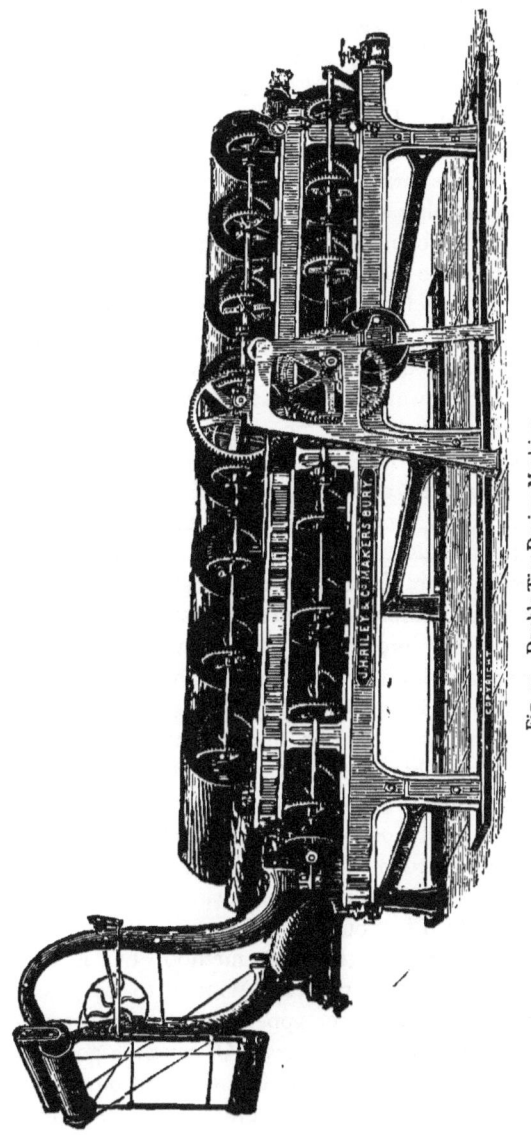

Fig. 74.—Double-Tier Drying Machine.

stimulated improvements by other machinists, and one of great merit is a Huddersfield make, very similar to that just described.

Cotton hosiery is raised by a machine fitted with india-rubber feed rollers, by which the risk of damaging the finest goods is reduced to a minimum. The regulation of the amount of nap or finish is effected by means of change wheels, so that the speed of the feed rollers may be varied as desired. In the cropping process of these fabrics, a peculiar knife is used, called the bayonet spiral, the sides of the blades being concave, which gives greater rigidity and adds considerably to the resistance of the knife when cropping. The blade when working for some time, instead of becoming blunt, tends to get sharper, and however much the diameter of the cutter is reduced, the concavity of the blade ensures a cutting edge being presented to the cloth, which, of course, obviates the necessity of grinding and saves a great loss of time.

The makers, a Sheffield firm, are able to construct these concave spiral cutters up to a width of 144 inches, and they can be adapted for all classes of goods, from the finest to the coarsest textures. Breakage of cutters leads to serious damage of cloths, but these spiral cutters, although hardened to the highest possible point, can be pulled out of shape without a fracture, they are so tenacious.

Drying.—The effects produced by different methods of drying vary greatly; there are drum-drying, tentering, and other machines, hot-air chambers, and open-air drying. Cloth dried on a cylinder, though becoming smoother, has the fibres flattened, and steaming afterwards, if the finish has been made with starch, flour, or gum, will cause these ingredients to swell, and a good effect will result by stoving the steamed cloth on rollers in a damp or cool place for a short time. Goods dried on the tenter are full to the feel; there are no flattened fibres or drawers out of position, and the cloth has the same appearance as if just out of the loom.

Hot or open-air drying is the best method, but is generally replaced by the tenter frame process. Goods to be calendered depend very much for their proper appearance upon the amount of friction and the pressure of the machines. Of course, a good finisher in every case has it in his power to produce the best possible effects by the use of fillings and finishing solutions, and sprinkling and steaming are very important agents for giving the best results.

A double-tier bevil-geared drying machine is shown in Fig. 74.

Damping.—This is a process that receives little or no attention, and is generally grossly neglected. However careful the finisher may have been in the previous operations, and although he has left nothing, as he imagines, undone to ensure success, he will often find himself confronted in his final effort with a mere rag, or, at all events, with a piece of cloth barely presentable, for if the damping is too little, the fabric will remain hard, rough, inflexible, and cannot be properly calendered. Further, it becomes too thick, is not pressed sufficiently, and altogether it has a miserable look all through But if it happens to be damped too much, it will be dull, without any body, and certain to *spot* in the calendering; the finishing agent will become weak or disappear entirely, and the goods will just look as if they had been washed. There is also the storing away in a place too *damp* or too *warm*, when mouldy spots begin to show, and both the strength of the cloth and its colour are injured.

These are very simple remarks to make, no doubt, and finishers may make light of them; but we may ask, Why are there such numbers of "rejects" and ruined goods before being sent out to purchasers?

Many old-fashioned manufacturers spray their goods whilst slowly winding on for the calender. The spraying, being done with a brush, is an inexpensive and very convenient method where expensive machines are not in use; and the orders are the same year in and year out. Dépierre says, "The best method of giving cloth the proper amount of moisture is to hang it for a time in a humid locality." There are many objections to such a process: in the first place, it would be too costly and irregular; stains would take place from the moisture falling in drops upon the fabric; and altogether the process would be barren of anything like a profitable undertaking. If damp salts are used in the finishing, then too much dampness is attracted before and after the finish, and, in any case, could only be used with fast colours.

Some damping machines have a rough stone roller revolving in a water trough, which delivers moisture to a brush; and this brush is made to revolve in a contrary direction to, and at a greater speed than, the stone roller.

The best damping machines known are those constructed on the injector system, as follows:—A number of pipes, perhaps twenty or thirty, dip into a water reservoir, the upper ends of which are pointed at an angle of from 40 to 60 degrees; larger pipes correspond with these, into which a current of air is driven and passes over the mouths of the pipes dipping into the water, and this rarefaction of the air causes the water to rise gradually and become vaporised. There is no dropping, and the perfect vapour moistens the cloth uniformly and thoroughly. This class of damping machinery will moisten about 24,000 yards in ten hours. After this the goods are rolled up on an iron roller and "beetled." In many places the old wood-faller beetle is still in use. This finish is a peculiar one, and has a full soft feel of a thread-like appearance, not obtainable by any other means. Fabrics intended for a beetle finish are rarely filled or weighted, very little being used in any case.

To be a good finisher, a man ought to be capable of finishing according to any recipé, and of setting his machinery precisely for all kinds and qualities of goods delivered to him and producing a perfect finishing effect. The extensive series of operations require the greatest care and no small amount of judgment, and the final result is of the utmost importance in the sale of the cloth.

INDEX.

	PAGE
ACETATE of Alumina, Preparation of	157
Alizarine Red Extract	150
Alkaline Leys	131
—— Vat	138
Arsenic in Prints	154
BALL Sizing	115
Ballooning	78
Beam Warping	129
Beetling	204
Bleach, Madder or Turkey	98
Bleaching	94
—— by Electricity	100
—— by Peroxide of Hydrogen	99
—— Final Operations in	97
—— Kier	95
—— Samples, Recipe for	100
—— Twaddle Test in	97
Bobbin Frame	40
Bobbins, Conditioning	79
Break	178
Bright Colours	131
Building of Mule Cops	64

INDEX.

	PAGE
CALCULATIONS, Carding	22
—— for Drafts and Speeds	36
—— for Mule Spinning	66
—— for Number of Healds	160
—— for Sectional Warping	126
—— for Warping	121
—— for Working for Slubbing and Roving Frames	48
Calender Finish	202
Calico Printing	145
—— Machine for	145
Card Grinding	25
—— Motion	18
Carders, Duties of	20
Carding of Cotton	16
—— Calculations	22
—— Remedies in	21
Cards, Clothing of	19
—— Setting the	20
—— Stripping of	21
Check Strap	171
Chemical Reactions	140
Clearer, Mote	110
Cloth Contraction	184
—— Injury to	154
—— Prevention of Unevenness in	168
—— Unlevel	178
Colour Dyeing	130
Colouring Stuffs	130
Colours	155
—— Bright	131
Combing of Cotton, Carding and	16
Conical Drum Winding	86
Conditioning Bobbins	79
Cop Dyeing	134
Cops, Building of Mule	64
Cotton, Carding and Combing of	16
—— Bleaching	94
—— Combing	26
—— Curtains and Laces	204

INDEX.

	PAGE
Cotton Fabrics, Finishing of	196
—— Industry	1
—— Natural Twist of	13
—— Opening	7
—— Preparing of	5
—— Sizing of	101
—— Test Value of	3
—— Varieties of	1
Counter-Faller	63
Cover	183
Covering, Roller	65
Cylinders, Main Speeds of	21

DAMPING	211
Damping Machine	109
Dark Blue Vats, Orange Discharge in	157
Dhootie Sizing	112
Dividend for Picks	166
Dobbies, Shedding Motions and	187
Doubling, Drawing and	30
—— Final Preparation for	39
—— Object in	30
—— Winding	87
Drafts and Speeds, Calculations for	36
Drawing and Doubling	30
—— Frame	34
—— —— Rules for	37
—— Object in	31
Drawing-in	159
Drying	210
Dyeing, Action of Indigo in	138
—— and Printing	145
—— Colour	130
—— Cops	134
—— Hank and Slubbing	136

INDEX.

	PAGE
ELECTRIC Stop Motion when Warp Yarns break	182
Electricity, Bleaching by	100
—— by Friction	33
Embossing	200
Extract Alizarine Red	150
—— Red	151
FABRIC, Padding and Topping the	149
Fabrics, Distinction in	191
Faller, Counter	63
Filling and Finishing Machine	111
Finish, Calender	202
—— Linen	199
—— Moire or Silk	200
Finishing of Cotton Fabrics	196
Fly or Bobbin Frame	40
Flyer, Combination of Movements of the	41
Flyer Frame	40
Frame, Drawing	34
—— Fly or Bobbin	40
—— Roving	44
—— Slubbing	44
GAS, Singeing by	200
Gassing Yarns	91
Grinding Cards	25
HAND Loom	165
Hand Mule	54
Hank Sizing	116
—— and Slubbing Dyeing	136
Healds, Calculation for Number of	160
—— Eccentricity of the	187
Hopper Feeder	7
Hot Vat	139
Hydrogen, Bleaching by Peroxide of	99

INDEX.

	PAGE
INDICATORS, Mule	69
Indigo Dyeing, Difficulties of	142
—— in Dyeing, Action of	138
—— Printing	152
—— Recovery of Spent	143
JACQUARD Machine	192
KIER, Bleaching	95
LACES, Cotton	204
Let-off Motions and Warp Tension	180
Leys, Alkaline	131
Linen Finish	199
Loom, Hand	165
Looming, Twisting and	163
Looms, Placing	163
—— Power	165
—— Varieties of Power	165
MACHINE for Calico Printing	145
—— for Damping	109
—— for Filling and Finishing	111
—— Jacquard	192
Machines, Caution necessary in Use of	5
Madder or Turkey Bleach	98
Managers, Advice to	48
Manganese, Preparation of Precipitated	157
Mill, Warping	120
Minders, Hints to	71
Mixings	4
—— Weight of	13
Moire or Silk Finish	200
Mote Clearer	110

	PAGE
Mule and its Structure	50
—— Carriage	61
—— —— Gaining of	53
—— Hand	54
—— Improved	57
—— Indicators	69
—— Lubricating the	55
—— Spinning, Calculations and Memoranda for	66
—— —— Economy in	56
—— —— Principal Points in	64
—— Yarn, Finding the Twist for	67
Mules, Working Capacity of	68

Nosing Motion, Self-acting . . . 59
Number of Healds, Calculation for the . 160

Opener . . . 8
Opening Cotton . . . 7
Orange Discharge in Dark Blue Vats . 157
Oxidising Print-work Chambers . 153

Padding and Topping the Fabric . 149
Pause 185
Peroxide of Hydrogen, Bleaching by 99
Pick, Scroll . . . 176
—— Time for the . 175
Picking Motion . . 170
—— —— Improved 179
Picks, Dividend for 166
Placing Looms . 163
Power Loom 165
—— Looms, Varieties of . 165
Practical Operations in Weaving . 194
Preventer, Trap . . 182

	PAGE
Printing Calico	145
—— Dyeing and	145
—— Machine for Calico	145
Print-work Chambers, Oxidising	153

QUADRANT . . . 62

RAISING	208
—— Reactions, Chemical	140
Recipes for Sizing	104
Recovery of Spent Indigo	143
Red, Alizarine Extract	150
—— Extract	151
Reds, Printing on Turkey	147
—— Turkey	206
—— Variety of	152
Reed Motion	178
Reeds	161
Reeling, Winding and	82
Resists	156
Rigging	199
Ring Spinning	73
—— —— Throstle and	72
Roller Covering	65
Rollers, Speed of	34
Roving Frame	44
Rovings, Condition of	22
—— Trial of	38
Rules for Drawing Frames	37
,, ,, Slasher Sizing	114

SCROLL Pick	176
Scutching	13
Sectional Warping	123
—— —— Calculations	126

Self-acting Nosing Motion	59
Setting the Cards	20
Shearing	207
Shedding Motions and Dobbies	187
Shuttle	172
Silk Finish, Moire or	200
Singeing Machine	200
Size, Preventing Putrefaction of	116
—— Quantity required	104
Sizing, Ball	115
—— Economics of	101
—— Hank	116
—— of Cotton	101
—— Process	102
—— —— Dhootie	112
—— Recipes for	104
—— Use of Tallow or Oil in	104
Slasher Sizing, Rules for	114
Slashing Operations	106
Slay	169
Sliver	31
Slubbing	38
—— and Roving Frames, Calculations for	48
—— Frame	44
Souring	97
Spent Indigo, Recovery of	143
Spindles	77
Spinning, Ring	73
—— Throstle	72
—— Throstle and Ring	72
Steeping	95
Stop Motion, Electric, when Warp Yarns break	182
Strap, Check	171
Stripping of Cards	21

TAKING-UP Motion	167
Tallow or Oil in Sizing, Use of	104

INDEX.

	PAGE
Tappets	176
Temples	186
Throstle and Ring Spinning . .	72
—— Spinning	72
Topping the Fabric, Padding and .	149
Trap Preventer	182
Treadles	173
Trial of Rovings	38
Turkey Bleach, Madder or . .	98
—— Reds	206
—— —— Printing on .	147
Twaddle Test in Bleaching .	97
Twist for Mule Yarn . .	67
Twisting	90
—— or Looming . . .	163

UNEVENNESS in Cloth, Prevention of .	168
Unlevel Cloth	178

VALUE of Cotton, Test of . . .	3
Vat, Alkaline . . .	138
Vat, Hot	139

WARP and Weft Yarns	54
Warp Tension	180
Warping, Beam	129
—— Calculations	121, 126
—— of Cotton	119
—— Mill	120
—— Object in	119
—— Sectional	123
Washing	94
Waste in Cotton Manufacture	93

INDEX.

	PAGE
Weaving, Practical Operations in	194
—— Preparatory Processes for	159
Weft Motion	177
—— Waste of	178
Winding and Reeling	82
—— Conical Drum	86
—— Doubling	87
YARNS, Gassing	91
Yarns, Quality of	3
—— Warp and Weft	54

Printed by Hazell, Watson, & Viney, Ld., London and Aylesbury.

Advertisements. i

Telegraphic Address: "Machinery, Manchester."
A B C CODE USED. 4TH EDITION.

ESTABLISHED 1814.

Manchester Royal Exchange: Tuesdays & Fridays, 1 to 3 p.m.
NO. 25 PILLAR.

CRIGHTON & SONS,

CASTLEFIELD IRON WORKS, MANCHESTER, ENGLAND.

Patentees and Makers of
COTTON, SILK, & WOOLLEN MACHINERY,
With the latest Improvements.

Sole Makers of Higgins' "Patent Express Frames."

SOLE LICENCEES FOR THE "PATENT DUST-SEPARATING CYCLONE,"
Dispensing with the necessity for Dust Chambers for Scutching Rooms.

Our Patent Automatic Hopper Feeder for Openers and Scutchers
is the most Reliable and Equal Distributor.

Cotton Gins (Complete Plants).
Bale Breaking, or Cotton Pulling and Mixing Machines.
Distributing Lattices for Cotton Mixing, etc.
Crighton's Patent Automatic Hopper Feeders for Openers, Scutchers, e .
Crighton's New Patent Openers, with or without "Feed Tables."
Crighton's New Patent "Exhaust" Crighton Openers.
Crighton's New Patent Combined Openers and Lap Machines, with or without "Exhaust."
Lap Machines, with Patent Leaf Extractors.
  "   "   "   "   "   "   and Cone Regulators.
  "   "   "   "   "   "   and "Piano Motion" Cone Regulators.
Derby Doublers of all Descriptions.
Grinding Machines of all Descriptions.—Horsfall's Grinders.
Carding Engines, with Rollers and Clearers.
  "   "   with Revolving Flats.
Drawing Frames, with Front and Back Stop Motions, Full Can Stop Motion, etc.
Slubbing, Intermediate, Roving, and Jack Frames, with Mason's Long and Short Collars, etc.
Proprietors (by assignment from W. Higgins and Sons) and Sole Makers of "Higgins' Patent Express" Slubbing, Intermediate, Roving, and Jack Frames.
Merino and Silk Roving Frames.—Silk Dandy Frames.
Improved Self-Acting Mules, Adapted for all Classes of Work.
Crighton's Improved Patent Self-Acting Twiners, with Movable Creels, or on the Mule Principle, with Stationary Creels.

Crighton's Patent Wool Cleaning and Opening Machine received the Silver Medal (Highest Award) at the Bradford Exhibition, 1882.

References to Spinners, Plans, and Estimates on Application.
Correspondence Solicited.

ETELY EQUIPPED WITH SPINNING AND MANUFACTURING MACHINERY IN ANY PART OF THE WORLD.

LORD BROTHERS,

MAKERS OF

COTTON OPENING, CLEANING, CARDING, SPINNING, & WEAVING MACHINERY.

CANAL STREET WORKS, TODMORDEN.

LORD BROTHERS are Makers of the following Machinery, off Newest Models, with all Latest Improvements:—

Patent "Cotton Pulling" or "Bale Breaking" and Mixing Machines.
Patent Automatic or Hopper Feeder for Openers and Scutchers.
Patent "Exhaust" Fans.
Patent "Exhaust" Openers.
Patent "Exhaust" Openers, combined with Scutchers and Lap Machines, for Drawing Cotton any distance up to 1000 Feet.
Patent Combined "Exhaust" Vertical-Cylinder Openers, with or without Scutchers and Lap Machines.
Patent Openers, with Cylinders and Beaters, Combined with Scutchers and Lap Machines.
Patent Scutchers, with Lap Machines.
Patent "Express" Cards.
Patent Improved "Piano" Regulators.
Patent "Revolving" Flat and Improved "Wellman" Flat Cards.
Single and Double Carding Engines, with Patent Setting Arrangements for Rollers, Clearers, Grinding Rollers and Mote Knife.
Drawing Frames, with Improved Stop Motions.
Slubbing Frames, with Patent Steps, Short or Long Collars, built to stand High Speeds.
Intermediate Frames, with   Do.   Do.
Roving Frames, with   Do.   Do.
Fine Jack Frames, with   Do.   Do.
Improved "High-speed Flyer" Throstle, with Patent Steel, New Patent "Self-Lubricating," or our Improved Ashworth's Cast-iron Collars.
Flyer Doubling Frames, on same Principle as High-speed Throstle, for every description of Thread or Lace Yarns.
Improved Rabbeth, Gravity or "Flexible," or other Ring Frames.
   "   Ring Doubling Frames.
Winding Frames.
Warping Mills.
Ball-sizing Machinery.
Warp-Drying Machines.
Beaming Machines.
Patent Looms, Single Shutter or Box.
Looms with Patent Positive "Let-off" Motion.
Patent Indigo Mills, etc., etc.

Millwrights and Brass and Iron Founders.

N.B.—LORD BROTHERS purchased the whole of the Patterns and Business of the late firm of John Elce & Co., Limited, and can execute orders for new or repairs for existing Machinery.

Advertisements.

Telegraphic Address:
"*Machinery, Manchester.*"
A B C CODE USED. 4TH EDITION.

ESTABLISHED 1814.

Manchester Royal Exchange:
Tuesdays & Fridays, 1 to 3 p.m.
NO. 25 PILLAR.

CRIGHTON & SONS,

CASTLEFIELD IRON WORKS, MANCHESTER, ENGLAND.

Patentees and Makers of
COTTON, SILK, & WOOLLEN MACHINERY,
With the latest Improvements.

Sole Makers of Higgins' "Patent Express Frames."

SOLE LICENCEES FOR THE "PATENT DUST-SEPARATING CYCLONE,"
Dispensing with the necessity for Dust Chambers for Scutching Rooms.

Our Patent Automatic Hopper Feeder for Openers and Scutchers
is the most Reliable and Equal Distributor.

Cotton Gins (Complete Plants).
Bale Breaking, or Cotton Pulling and Mixing Machines.
Distributing Lattices for Cotton Mixing, etc.
Crighton's Patent Automatic Hopper Feeders for Openers, Scutchers, etc.
Crighton's New Patent Openers, with or without "Feed Tables."
Crighton's New Patent "Exhaust" Crighton Openers.
Crighton's New Patent Combined Openers and Lap Machines, with or without "Exhaust."
Lap Machines, with Patent Leaf Extractors.
" " " " " " and Cone Regulators.
" " " " " " and "Piano Motion" Cone Regulators.
Derby Doublers of all Descriptions.
Grinding Machines of all Descriptions.—Horsfall's Grinders.
Carding Engines, with Rollers and Clearers.
" " with Revolving Flats.
Drawing Frames, with Front and Back Stop Motions, Full Can Stop Motion, etc.
Slubbing, Intermediate, Roving, and Jack Frames, with Mason's Long and Short Collars, etc.
Proprietors (by assignment from W. Higgins and Sons) and Sole Makers of "Higgins' Patent Express" Slubbing, Intermediate, Roving, and Jack Frames.
Merino and Silk Roving Frames.—Silk Dandy Frames.
Improved Self-Acting Mules, Adapted for all Classes of Work.
Crighton's Improved Patent Self-Acting Twiners, with Movable Creels, or on the Mule Principle, with Stationary Creels.

Crighton's Patent Wool Cleaning and Opening Machine received the Silver Medal (Highest Award) at the Bradford Exhibition, 1882.

References to Spinners, Plans, and Estimates on Application.
Correspondence Solicited.

MILLS COMPLETELY EQUIPPED WITH SPINNING AND MANUFACTURING MACHINERY IN ANY PART OF THE WORLD.

LORD BROTHERS,

MAKERS OF

COTTON OPENING, CLEANING, CARDING, SPINNING, & WEAVING MACHINERY.

CANAL STREET WORKS, TODMORDEN.

LORD BROTHERS are Makers of the following Machinery, off Newest Models, with all Latest Improvements:—

Patent "Cotton Pulling" or "Bale Breaking" and Mixing Machines.
Patent Automatic or Hopper Feeder for Openers and Scutchers.
Patent "Exhaust" Fans.
Patent "Exhaust" Openers.
Patent "Exhaust" Openers, combined with Scutchers and Lap Machines, for Drawing Cotton any distance up to 1000 Feet.
Patent Combined "Exhaust" Vertical-Cylinder Openers, with or without Scutchers and Lap Machines.
Patent Openers, with Cylinders and Beaters, Combined with Scutchers and Lap Machines.
Patent Scutchers, with Lap Machines.
Patent "Express" Cards.
Patent Improved "Piano" Regulators.
Patent "Revolving" Flat and Improved "Wellman" Flat Cards.
Single and Double Carding Engines, with Patent Setting Arrangements for Rollers, Clearers, Grinding Rollers and Mote Knife.
Drawing Frames, with Improved Stop Motions.
Slubbing Frames, with Patent Steps, Short or Long Collars, built to stand High Speeds.

Intermediate Frames, with	Do.	Do.
Roving Frames, with	Do.	Do.
Fine Jack Frames, with	.	Do.

Improved "High-speed Flyer" Throstle, with Patent Steel, New Patent "Self-Lubricating," or our Improved Ashworth's Cast-iron Collars.
Flyer Doubling Frames, on same Principle as High-speed Throstle, for every description of Thread or Lace Yarns.
Improved Rabbeth, Gravity or "Flexible," or other Ring Frames.
 ,,   Ring Doubling Frames.
Winding Frames.
Warping Mills.
Ball-sizing Machinery.
Warp-Drying Machines.
Beaming Machines.
Patent Looms, Single Shutter or Box.
Looms with Patent Positive "Let-off" Motion.
Patent Indigo Mills, etc., etc.

Millwrights and Brass and Iron Founders.

N.B.—LORD BROTHERS purchased the whole of the Patterns and Business of the late firm of John Elce & Co., Limited, and can execute orders for new or repairs for existing Machinery.

BROOKS & DOXEY,

(Late *SAMUEL BROOKS*),

Cotton, Wool, Worsted, etc., Machinists, MANCHESTER.

MAKERS OF

Preparation and Spinning Machinery

For COTTON, etc., etc., including

Carding Engines, Drawing Frames, Slubbing, Intermediate & Roving Frames.

> THESE Machines contain many Patented Inventions of great value for increasing the production and improving the quality of the yarn.

Ring Spinning Frames *for Warp and Weft, etc.*

Ring and Flyer Doubling Frames *for all classes of Yarn.*

Hill & Brown Patent Winding Frame *for Winding Yarn on to Paper Tubes, dispensing with Two-Headed Bobbins.*

Upright Spindle Winding Frames, Cop Bobbin Reels, Bundling Presses, etc.

Patent Waste Picking Machine *for taking the hard threads from Underclearer Waste of Mules and other Spinning Frames.*

"Flyer" COTTON or WORSTED Spinning and Doubling Frames Converted (at reasonable cost) into "RING" System. Production very much increased—in some cases treble that of the old flyer frames before being converted. Results almost equal to new frames. Thousands of spindles can be shown at work successfully.

ALSO, SOLE MAKERS OF THE

"American Standard Ring Traveller"

For Spinning and Twisting, in Steel or Composition. STEEL FLYERS.

The Travellers are made from the Best Selected Material, under the superintendence of a most experienced Manager. A Large and Expensive Plant, with the most accurately working Machinery for producing every Traveller exactly to Micrometer Gauge, has been put down, so that customers may rely upon the Greatest Exactness in the Standard Weights and Numbers of each size of Traveller.

Price Lists on Application. Trial Orders Solicited. Large Stocks always on Hand.

SPECIAL STAINLESS GREASE FOR DOUBLING RINGS.

Re **JOHN TATHAM & SONS,** *Limited, Rochdale.*

We beg to intimate that we have purchased the Good-will of the Business, the Order Books, and the whole of the Wood and Iron Patterns of the above-named firm, for the construction of the various machines used in the processes of Opening, Teasing, Carding, Combing, Drawing, Condensing, Spinning, Winding, Doubling, Twisting, Warping, and Reeling, the Fibres of Waste Wool, Worsted, Cotton, Silk, Flax, Tow, and Asbestos. Orders for New Machines or Sundries will receive prompt attention.

Communications to be addressed to **BROOKS & DOXEY,** Union Iron Works, West Gorton, MANCHESTER

Union Iron Works, West Gorton. | Junction Iron Works, Newton Heath.

Advertisements.

Telegraphic Address:
"ASA, OLDHAM."

Telephone:
No. 7, OLDHAM.

ASA LEES AND CO,
LIMITED,
Soho Iron Works,
OLDHAM.

MANCHESTER OFFICE (*Open Tuesdays and Fridays*):
27, HOPWOOD AVENUE.

CONSTRUCTORS OF

ALL KINDS OF MACHINERY

FOR

Preparing, Spinning, and Doubling Cotton and Wool.

SOLE AGENTS FOR THE CONTINENT OF EUROPE:—
BAERLEIN & CO.,
12, Blackfriars Street, Salford, MANCHESTER,

To whom all Communications relating to Continental Business should be addressed.

AGENTS FOR INDIA:—
BRADBURY, BRADY & Co., Bruce Lane, Fort, Bombay.

ELECTRIC LIGHTING.

THE
Manchester Edison Swan Company, Ltd.,
MANUFACTURERS AND MERCHANTS,
TEMPLE CHAMBERS, 33, BRAZENNOSE STREET,
MANCHESTER.

Secretary: JOHN EDWIN SHARPLES.
Showrooms and Stores: Victoria Street, Manchester.
Works: 27, Little John Street, Deansgate, Manchester.

CONTRACTORS
For House and Mill Lighting, Central Stations, etc.

Apply for Price Lists of
INCANDESCENT AND ARC LAMPS,
ELECTRICAL PLANT AND MATERIAL OF EVERY KIND.   ESTIMATES FREE.

GEO. THOMAS & CO.,
TELEGRAPHIC ADDRESS:
"SAMOHT, MANCHESTER."
NATIONAL TELEPHONE, NO. 1183.

72A, DEANSGATE,
MANCHESTER.

Engineers and Exporters of all Classes of Machinery.

SPECIALITIES :

	PRICE		
	£	s.	d.
Patent Universal Yarn Assorting Balances, for Indicating the Counts of Yarn in Small Lengths, or Bits of Cloth . .	1	10	0
Patent Fibre (Staple) Testing Machine .	9	0	0
Improved Patent Spindle Tachometer, for Showing at a Glance (without Watch) No. of Revolutions per min., up to 12,000	3	5	0
Patent "Friction" Oil Tester . . .	25	0	0
Patent "Lever" Pattern Cutting Machine	£5 to £10		
Patent "Samoht" Self-closing Tap . .	0	6	0

PLEASE WRITE FOR CIRCULARS.

LEOPOLD CASSELLA & CO.,
Frankfort-on-the-Main, GERMANY.

BRANCH:
MANUFACTURE LYONNAISE DE MATIÈRES COLORANTES,
LYONS, FRANCE.

MANUFACTURERS OF ALL SORTS OF

ANILINE DYE-STUFFS.

Proprietors of many valuable Patents, as, for instance

FOR THE

CELEBRATED DIAMINE COLOURS,

all dyeing cotton direct without a mordant;

ALSO FOR

Naphtol Blacks, Naphtylamine Blacks, Anthracite Blacks, Naphtol Green, Thioflavine, Indazine, Brilliant Scarlets, Brilliant Croceines, Cyanole, Formyl Violets, Metaphenylene Blues, New Methylene Blues, etc., etc.

MANUFACTURERS OF

ALL SORTS OF DYE-STUFFS for WOOL, SILK, and COTTON DYEING and PRINTING, for LINEN, LEATHER, JUTE, STRAW, etc.

AGENTS IN GREAT BRITAIN:

PRONK, DAVIS, & CO., 9, Savage Gardens, Tower Hill, London, E.C.
and 19, Temple Street, Birmingham.
SCHOELLES & BROWN, 32, Queen Street, Albert Square, Manchester.
S. A. LIEBERT & CO., 8-10, Prince's Square, Glasgow.
BEVER & WOLF, 46, Vicar Lane, Bradford, Yorks.
CHARLES FORTH & SON, New Basford, Nottingham.

Samples and prices sent on application to any of these firms.

Advertisements. vii

JAMES TOMLINSON, Soho Works, Rochdale.

Raising or Napping Machine

ON THE

MOSER SYSTEM

FOR

FLANNELLETTES, BLANKETS, COTTON and WOOLLEN GOODS, of all Kinds, etc.

(See page 208.)

NEW AND IMPROVED MACHINE, WITH 24 ROLLERS.

TEXTILE INDUSTRIES
AND
JOURNAL OF FABRICS,
10, ANN PLACE, BRADFORD, YORKS.
AND AT
MANCHESTER.

PUBLISHED 12TH OF EACH MONTH.

Ten Shillings per Annum (Post Free) in advance.

THE JOURNAL CONTAINS

VALUABLE INFORMATION AND PRACTICAL ARTICLES ON EVERY BRANCH OF THE TEXTILE TRADES,

Of which this Volume, "COTTON MANUFACTURE," is a Specimen.

The Journal also contains **Original Designs** for every class of Fabric for Ladies' and Gentlemen's wear, with

WOVEN SAMPLES.

Also all classes of

DESIGNS FOR ORNAMENTAL FABRICS,

such as are not otherwise published, except in expensive works.

The first Journal to give **Woven Samples** with **Designs** and **Particulars** for Weaving them. Every Manufacturer, Mill Manager, Overlooker, and Textile Student should subscribe to this Journal.

PATENT DEPARTMENT.

The Proprietors of this Journal are registered Patent Agents, and are especially qualified to act as

PATENT AGENTS TO THE TEXTILE TRADES,

And to execute all work connected with Patents, Designs, and Trade Marks.

CORRESPONDENCE INVITED.   ADVICE GRATIS.

ADDRESS:—

H. & R. T. LORD, 10, Ann Place, Bradford,
YORKSHIRE.

Advertisements.

Telegraphic Address, "HETH, MANCHESTER."   Telephone No. 230.

JOHN HETHERINGTON & SONS, Ltd.,
MANCHESTER.
ALSO PROPRIETORS OF CURTIS, SONS & CO., PHŒNIX WORKS.

PATENT SELF-ACTING MULE.

WE beg to draw attention to the following improvements in our new **SELF-ACTING MULE**, which, during the time that it has been in the market, have brought it rapidly into favour, and enabled us to secure a very large number of orders.

The **HEADSTOCK** has been entirely remodelled, all the parts therein have been strengthened, it is bolted to strong foundation plates which carry the copping-motion and main slips, and all the parts are so arranged as to allow of their being taken off without displacing others, which facilitates the getting at and adjusting of the motions.

The **RIMSHAFT** is made 1¾ inches diameter of steel, with the bosses forged on; it can be run at any speed up to 1000 revolutions per minute, and we would call attention to the advantages of being able to run the Rimshaft at as quick a speed as is easily attainable from the ordinary line shaft, as it greatly facilitates the satisfactory working of the mule; the ordinary driving pulleys are 16 inches diameter, but this can be easily increased if desired; our usual tin-roller pulley is 12 inches diameter, but we can put in a 14-inch if desired, as the Rimshaft is run so quickly.

The **RIMBAND CARRIER PULLEYS** are 13 inches in diameter. This increased size, compared with those ordinarily used, greatly reduces the wear of the rimbands, and as they run at a correspondingly much lower speed and upon steel centres, the lubrication is effected much better, and the risk of fire is greatly reduced in this, one of the most dangerous parts of the Headstock.

The **BACKING-OFF FRICTION CONE** is 19 inches diameter, giving increased power to this important movement; the lever for putting it in and out of gear passes on both sides of it, so that it is always kept square to its work.

The **MAIN DRIVING BELT** controls only the turning of the spindles and the rollers, and the outward movement of the carriage; the other motions, such as backing-off and taking-in, are driven by the pulley upon the ordinary counter-shaft, working into a grooved pulley on the end of the shaft, which comes on the front part of the Headstock. The grooved pulley used is an ordinary rimband pulley, which is easily changed, so that the speed of the various motions can be regulated as required. A special Band-tightening Apparatus is used which comes on the slack side of the band, so that the slack is taken up as required in a very simple manner.

The **PINIONS** for the twist are spurs; they are made much larger in diameter than those usually employed, so as to obtain a smaller amount of change and a nearer approximation to the counts of yarn required. The change draft pinion is also of large size.

NO ANTAGONISTIC MOVEMENTS can be put into operation at the same time, as the changes are so arranged and connected that the one must be out of gear before the other is capable of acting. The cam-shaft has all the cams made solid with it, and is driven by a friction cone controlled in the same manner as described for the backing-off friction.

The **STRAP LEVER** is a separate one, and is connected only to the controlling levers by springs, so that the force by which the belt is moved from one pulley to the other can be regulated and controlled by the strength of the spring, thus saving considerable wear and tear of the belts; the outward movement of the carriage is utilised to bring the belt on the loose pulley, and the levers are so arranged that the carriage can begin to control this movement in any position within 12 inches of its outward run; when a twist motion is used, this, of course, is not employed. The inward run of the carriage, immediately before arriving at its termination, is in a similar way arranged to put the belt from the loose to the fast pulley. In case the carriage is run in slowly, the winding-on is not interfered with, and the ends are not broken as in other mules.

The **REGULATING (STRAPPING OR GOVERNING) MOTION** is made in such a way that a very slight variation in the position of the fallers brings it into gear, and there is a cam which again puts it out of gear, without any strain upon the counter-faller.

The **FULL-COP STOP MOTION**, if ordered, stops the mule when the cops of any desired length are full; the cops are then all the same size. This motion is particularly suitable for weft mules.

The **IMPROVED BACKING-OFF CAM** can be so adjusted that the faller follows the yarn as it is uncoiled from the spindles, and there is also a backing-off chain tightening motion which is connected with the builder plates.

The two ordinary **SCROLL BANDS** for taking in the carriage we are making in one continuous length, and we use a patent tightening apparatus by which the same strain is brought upon each end, and there is also facility for giving out as much slack band as may be required for piecing-up or re-knotting, and so making it serve a much longer time. It is impossible to enumerate in a circular all the improvements introduced, but we invite friends to pay a visit to our works, where we can explain in detail what are the merits of our construction. The production is the largest, and the cost of repairs lower than any Mules working; and in the opinion of those using them they take less power to drive, consequent on the improvements introduced. We shall be happy to show intending purchasers these Mules working at several Mills in and about Manchester, Oldham, Stockport, and Heywood.

Also Makers of Bale Breakers, Automatic Feeders, Crighton Openers, Scutchers, Revolving Flat Carding Engines, Drawing Frame with Metallic Rollers, Ribbon Lap Machine, Heilmann Combing Machine, all kinds of Fly Frames, Ring Spinning Throstles, Ring Doubling Frames, Patent quick traverse Winding Frames, together with Bundling Presses, Cop and Bobbin Reels, and all kinds of Woollen and Worsted Machinery.

USEFUL BOOKS for COTTON MANUFACTURERS.

Second Edition, Revised. Crown 8vo, Cloth, price 7s. 6d.

THE MANUAL OF COLOURS AND DYE-WARES:

Their Properties, Applications, Valuations, Impurities, and Sophistications. For the use of Dyers, Printers, Drysalters, Brokers, etc.

By J. W. SLATER.

"A complete encyclopædia of the *materia tinctoria*. The information given respecting each article is full and precise, and the methods of determining the value of articles such as these, so liable to sophistication, are given with clearness, and are practical as well as valuable."—*Chemist and Druggist.*

Fourth Edition, Revised and Enlarged. Demy 8vo, 250 pages, Strongly Bound, price 6s.

FACTORY ACCOUNTS:

Their Principles and Practice. A Handbook for Accountants and Manufacturers, with Appendices on the Nomenclature of Machine Details; the Income Tax Acts; the Rating of Factories; Fire and Boiler Insurance; the Factory and Workshop Acts, etc.; including also a Glossary of Terms and a large number of Specimen Rulings.

By EMILE GARCKE and J. M. FELLS.

"A very interesting description of the requirements of Factory Accounts. . . . The principle of assimilating the Factory Accounts to the general commercial books is one which we thoroughly agree with."—*Accountants' Journal.*

Third Edition, Revised and Improved. Octavo, Strongly Bound, price 18s.

THE NUMBER, WEIGHT, AND FRACTIONAL CALCULATOR.

Containing upwards of 250,000 Separate Calculations, showing at a glance the value of 422 different rates, ranging from $\frac{1}{16}$th of a Penny to 20s. each, or per cwt., and £20 per ton, of any number of articles consecutively, from 1 to 470; any number of cwts., qrs., and lbs., from 1 cwt. to 470 cwts.; any number of tons, cwts., qrs., and lbs., from 1 to 1000 tons.

By WILLIAM CHADWICK, Public Accountant.

"It is as easy of reference for any answer, or any number of answers, as a dictionary, and the references are even more quickly made. For making up accounts or estimates the book must prove invaluable to all who have any considerable quantity of calculations, involving price and measure in any combination, to do."—*Engineer.*

☞ **For other Works for Manufacturers, see full Catalogue.**

LONDON: CROSBY LOCKWOOD & SON, 7, STATIONERS' HALL COURT, E.C.

7, Stationers' Hall Court, London, E.C.
October, 1895.

A CATALOGUE OF BOOKS

INCLUDING NEW AND STANDARD WORKS IN

ENGINEERING: CIVIL, MECHANICAL AND MARINE;
ELECTRICITY AND ELECTRICAL ENGINEERING;
MINING, METALLURGY; ARCHITECTURE,
BUILDING, INDUSTRIAL AND DECORATIVE ARTS;
SCIENCE, TRADE AND MANUFACTURES;
AGRICULTURE, FARMING, GARDENING;
AUCTIONEERING, VALUING AND ESTATE AGENCY
LAW AND MISCELLANEOUS.

PUBLISHED BY

CROSBY LOCKWOOD & SON.

MECHANICAL ENGINEERING, etc.

D. K. Clark's Pocket-Book for Mechanical Engineers.
THE MECHANICAL ENGINEER'S POCKET-BOOK OF TABLES, FORMULÆ, RULES, AND DATA: A Handy Book of Reference for Daily Use in Engineering Practice. By D. KINNEAR CLARK, M. Inst. C.E., Author of "Railway Machinery," "Tramways," &c. Second Edition, Revised and Enlarged, Small 8vo, 700 pages, 9s. bound in flexible leather covers, with rounded corners and gilt edges.

SUMMARY OF CONTENTS.

MATHEMATICAL TABLES.—MEASUREMENT OF SURFACES AND SOLIDS.—ENGLISH WEIGHTS AND MEASURES.—FRENCH METRIC WEIGHTS AND MEASURES.—FOREIGN WEIGHTS AND MEASURES.—MONEYS.—SPECIFIC GRAVITY, WEIGHT AND VOLUME.—MANUFACTURED METALS.—STEEL PIPES.—BOLTS AND NUTS.—SUNDRY ARTICLES IN WROUGHT AND CAST IRON, COPPER, BRASS, LEAD, TIN, ZINC.—STRENGTH OF MATERIALS.—STRENGTH OF TIMBER.—STRENGTH OF CAST IRON.—STRENGTH OF WROUGHT IRON.—STRENGTH OF STEEL.—TENSILE STRENGTH OF COPPER, LEAD, ETC.—RESISTANCE OF STONES AND OTHER BUILDING MATERIALS.—RIVETED JOINTS IN BOILER PLATES.—BOILER SHELLS.—WIRE ROPES AND HEMP ROPES.—CHAINS AND CHAIN CABLES.—FRAMING.—HARDNESS OF METALS, ALLOYS AND STONES.—LABOUR OF ANIMALS.—MECHANICAL PRINCIPLES.—GRAVITY AND FALL OF BODIES.—ACCELERATING AND RETARDING FORCES.—MILL GEARING, SHAFTING, &c.—TRANSMISSION OF MOTIVE POWER.—HEAT.—COMBUSTION: FUELS.—WARMING, VENTILATION, COOKING STOVES.—STEAM.—STEAM ENGINES AND BOILERS.—RAILWAYS.—TRAMWAYS.—STEAM SHIPS.—PUMPING STEAM ENGINES AND PUMPS.—COAL GAS, GAS ENGINES, &c.—AIR IN MOTION.—COMPRESSED AIR.—HOT AIR ENGINES.—WATER POWER.—SPEED OF CUTTING TOOLS.—COLOURS.—ELECTRICAL ENGINEERING.

*** OPINIONS OF THE PRESS.

"Mr. Clark manifests what is an innate perception of what is likely to be useful in a pocket-book, and he is really unrivalled in the art of condensation. Very frequently we find the information on a given subject is supplied by giving a summary description of an experiment, and a statement of the results obtained. There is a very excellent steam table, occupying five-and-a-half pages; and there are rules given for several calculations, which rules cannot be found in other pocket-books, as, for example, that on page 497, for getting at the quantity of water in the shape of priming in any known weight of steam. It is very difficult to hit upon any mechanical engineering subject concerning which this work supplies no information, and the excellent index at the end adds to its utility. In one word, it is an exceedingly handy and efficient tool, possessed of which the engineer will be saved many a wearisome calculation, or yet more wearisome hunt through various text-books and treatises, and, as such, we can heartily recommend it to our readers, who must not run away with the idea that Mr. Clark's Pocket-book is only Molesworth in another form. On the contrary, each contains what is not to be found in the other; and Mr Clark takes more room and deals at more length with many subjects than Molesworth possibly could."—*The Engineer.*

"It would be found difficult to compress more matter within a similar compass, or produce a book of 650 pages which should be more compact or convenient for pocket reference. . . . Will be appreciated by mechanical engineers of all classes."—*Practical Engineer.*

"Just the kind of work that practical men require to have near to them."—*English Mechanic.*

MR. HUTTON'S PRACTICAL HANDBOOKS.

Handbook for Works' Managers.

THE WORKS' MANAGER'S HANDBOOK OF MODERN RULES, TABLES, AND DATA. For Engineers, Millwrights, and Boiler Makers; Tool Makers, Machinists, and Metal Workers; Iron and Brass Founders, &c. By W. S. HUTTON, Civil and Mechanical Engineer, Author of "The Practical Engineer's Handbook." Fifth Edition, carefully Revised, with Additions. In One handsome Volume, medium 8vo, price 15s. strongly bound.

☞ *The Author having compiled Rules and Data for his own use in a great variety of modern engineering work, and having found his notes extremely useful, decided to publish them—revised to date—believing that a practical work, suited to the* DAILY REQUIREMENTS OF MODERN ENGINEERS, *would be favourably received.*

*** OPINIONS OF THE PRESS.

"Of this edition we may repeat the appreciative remarks we made upon the first and third. Since the appearance of the latter very considerable modifications have been made, although the total number of pages remains almost the same. It is a very useful collection of rules, tables, and workshop and drawing office data."—*The Engineer,* May 10, 1895.

"The author treats every subject from the point of view of one who has collected workshop notes for application in workshop practice, rather than from the theoretical or literary aspect. The volume contains a great deal of that kind of information which is gained only by practical experience, and is seldom written in books."—*The Engineer,* June 5, 1885.

"The volume is an exceedingly useful one, brimful with engineers' notes, memoranda, and rules, and well worthy of being on every mechanical engineer's bookshelf."—*Mechanical World.*

"The information is precisely that likely to be required in practice. . . . The work forms a desirable addition to the library not only of the works' manager, but of anyone connected with general engineering."—*Mining Journal.*

"A formidable mass of facts and figures, readily accessible through an elaborate index. . . . Such a volume will be found absolutely necessary as a book of reference in all sorts of 'works' connected with the metal trades."—*Ryland's Iron Trades Circular.*

"Brimful of useful information, stated in a concise form, Mr. Hutton's books have met a pressing want among engineers. The book must prove extremely useful to every practical man possessing a copy."—*Practical Engineer.*

New Manual for Practical Engineers.

THE PRACTICAL ENGINEER'S HANDBOOK, Comprising a Treatise on Modern Engines and Boilers, Marine, Locomotive, and Stationary. And containing a large collection of Rules and Practical Data relating to recent Practice in Designing and Constructing all kinds of Engines, Boilers, and other Engineering work. The whole constituting a comprehensive Key to the Board of Trade and other Examinations for Certificates of Competency in Modern Mechanical Engineering. By WALTER S. HUTTON, Civil and Mechanical Engineer, Author of "The Works' Manager's Handbook for Engineers," &c. With upwards of 370 Illustrations. Fourth Edition, Revised, with Additions. Medium 8vo, nearly 500 pp., price 18s. strongly bound.

☞ *This work is designed as a companion to the Author's* "WORKS' MANAGER'S HANDBOOK." *It possesses many new and original features, and contains, like its predecessor, a quantity of matter not originally intended for publication, but collected by the Author for his own use in the construction of a great variety of* MODERN ENGINEERING WORK.

The information is given in a condensed and concise form, and is illustrated by upwards of 370 Woodcuts; and comprises a quantity of tabulated matter of great value to all engaged in designing, constructing, or estimating for ENGINES, BOILERS, *and* OTHER ENGINEERING WORK.

*** OPINIONS OF THE PRESS.

"We have kept it at hand for several weeks, referring to it as occasion arose, and we have not on a single occasion consulted its pages without finding the information of which we were in quest."—*Athenæum.*

"A thoroughly good practical handbook, which no engineer can go through without learning something that will be of service to him."—*Marine Engineer.*

"An excellent book of reference for engineers, and a valuable text-book for students of engineering."—*Scotsman.*

"This valuable manual embodies the results and experience of the leading authorities on mechanical engineering."—*Building News.*

"The author has collected together a surprising quantity of rules and practical data, and has shown much judgment in the selections he has made. . . . There is no doubt that this book is one of the most useful of its kind published, and will be a very popular compendium."—*Engineer.*

"A mass of information, set down in simple language, and in such a form that it can be easily referred to at any time. The matter is uniformly good and well chosen, and is greatly elucidated by the illustrations. The book will find its way on to most engineers' shelves, where it will rank as one of the most useful books of reference."—*Practical Engineer.*

"Full of useful information, and should be found on the office shelf of all practical engineers."—*English Mechanic.*

MR. HUTTON'S PRACTICAL HANDBOOKS—continued.

Practical Treatise on Modern Steam-Boilers.

STEAM BOILER CONSTRUCTION. A Practical Handbook for Engineers, Boiler-Makers, and Steam Users. Containing a large Collection of Rules and Data relating to Recent Practice in the Design, Construction, and Working of all Kinds of Stationary, Locomotive, and Marine Steam-Boilers. By WALTER S. HUTTON, Civil and Mechanical Engineer, Author of "The Works' Manager's Handbook," "The Practical Engineer's Handbook," &c. With upwards of 300 Illustrations. Second Edition, medium 8vo, 18s. cloth.

☞ THIS WORK *is issued in continuation of the Series of Handbooks written by the Author, viz:*—"THE WORKS' MANAGER'S HANDBOOK" *and* "THE PRACTICAL ENGINEER'S HANDBOOK," *which are so highly appreciated by Engineers for the practical nature of their information; and is consequently written in the same style as those works.*

The Author believes that the concentration, in a convenient form for easy reference, of such a large amount of thoroughly practical information on Steam-Boilers, will be of considerable service to those for whom it is intended, and he trusts the book may be deemed worthy of as favourable a reception as has been accorded to its predecessors.

*** OPINIONS OF THE PRESS.

"Every detail, both in boiler design and management, is clearly laid before the reader. The volume shows that boiler construction has been reduced to the condition of one of the most exact sciences; and such a book is of the utmost value to the *fin de siècle* Engineer and Works' Manager."—*Marine Engineer.*

"There has long been room for a modern handbook on steam boilers; there is not that room now, because Mr. Hutton has filled it. It is a thoroughly practical book for those who are occupied in the construction, design, selection, or use of boilers."—*Engineer.*

"The book is of so important and comprehensive a character that it must find its way into the libraries of every one interested in boiler using or boiler manufacture if they wish to be thoroughly informed. We strongly recommend the book for the intrinsic value of its contents."—*Machinery Market.*

"The value of this book can hardly be over-estimated. The author's rules, formulæ, &c., are all very fresh, and it is impossible to turn to the work and not find what you want. No practical engineer should be without it."—*Colliery Guardian.*

Hutton's "Modernised Templeton."

THE PRACTICAL MECHANICS' WORKSHOP COMPANION. Comprising a great variety of the most useful Rules and Formulæ in Mechanical Science, with numerous Tables of Practical Data and Calculated Results for Facilitating Mechanical Operations. By WILLIAM TEMPLETON, Author of "The Engineer's Practical Assistant," &c. &c. Seventeenth Edition, Revised, Modernised, and considerably Enlarged by WALTER S. HUTTON, C.E., Author of "The Works' Manager's Handbook," "The Practical Engineer's Handbook," &c. Fcap. 8vo, nearly 500 pp., with 8 Plates and upwards of 250 Illustrative Diagrams, 6s. strongly bound for workshop or pocket wear and tear.

*** OPINIONS OF THE PRESS.

"In its modernised form Hutton's 'Templeton' should have a wide sale, for it contains much valuable information which the mechanic will often find of use, and not a few tables and notes which he might look for in vain in other works. This modernised edition will be appreciated by all who have learned to value the original editions of 'Templeton.'"—*English Mechanic.*

"It has met with great success in the engineering workshop, as we can testify; and there are great many men who, in a great measure, owe their rise in life to this little book."—*Building News.*

"This familiar text-book—well known to all mechanics and engineers—is of essential service to the every-day requirements of engineers, millwrights, and the various trades connected with engineering and building. The new modernised edition is worth its weight in gold."—*Building News.* (Second Notice.)

"This well-known and largely-used book contains information, brought up to date, of the sort so useful to the foreman and draughtsman. So much fresh information has been introduced as to constitute it practically a new book. It will be largely used in the office and workshop."—*Mechanical World.*

"The publishers wisely entrusted the task of revision of this popular, valuable, and useful book to Mr. Hutton than whom a more competent man they could not have found."—*Iron.*

Templeton's Engineer's and Machinist's Assistant.

THE ENGINEER'S, MILLWRIGHT'S, AND MACHINIST'S PRACTICAL ASSISTANT. A collection of Useful Tables, Rules, and Data. By WILLIAM TEMPLETON. Seventh Edition, with Additions. 18mo, 2s. 6d. cloth.

*** OPINIONS OF THE PRESS.

"Occupies a foremost place among books of this kind. A more suitable present to an apprentice to any of the mechanical trades could not possibly be made."—*Building News.*

"A deservedly popular work. It should be in the 'drawer' of every mechanic."—*English Mechanic.*

Foley's Office Reference Book for Mechanical Engineers.

THE MECHANICAL ENGINEER'S REFERENCE BOOK, for Machine and Boiler Construction. In Two Parts. Part I. GENERAL ENGINEERING DATA. Part II. BOILER CONSTRUCTION. With 51 Plates and numerous Illustrations. By NELSON FOLEY. M.I.N.A. Second Edition, Revised throughout and much Enlarged. Folio, £3 3s. net, half-bound. [*Just published.*

SUMMARY OF CONTENTS.

PART I.

MEASURES.
CIRCUMFERENCES AND AREAS, &c., SQUARES, CUBES, FOURTH POWERS.
SQUARE AND CUBE ROOTS.
SURFACE OF TUBES.
RECIPROCALS.
LOGARITHMS.
MENSURATION.
SPECIFIC GRAVITIES AND WEIGHTS.
WORK AND POWER.
HEAT.
COMBUSTION.
EXPANSION AND CONTRACTION.
EXPANSION OF GASES.
STEAM.
STATIC FORCES.
GRAVITATION AND ATTRACTION.
MOTION AND COMPUTATION OF RESULTING FORCES.
ACCUMULATED WORK.
CENTRE AND RADIUS OF GYRATION.
MOMENT OF INERTIA.
CENTRE OF OSCILLATION.
ELECTRICITY.
STRENGTH OF MATERIALS.
ELASTICITY.
TEST SHEETS OF METALS.
FRICTION.
TRANSMISSION OF POWER.
FLOW OF LIQUIDS.
FLOW OF GASES.
AIR PUMPS, SURFACE CONDENSERS, &c.
SPEED OF STEAMSHIPS.
PROPELLERS.
CUTTING TOOLS.
FLANGES.
COPPER SHEETS AND TUBES.
SCREWS, NUTS, BOLT HEADS, &c.
VARIOUS RECIPES AND MISCELLANEOUS MATTER.

WITH DIAGRAMS FOR VALVE-GEAR, BELTING AND ROPES, DISCHARGE AND SUCTION PIPES, SCREW PROPELLERS, AND COPPER PIPES.

PART II.

TREATING OF, POWER OF BOILERS.
USEFUL RATIOS.
NOTES ON CONSTRUCTION.
CYLINDRICAL BOILER SHELLS.
CIRCULAR FURNACES.
FLAT PLATES.
STAYS.
GIRDERS.
SCREWS.
HYDRAULIC TESTS.
RIVETING.
BOILER SETTING, CHIMNEYS, AND MOUNTINGS.
FUELS, &c.
EXAMPLES OF BOILERS AND SPEEDS OF STEAMSHIPS.
NOMINAL AND NORMAL HORSE POWER.

WITH DIAGRAMS FOR ALL BOILER CALCULATIONS AND DRAWINGS OF MANY VARIETIES OF BOILERS.

*** OPINIONS OF THE PRESS.

"The book is one which every mechanical engineer may, with advantage to himself, add to his library."—*Industries.*

"Mr. Foley is well fitted to compile such a work. . . . The diagrams are a great feature of the work. . . . Regarding the whole work, it may be very fairly stated that Mr. Foley has produced a volume which will undoubtedly fulfil the desire of the author and become indispensable to all mechanical engineers."—*Marine Engineer.*

"We have carefully examined this work, and pronounce it a most excellent reference book for the use of marine engineers."—*Journal of American Society of Naval Engineers.*

"A veritable monument of industry on the part of Mr. Foley, who has succeeded in producing what is simply invaluable to the engineering profession."—*Steamship.*

Coal and Speed Tables.

A POCKET BOOK OF COAL AND SPEED TABLES, for Engineers and Steam-users. By NELSON FOLEY, Author of "The Mechanical Engineer's Reference Book." Pocket-size, 3s. 6d. cloth.

"These tables are designed to meet the requirements of every-day use; they are of sufficient scope for most practical purposes, and may be commended to engineers and users of steam."—*Iron.*

"This pocket-book well merits the attention of the practical engineer. Mr. Foley has compiled a very useful set of tables, the information contained in which is frequently required by engineers, coal consumers, and users of steam."—*Iron and Coal Trades Review.*

Steam Engine.

TEXT-BOOK ON THE STEAM ENGINE. With a Supplement on GAS ENGINES, and PART II. ON HEAT ENGINES. By T. M. GOODEVE, M.A., Barrister-at-Law, Professor of Mechanics at the Royal College of Science, London; Author of "The Principles of Mechanics," "The Elements of Mechanism," &c. Twelfth Edition, Enlarged. Crown 8vo, 6s. cloth.

"Professor Goodeve has given us a treatise on the steam engine, which will bear comparison with anything written by Huxley or Maxwell, and we can award it no higher praise."—*Engineer.*

"Mr. Goodeve's text-book is a work of which every young engineer should possess himself."—*Mining Journal.*

Gas Engines.

ON GAS ENGINES. With Appendix describing a Recent Engine with Tube Igniter. By T. M. GOODEVE, M.A. Crown 8vo, 2s. 6d. cloth.

"Like all Mr. Goodeve's writings, the present is no exception in point of general excellence. It is a valuable little volume."—*Mechanical World.*

Steam Engine Design.

A HANDBOOK ON THE STEAM ENGINE, with especial Reference to Small and Medium-sized Engines. For the Use of Engine Makers, Mechanical Draughtsmen, Engineering Students, and Users of Steam Power. By HERMAN HAEDER, C.E. English Edition, Re-edited by the Author from the Second German Edition, and Translated, with considerable Additions and Alterations, by H. H. P. POWLES, A.M.I.C.E., M.I.M.E. With nearly 1,100 Illustrations. Crown 8vo, 9s. cloth.

"A perfect encyclopædia of the steam engine and its details, and one which must take a permanent place in English drawing-offices and workshops."—*A Foreman Pattern-maker.*

"This is an excellent book, and should be in the hands of all who are interested in the construction and design of medium-sized stationary engines. . . . A careful study of its contents and the arrangement of the sections leads to the conclusion that there is probably no other book like it in this country. The volume aims at showing the results of practical experience, and it certainly may claim a complete achievement of this idea."—*Nature.*

"There can be no question as to its value. We cordially commend it to all concerned in the design and construction of the steam engine."—*Mechanical World.*

Steam Boilers.

A TREATISE ON STEAM BOILERS: Their Strength, Construction, and Economical Working. By R. WILSON, C.E. Fifth Edition. 12mo, 6s. cloth.

"The best treatise that has ever been published on steam boilers."—*Engineer.*

"The author shows himself perfect master of his subject, and we heartily recommend all employing steam power to possess themselves of the work."—*Ryland's Iron Trade Circular.*

Boiler Chimneys.

BOILER AND FACTORY CHIMNEYS: Their Draught-Power and Stability. With a Chapter on *Lightning Conductors.* By ROBERT WILSON, A.I.C.E., Author of "A Treatise on Steam Boilers," &c. Second Edition. Crown 8vo, 3s. 6d. cloth.

"A valuable contribution to the literature of scientific building."—*The Builder.*

Boiler Making.

THE BOILER-MAKER'S READY RECKONER AND ASSISTANT. With Examples of Practical Geometry and Templating, for the Use of Platers, Smiths, and Riveters. By JOHN COURTNEY, Edited by D. K. CLARK, M.I.C.E. Third Edition, 480 pp., with 140 Illustrations. Fcap. 8vo, 7s. half-bound.

"No workman or apprentice should be without this book."—*Iron Trade Circular.*

Locomotive Engine Development.

THE LOCOMOTIVE ENGINE AND ITS DEVELOPMENT. A Popular Treatise on the Gradual Improvements made in Railway Engines between 1803 and 1894. By CLEMENT E. STRETTON, C.E., Author of "Safe Railway Working," &c. Third Edition, Revised and Enlarged. With 95 Illustrations. Crown 8vo, 2s. 6d. cloth. [*Just published.*

"Students of railway history and all who are interested in the evolution of the modern locomotive will find much to attract and entertain in this volume." - *The Times.*

"The author of this work is well known to the railway world, and no one, probably, has a better knowledge of the history and development of the locomotive. The volume before us should be of value to all connected with the railway system of this country."—*Nature.*

Estimating for Engineering Work, &c.

ENGINEERING ESTIMATES, COSTS, AND ACCOUNTS: A Guide to Commercial Engineering. With numerous Examples of Estimates and Costs of Millwright Work, Miscellaneous Productions, Steam Engines and Steam Boilers; and a Section on the Preparation of Costs Accounts. By A GENERAL MANAGER. Demy 8vo, 12s. cloth.

"This is an excellent and very useful book, covering subject-matter in constant requisition in every factory and workshop. . . . . The book is invaluable, not only to the young engineer, but also to the estimate department of every works."—*Builder.*

"We accord the work unqualified praise. The information is given in a plain, straightforward manner, and bears throughout evidence of the intimate practical acquaintance of the author with every phrase of commercial engineering."—*Mechanical World.*

Boiler Making.

PLATING AND BOILER MAKING: A Practical Handbook for Workshop Operations. By JOSEPH G. HORNER, A.M.I.M.E. ("A Foreman Pattern Maker"), Author of "Pattern Making," &c. 380 pages, with 338 Illustrations. Crown 8vo, 7s. 6d. cloth. [*Just published.*

"A thoroughly practical, plainly-written treatise. The volume merits commendation. The author's long experience enables him to write with full knowledge of his subject."—*Glasgow Herald.*

Engineering Construction.

PATTERN-MAKING : A Practical Treatise, embracing the Main Types of Engineering Construction and including Gearing, both Hand and Machine-made, Engine Work, Sheaves and Pulleys, Pipes and Columns, Screws, Machine Parts, Pumps and Cocks, the Moulding of Patterns in Loam and Greensand, &c.; together with the methods of Estimating the weight of Castings ; to which is added an Appendix of Tables for Workshop Reference. By JOSEPH G. HORNER, A.M.I.M.E. ("Foreman Pattern Maker"), Author of "Plating and Boiler Making," &c. Second Edition, thoroughly Revised and much Enlarged. With upwards of 450 Illustrations. Crown 8vo, 7s. 6d. cloth.

"A well-written technical guide, evidently written by a man who understands and has practised what he has written about. . . . . We cordially recommend it to engineering students, young journeymen, and others desirous of being initiated into the mysteries of pattern-making."—*Builder.*

"More than 370 illustrations help to explain the text, which is, however, always clear and explicit, thus rendering the work an excellent *vade mecum* for the apprentice who desires to become master of his trade."—*English Mechanic.*

Dictionary of Mechanical Engineering Terms.

LOCKWOOD'S DICTIONARY OF TERMS USED IN THE PRACTICE OF MECHANICAL ENGINEERING, embracing those current in the Drawing Office, Pattern Shop, Foundry, Fitting, Turning, Smiths', and Boiler Shops, &c. &c. Comprising upwards of 6,000 Definitions. Edited by JOSEPH G. HORNER, A.M.I.M.E. ("Foreman Pattern Maker"), Author of "Pattern Making," &c. Second Edition, Revised, with Additions. Crown 8vo, 7s. 6d. cloth.

"Just the sort of handy dictionary required by the various trades engaged in mechanical engineering. The practical engineering pupil will find the book of great value in his studies, and every foreman engineer and mechanic should have a copy."—*Building News.*

"One of the most useful books which can be presented to a mechanic or student."—*English Mechanic.*

"Not merely a dictionary, but, to a certain extent, also a most valuable guide. It strikes us as a happy idea to combine with a definition of the phrase useful information on the subject of which it treats."—*Machinery Market.*

Mill Gearing.

TOOTHED GEARING : A Practical Handbook for Offices and Workshops. By JOSEPH HORNER, A.M.I.M.E. ("Foreman Pattern Maker"), Author of "Pattern Making," &c. With 184 Illustrations. Crown 8vo, 6s. cloth.

SUMMARY OF CONTENTS.

CHAP. I. PRINCIPLES.— II. FORMATION OF TOOTH PROFILES.—III. PROPORTIONS OF TEETH.—IV. METHODS OF MAKING TOOTH FORMS.— V. INVOLUTE TEETH —VI. SOME SPECIAL TOOTH FORMS.—VII. BEVEL WHEELS.—VIII. SCREW GEARS. — IX. WORM GEARS.— X. HELICAL WHEELS.--XI. SKEW BEVELS.—XII. VARIABLE AND OTHER GEARS.—XIII. DIAMETRICAL PITCH. — XIV. THE ODONTOGRAPH.— XV. PATTERN GEARS.—XVI. MACHINE MOULDING GEARS.—XVII. MACHINE CUT GEARS.—XVIII. PROPORTION OF WHEELS.

"We must give the book our unqualified praise for its thoroughness of treatment and we can heartily recommend it to all interested as the most practical book on the subject yet written."—*Mechanical World.*

Fire Engineering.

FIRES, FIRE-ENGINES, AND FIRE-BRIGADES. With a History of Fire-Engines, their Construction, Use, and Management ; Remarks on Fire-Proof Buildings, and the Preservation of Life from Fire ; Statistics of the Fire Appliances in English Towns ; Foreign Fire Systems ; Hints on Fire-Brigades, &c. &c. By CHARLES F. T. YOUNG, C.E. With numerous Illustrations, 544 pp., demy 8vo, £1 4s. cloth.

"To such of our readers as are interested in the subject of fires and fire apparatus, we can most heartily commend this book. It is really the only English work we now have upon the subject."—*Engineering.*

"It displays much evidence of careful research, and Mr. Young has put his facts neatly together. His acquaintance with the practical details of the construction of steam fire engines, old and new, and the conditions with which it is necessary they should comply, is accurate and full."—*Engineer.*

Stone-working Machinery.

STONE-WORKING MACHINERY, and the Rapid and Economical Conversion of Stone. With Hints on the Arrangement and Management of Stone Works. By M. POWIS BALE, M.I.M.E. With Illustrations. Crown 8vo, 9s.

"The book should be in the hands of every mason or student of stonework."—*Colliery Guardian.*
"A capital handbook for all who manipulate stone for building or ornamental purposes."—*Machinery Market.*

Pump Construction and Management.

PUMPS AND PUMPING: A Handbook for Pump Users. Being Notes on Selection, Construction, and Management. By M. POWIS BALE, M.I.M.E., Author of "Woodworking Machinery," "Saw Mills," &c. Second Edition, Revised. Crown 8vo, 2s. 6d. cloth.

"The matter is set forth as concisely as possible. In fact, condensation rather than diffuseness has been the author's aim throughout; yet he does not seem to have omitted anything likely to be of use."—*Journal of Gas Lighting.*
"Thoroughly practical and simply and clearly written."—*Glasgow Herald.*

Milling Machinery, &c.

MILLING MACHINES AND PROCESSES: A Practical Treatise on Shaping Metals by Rotary Cutters. Including Information on Making and Grinding the Cutters. By PAUL N. HASLUCK, Author of "Lathe-Work," "Handybooks for Handicrafts," &c. With upwards of 300 Engravings, including numerous Drawings by the Author. Large crown 8vo, 352 pages, 12s. 6d. cloth.

"A new departure in engineering literature. . . . We can recommend this work to all interested in milling machines; it is what it professes to be—a practical treatise."—*Engineer.*
"A capital and reliable book which will no doubt be of considerable service both to those who are already acquainted with the process as well as to those who contemplate its adoption."—*Industries.*

Turning.

LATHE-WORK: A Practical Treatise on the Tools, Appliances, and Processes employed in the Art of Turning. By PAUL N. HASLUCK. Fifth Edition. Crown 8vo, 5s. cloth.

"Written by a man who knows not only how work ought to be done, but who also knows how to do it, and how to convey his knowledge to others. To all turners this book would be valuable."—*Engineering.*
"We can safely recommend the work to young engineers. To the amateur it will simply be invaluable. To the student it will convey a great deal of useful information."—*Engineer.*

Screw-Cutting.

SCREW THREADS: And Methods of Producing Them. With numerous Tables and complete Directions for using Screw-Cutting Lathes. By PAUL N. HASLUCK, Author of "Lathe-Work," &c. With Seventy-four Illustrations. Third Edition, Revised and Enlarged. Waistcoat-pocket size, 1s. 6d. cloth.

"Full of useful information, hints and practical criticism. Taps, dies, and screwing tools generally are illustrated and their action described."—*Mechanical World.*
"It is a complete compendium of all the details of the screw-cutting lathe; in fact a *multum-in-parvo* on all the subjects it treats upon."—*Carpenter and Builder.*

Smith's Tables for Mechanics, &c.

TABLES, MEMORANDA, AND CALCULATED RESULTS, FOR MECHANICS, ENGINEERS, ARCHITECTS, BUILDERS, &C. Selected and Arranged by FRANCIS SMITH. Fifth Edition. thoroughly Revised and Enlarged, with a New Section of ELECTRICAL TABLES, FORMULÆ, & MEMORANDA. Waistcoat-pocket size, 1s. 6d. limp leather.

"It would, perhaps, be difficult to make a small pocket-book selection of notes and formulæ to suit ALL engineers as it would be to make a universal medicine; but Mr. Smith's waistcoat-pocket collection may be looked upon as a successful attempt."—*Engineer.*
"The best example we have ever seen of 270 pages of useful matter packed into the dimensions of a card-case."—*Building News.*   "A veritable pocket treasury of knowledge."—*Iron.*

French-English Glossary for Engineers, &c.

A POCKET GLOSSARY OF TECHNICAL TERMS: ENGLISH-FRENCH, FRENCH-ENGLISH; with Tables suitable for the Architectural, Engineering, Manufacturing, and Nautical Professions. By JOHN JAMES FLETCHER, Engineer and Surveyor. Second Edition, Revised and Enlarged, 200 pp. Waistcoat-pocket size, 1s. 6d. limp leather.

"It is a very great advantage for readers and correspondents in France and England to have so large a number of the words relating to engineering and manufacturers collected in a liliputian volume. The little book will be useful both to students and travellers."—*Architect.*
"The glossary of terms is very complete, and many of the Tables are new and well arranged. We cordially commend the book."—*Mechanical World.*

Year-Book of Engineering Formulæ, &c.

THE ENGINEER'S YEAR-BOOK FOR 1895. Comprising Formulæ Rules, Tables, Data and Memoranda in Civil, Mechanical, Electrical, Marine and Mine Engineering. By H. R. KEMPE, A.M.Inst.C.E., M.I.E.E., Technical Officer of the Engineer-in-Chief's Office, General Post Office, London, Author of "A Handbook of Electrical Testing," "The Electrical Engineer's Pocket-Book," &c. With 750 Illustrations, specially Engraved for the work. Crown 8vo, 650 pages, 8s. leather. [*Just published.*]

"Represents an enormous quantity of work, and forms a desirable book of reference."—*The Engineer.*

"The volume is distinctly in advance of most similar publications in this country."—*Engineering.*

"This valuable and well-designed book of reference meets the demands of all descriptions of engineers."—*Saturday Review.*

"Teems with up-to-date information in every branch of engineering and construction."—*Building News.*

"The needs of the engineering profession could hardly be supplied in a more admirable, complete and convenient form. To say that it more than sustains all comparisons is praise of the highest sort, and that may justly be said of it."—*Mining Journal.*

"There is certainly room for the new comer, which supplies explanations and directions, as well as formulæ and tables. It deserves to become one of the most successful of the technical annuals."—*Architect.*

"Brings together with great skill all the technical information which an engineer has to use day by day. It is in every way admirably equipped, and is sure to prove successful."—*Scotsman.*

"The up-to-dateness of Mr. Kempe's compilation is a quality that will not be lost on the busy people for whom the work is intended."—*Glasgow Herald.*

Portable Engines.

THE PORTABLE ENGINE: ITS CONSTRUCTION AND MANAGEMENT: A Practical Manual for Owners and Users of Steam Engines generally. By WILLIAM DYSON WANSBROUGH. With 90 Illustrations. Crown 8vo, 3s. 6d. cloth.

"This is a work of value to those who use steam machinery. . . . Should be read by every one who has a steam engine, on a farm or elsewhere."—*Mark Lane Express.*

"We cordially commend this work to buyers and owners of steam engines, and to those who have to do with their construction or use."—*Timber Trades Journal.*

"Such a general knowledge of the steam-engine as Mr. Wansbrough furnishes to the reader should be acquired by all intelligent owners and others who use the steam engine."—*Building News.*

"An excellent text-book of this useful form of engine. The 'Hints to Purchasers' contain a good deal of common-sense and practical wisdom."—*English Mechanic.*

Iron and Steel.

"IRON AND STEEL": A Work for the Forge, Foundry, Factory, and Office. Containing ready, useful, and trustworthy Information for Ironmasters and their Stock-takers; Managers of Bar, Rail, Plate, and Sheet Rolling Mills; Iron and Metal Founders; Iron Ship and Bridge Builders; Mechanical, Mining, and Consulting Engineers; Architects, Contractors, Builders, and Professional Draughtsmen. By CHARLES HOARE, Author of "The Slide Rule," &c. Eighth Edition, Revised throughout and considerably Enlarged. 32mo, 6s. leather.

"For comprehensiveness the book has not its equal."—*Iron.*

"One of the best of the pocket books."—*English Mechanic.*

"We cordially recommend this book to those engaged in considering the details of all kinds of iron and steel works."—*Naval Science.*

Elementary Mechanics.

CONDENSED MECHANICS. A Selection of Formulæ, Rules, Tables, and Data for the Use of Engineering Students, Science Classes, &c. In accordance with the Requirements of the Science and Art Department. By W. G. CRAWFORD HUGHES, A.M.I.C.E. Crown 8vo, 2s. 6d. cloth.

"The book is well fitted for those who are either confronted with practical problems in their work, or are preparing for examination and wish to refresh their knowledge by going through their formulæ again."—*Marine Engineer.*

"It is well arranged, and meets the wants of those for whom it is intended."—*Railway News.*

Steam.

THE SAFE USE OF STEAM. Containing Rules for Unprofessional Steam-users. By an ENGINEER. Sixth Edition. Sewed, 6d.

"If steam-users would but learn this little book by heart, boiler explosions would become sensations by their rarity."—*English Mechanic.*

Warming.

HEATING BY HOT WATER; with Information and Suggestions on the best Methods of Heating Public, Private and Horticultural Buildings. By WALTER JONES. Second Edition. With 96 Illustrations, crown 8vo, 2s. 6d. net.

"We confidently recommend all interested in heating by hot water to secure a copy of this valuable little treatise."—*The Plumber and Decorator.*

CIVIL ENGINEERING, SURVEYING, etc.

Water Supply and Water-Works.

THE WATER SUPPLY OF TOWNS AND THE CONSTRUCTION OF WATER-WORKS: A Practical Treatise for the Use of Engineers and Students of Engineering. By W. K. BURTON, A.M.Inst.C.E., Professor of Sanitary Engineering in the Imperial University, Tokyo, Japan, and Consulting Engineer to the Tokyo Water-Works. With an Appendix on THE EFFECTS OF EARTHQUAKES ON WATER-WORKS, by Professor JOHN MILNE, F.R.S. With numerous Plates and Illustrations. Super-royal 8vo, 25s. buckram. [*Just published.*

CONTENTS.

CHAP.
- I.—INTRODUCTORY.
- II.—DIFFERENT QUALITIES OF WATER.
- III.—QUANTITY OF WATER TO BE PROVIDED.
- IV.—ON ASCERTAINING WHETHER A PROPOSED SOURCE OF SUPPLY IS SUFFICIENT.
- V.—ON ESTIMATING THE STORAGE CAPACITY REQUIRED TO BE PROVIDED.
- VI.—CLASSIFICATION OF WATERWORKS.
- VII.—IMPOUNDING RESERVOIRS.
- VIII.—EARTHWORK DAMS.
- IX.—MASONRY DAMS.
- X.—THE PURIFICATION OF WATER.
- XI.—SETTLING RESERVOIRS.
- XII.—SAND FILTRATION.
- XIII.—PURIFICATION OF WATER BY ACTION OF IRON—SOFTENING OF WATER BY ACTION OF LIME—NATURAL FILTRATION.

CHAP.
- XIV.—SERVICE OR CLEAN WATER RESERVOIRS—WATER TOWERS—STAND PIPES.
- XV.—THE CONNECTION OF SETTLING RESERVOIRS, FILTER BEDS AND SERVICE RESERVOIRS.
- XVI.—PUMPING MACHINERY.
- XVII.—FLOW OF WATER IN CONDUITS—PIPES AND OPEN CHANNELS.
- XVIII.—DISTRIBUTION SYSTEMS.
- XIX.—SPECIAL PROVISIONS FOR THE EXTINCTION OF FIRE.
- XX.—PIPES FOR WATERWORKS.
- XXI.—PREVENTION OF WASTE OF WATER.
- XXII.—VARIOUS APPLICATIONS USED IN CONNECTION WITH WATERWORKS.

APPENDIX. By PROF. JOHN MILNE, F.R.S.—CONSIDERATIONS CONCERNING THE PROBABLE EFFECTS OF EARTHQUAKES ON WATER-WORKS, AND THE SPECIAL PRECAUTIONS TO BE TAKEN IN EARTHQUAKE COUNTRIES.

"The chapter upon filtration of water is very complete, and the details of construction well illustrated. . . . The work should be specially valuable to civil engineers engaged in work in Japan, but the interest is by no means confined to that locality."—*Engineer.*

"It is with great pleasure that we chronicle an addition to the literature of this important branch of engineering, and we congratulate the author upon the practical common sense shown in the preparation of this work. . . . The plates and diagrams have evidently been prepared with great care, and cannot fail to be of great assistance to the student."—*Builder.*

"The whole art of waterworks construction is dealt with in a clear and comprehensive fashion in this handsome volume. . . . Mr. Burton's practical treatise shows in all its sections the fruit of independent study and individual experience. It is largely based upon his own practice in the branch of engineering of which it treats."—*Saturday Review.*

The Water-Supply of Cities and Towns.

A COMPREHENSIVE TREATISE ON THE WATER-SUPPLY OF CITIES AND TOWNS. By WILLIAM HUMBER, A.-M. Inst. C.E., and M. Inst. M.E., Author of "Cast and Wrought Iron Bridge Construction," &c. &c. Illustrated with 50 Double Plates, 1 Single Plate, Coloured Frontispiece, and upwards of 250 Woodcuts, and containing 400 pages of Text. Imp. 4to, £6 6s. elegantly and substantially half-bound in morocco.

LIST OF CONTENTS.

- I. HISTORICAL SKETCH OF SOME OF THE MEANS THAT HAVE BEEN ADOPTED FOR THE SUPPLY OF WATER TO CITIES AND TOWNS.—II. WATER AND THE FOREIGN MATTER USUALLY ASSOCIATED WITH IT.—III. RAINFALL AND EVAPORATION.—IV. SPRINGS AND THE WATER-BEARING FORMATIONS OF VARIOUS DISTRICTS.—V. MEASUREMENT AND ESTIMATION OF THE FLOW OF WATER.—VI. ON THE SELECTION OF THE SOURCE OF SUPPLY.—VII. WELLS.—VIII. RESERVOIRS.—IX. THE PURIFICATION OF WATER.—X. PUMPS.—XI. PUMPING MACHINERY.—XII. CONDUITS.—XIII. DISTRIBUTION OF WATER.—XIV. METERS SERVICE PIPES, AND HOUSE FITTINGS.—XV. THE LAW AND ECONOMY OF WATER WORKS.—XVI. CONSTANT AND INTERMITTENT SUPPLY.—XVII. DESCRIPTION OF PLATES.—APPENDICES, GIVING TABLES OF RATES OF SUPPLY, VELOCITIES, &c. &c., TOGETHER WITH SPECIFICATIONS OF SEVERAL WORKS ILLUSTRATED, AMONG WHICH WILL BE FOUND: ABERDEEN, BIDEFORD, CANTERBURY, DUNDEE, HALIFAX, LAMBETH, ROTHERHAM, DUBLIN, AND OTHERS.

"The most systematic and valuable work upon water supply hitherto produced in English, or in any other language. . . . Mr. Humber's work is characterised almost throughout by an exhaustiveness much more distinctive of French and German than of English technical treatises."—*Engineer.*

"We can congratulate Mr. Humber on having been able to give so large an amount of information on a subject so important as the water supply of cities and towns. The plates, fifty in number, are mostly drawings of executed works, and alone would have commanded the attention of every engineer whose practice may lie in this branch of the profession."—*Builder.*

Water Supply.

RURAL WATER SUPPLY. A Practical Handbook on the Supply of Water and Construction of Waterworks for Small Conntry Districts. By ALLAN GREENWELL, A.M.I.C.E., and W. T. CURRY, A.M.I.C.E., F.G.S. With Illustrations. Crown 8vo, 5s. cloth. *[Just ready.*

Hydraulic Tables.

HYDRAULIC TABLES, CO-EFFICIENTS, AND FORMULÆ for Finding the Discharge of Water from Orifices, Notches, Weirs, Pipes, and Rivers. With New Formulæ, Tables, and General Information on Rain-fall, Catchment-Basins, Drainage, Sewerage, Water Supply for Towns and Mill Power. By JOHN NEVILLE, Civil Engineer, M.R.I.A. Third Edition, carefully revised, with considerable Additions. Numerous Illustrations. Crown 8vo, 14s. cloth.

"Alike valuable to students and engineers in practice; its study will prevent the annoyance of avoidable failures, and assist them to select the readiest means of successfully carrying out any given work connected with hydraulic engineering."—*Mining Journal.*

"It is, of all English books on the subject, the one nearest to completeness . . . . From the good arrangement of the matter, the clear explanations and abundance of formulæ, the carefully calculated tables, and, above all, the thorough acquaintance with both theory and construction, which is displayed from first to last, the book will be found to be an acquisition."—*Architect.*

Hydraulics.

HYDRAULIC MANUAL. Consisting of Working Tables and Explanatory Text. Intended as a Guide in Hydraulic Calculations and Field Operations. By LOWIS D'A. JACKSON, Author of "Aid to Survey Practice," "Modern Metrology," &c. Fourth Edition, Enlarged. Large crown 8vo, 16s. cloth.

"The author has had a wide experience in hydraulic engineering and has been a careful observer of the facts which have come under his notice, and from the great mass of material at his command he has constructed a manual which may be accepted as a trustworthy guide to this branch of the engineer's profession. We can heartily recommend this volume to all who desire to be acquainted with the latest development of this important subject."—*Engineering.*

"The standard work in this department of mechanics."—*Scotsman.*

"The most useful feature of this work is its freedom from what is superannuated, and its thorough adoption of recent experiments; the text is in fact in great part a short account of the great modern experiments."—*Nature.*

Water Storage, Conveyance, and Utilisation.

WATER ENGINEERING: A Practical Treatise on the Measurement, Storage, Conveyance, and Utilisation of Water for the Supply of Towns, for Mill Power, and for other Purposes. By CHARLES SLAGG, A.-M.Inst.C.E., Author of "Sanitary Work in the Smaller Towns, and in Villages," &c. Second Edition, with numerous Illustrations. Crown 8vo, 7s. 6d. cloth.

"As a small practical treatise on the water supply of towns, and on some applications of waterpower, the work is in many respects excellent."—*Engineering.*

"The author has collated the results deduced from the experiments of the most eminent authorities, and has presented them in a compact and practical form, accompanied by very clear and detailed explanations. . . . The application of water as a motive power is treated very carefully and exhaustively."—*Builder.*

"For anyone who desires to begin the study of hydraulics with a consideration of the practical applications of the science there is no better guide."—*Architect.*

Drainage.

ON THE DRAINAGE OF LANDS, TOWNS, AND BUILDINGS. By G. D. DEMPSEY, C.E., Author of "The Practical Railway Engineer," &c. Revised, with large Additions on RECENT PRACTICE IN DRAINAGE ENGINEERING, by D. KINNEAR CLARK, M.Inst. C.E., Author of "Tramways: their Construction and Working," "A Manual of Rules, Tables, and Data for Mechanical Engineers," &c. Second Edition, Corrected. Fcap. 8vo, 5s. cloth.

"The new matter added to Mr. Dempsey's excellent work is characterised by the comprehensive grasp and accuracy of detail for which the name of Mr. D. K. Clark is a sufficient voucher."—*Athenæum.*

"As a work on recent practice in drainage engineering, the book is to be commended to all who are making that branch of engineering science their special study."—*Iron.*

"A comprehensive manual on drainage engineering, and a useful introduction to the student."—*Building News.*

River Engineering.

RIVER BARS: The Causes of their Formation, and their Treatment by "Induced Tidal Scour;" with a Description of the Successful Reduction by this Method of the Bar at Dublin. By I. J. MANN, Assist. Eng. to the Dublin Port and Docks Board. Royal 8vo, 7s. 6d. cloth.

"We recommend all interested in harbour works—and, indeed, those concerned in the improvements of rivers generally—to read Mr. Mann's interesting work on the treatment of river bars."—*Engineer.*

Tramways and their Working.

TRAMWAYS : THEIR CONSTRUCTION AND WORKING. Embracing a Comprehensive History of the System ; with an exhaustive Analysis of the Various Modes of Traction, including Horse Power, Steam, Cable Traction, Electric Traction, &c. ; a Description of the Varieties of Rolling Stock ; and ample Details of Cost and Working Expenses. New Edition, Thoroughly Revised, and Including the Progress recently made in Tramway Construction, &c. &c. By D. KINNEAR CLARK, M.Inst. C.E. With 400 Illustrations. 8vo, 780 pages. Price 28s., buckram. [*Just published.*

"Although described as a new edition, this book is really a new one, a large part of it, which covers historical ground, having been re-written and amplified ; while the parts which relate to all that has been done since 1882 appear in this edition only. It is sixteen years since the first edition appeared, and twelve years since the supplementary volume to the first book was published. After a lapse, then, of twelve years, it is obvious that the author has at his disposal a vast quantity of descriptive and statistical information, with which he may, and has, produced a volume of great value to all interested in tramway construction and working. The new volume is one which will rank, among tramway engineers and those interested in tramway working, with his world-famed book on railway machinery."—*The Engineer,* March 8, 1895.

"An exhaustive and practical work on tramways, in which the history of this kind of locomotion, and a description and cost of the various modes of laying tramways, are to be found."—*Building News.*

"The best form of rails, the best mode of construction, and the best mechanical appliances, are so fairly indicated in the work under review that any engineer about to construct a tramway will be enabled at once to obtain the practical information which will be of most service to him."—*Athenæum.*

"Of this work we spoke in terms of deservedly high praise on its first appearance. . . . The work is a standard one, and constitutes really all that an engineer about to construct a tramway would need. Mr. Clark has the very highest reputation, both for soundness and for his tact in imparting instruction. Of this work it is impossible to speak too highly."—*Colliery Guardian.*

Student's Text-Book on Surveying.

PRACTICAL SURVEYING : A Text-Book for Students preparing for Examinations or for Survey-work in the Colonies. By GEORGE W. USILL, A.M.I.C.E., Author of "The Statistics of the Water Supply of Great Britain." With 4 Lithographic Plates and upwards of 330 Illustrations. Third Edition, Revised and Enlarged. Including Tables of Natural Sines, Tangents, Secants, &c. Crown 8vo, 7s. 6d. cloth ; or, on THIN PAPER, bound in limp leather, gilt edges, rounded corners, for pocket use, price 12s. 6d.

"The best forms of instruments are described as to their construction, uses and modes of employment, and there are innumerable hints on work and equipment such as the author, in his experience as surveyor, draughtsman and teacher, has found necessary, and which the student in his inexperience will find most serviceable."—*Engineer.*

"The latest treatise in the English language on surveying, and we have no hesitation in saying that the student will find it a better guide than any of its predecessors. . . . Deserves to be recognised as the first book which should be put in the hands of a pupil of Civil Engineering, and every gentleman of education who sets out for the Colonies would find it well to have a copy."—*Architect.*

Survey Practice.

AID TO SURVEY PRACTICE : for Reference in Surveying, Level- ling, and Setting-out ; and in Route Surveys of Travellers by Land and Sea. With Tables, Illustrations, and Records. By LOWIS D'A. JACKSON, A.M.I.C.E., Author of "Hydraulic Manual," "Modern Metrology," &c. Second Edition, Enlarged. Large crown 8vo, 12s. 6d. cloth.

"Mr. Jackson has produced a valuable *vade-mecum* for the surveyor. We can recommend this book as containing an admirable supplement to the teaching of the accomplished surveyor."—*Athenæum.*

"As a text-book we should advise all surveyors to place it in their libraries, and study well the matured instructions afforded in its pages."—*Colliery Guardian.*

"The author brings to his work a fortunate union of theory and practical experience which, aided by a clear and lucid style of writing, renders the book a very useful one."—*Builder.*

Field-Book for Engineers.

THE ENGINEER'S, MINING SURVEYOR'S, AND CONTRACTOR'S FIELD-BOOK. Consisting of a Series of Tables, with Rules, Explanations of Systems, and use of Theodolite for Traverse Surveying and Plotting the Work with minute accuracy by means of Straight Edge and Set Square only ; Levelling with the Theodolite, Casting-out and Reducing Levels to Datum, and Plotting Sections in the ordinary manner ; Setting-out Curves with the Theodolite by Tangential Angles and Multiples with Right and Left-hand Readings of the Instrument ; Setting-out Curves without Theodolite on the System of Tangential Angles by Sets of Tangents and Offsets ; and Earthwork Tables to 80 feet deep, calculated for every 6 inches in depth. By W. DAVIS HASKOLL, C.E. With numerous Woodcuts. Fourth Edition, Enlarged. Crown 8vo, 12s. cloth.

"The book is very handy ; the separate tables of sines and tangents to every minute will make i useful for many other purposes, the genuine traverse tables existing all the same."—*Athenæum.*

"Every person engaged in engineering field operations will estimate the importance of such a work and the amount of valuable time which will be saved by reference to a set of reliable tables prepared with the accuracy and fulness of those given in this volume."—*Railway News.*

Surveying, Land and Marine.

LAND AND MARINE SURVEYING, in Reference to the Preparation of Plans for Roads and Railways; Canals, Rivers, Towns' Water Supplies; Docks and Harbours. With Description and Use of Surveying Instruments. By W. DAVIS HASKOLL, C.E., Author of "Bridge and Viaduct Construction," &c. Second Edition, Revised, with Additions. Large crown 8vo, 9s. cloth.

"This book must prove of great value to the student. We have no hesitation in recommending it, feeling assured that it will more than repay a careful study."—*Mechanical World.*
"A most useful and well arranged book for the aid of a student. We can strongly recommend it as a carefully-wr tten and valuable text-book. It enjoys a well-deserved repute among surveyors."—*Builder.*
"This volume cannot fail to prove of the utmost practical utility. It may be safely recommended to all students who aspire to become clean and expert surveyors."—*Mining Journal.*

Levelling.

A TREATISE ON THE PRINCIPLES AND PRACTICE OF LEVELLING. Showing its Application to purposes of Railway and Civil Engineering in the Construction of Roads; with Mr. TELFORD'S Rules for the same. By FREDERICK W. SIMMS, F.G.S., M. Inst. C.E. Seventh Edition, with the addition of LAW's Practical Examples for Setting-out Railway Curves, and TRAUTWINE'S Field Practice of Laying-out Circular Curves. With 7 Plates and numerous Woodcuts, 8vo, 8s. 6d. cloth. *** TRAUTWINE on CURVES may be had separate, 5s.

"The text-book on levelling in most of our engineering schools and colleges."—*Engineer.*
"The publishers have rendered a substantial service to the profession, especially to the younger members, by bringing out the present edition of Mr. Simms's useful work."—*Engineering.*

Trigonometrical Surveying.

AN OUTLINE OF THE METHOD OF CONDUCTING A TRIGONOMETRICAL SURVEY, for the Formation of Geographical and Typographical Maps and Plans, Military Reconnaissance, LEVELLING, &c., with Useful Problems, Formulæ, and Tables. By Lieut.-General FROME, R.E. Fourth Edition, Revised and partly Re-written by Major-General Sir CHARLES WARREN, G.C.M.G., R.E. With 19 Plates and 115 Woodcuts, royal 8vo, 16s. cloth.

"No words of praise from us can strengthen the position so well and so steadily maintained by this work. Sir Charles Warren has revised the entire work, and made such additions as were necessary to bring every portion of the contents up to the present date."—*Broad Arrow.*

Curves, Tables for Setting-out.

TABLES OF TANGENTIAL ANGLES AND MULTIPLES FOR SETTING-OUT CURVES from 5 to 200 Radius. By A. BEAZELEY, M.Inst.C.E. 4th Edition. Printed on 48 Cards, and sold in a cloth box, waistcoat-pocket size, 3s. 6d.

"Each table is printed on a small card, which, being placed on the theodolite, leaves the hands free to manipulate the instrument—no small advantage as regards the rapidity of work."—*Engineer.*
"Very handy: a man may know that all his day's work must fall on two of these cards, which he puts into his own card-case, and leaves the rest behind."—*Athenæum.*

Earthwork.

EARTHWORK TABLES. Showing the Contents in Cubic Yards of Embankments, Cuttings, &c., of Heights or Depths up to an average of 80 feet. By JOSEPH BROADBENT, C.E., and FRANCIS CAMPIN, C.E. Crown 8vo, 5s. cloth.

"The way in which accuracy is attained, by a simple division of each cross section into three elements, two in which are constant and one variable, is ingenious."—*Athenæum.*

Earthwork, Measurement of.

A MANUAL ON EARTHWORK. By ALEX. J. S. GRAHAM, C.E. With numerous Diagrams. Second Edition. 18mo, 2s. 6d. cloth.

Tunnelling.

PRACTICAL TUNNELLING. Explaining in detail the Setting-out of the Works, Shaft-sinking, and Heading-driving, Ranging the Lines and Levelling under Ground, Sub-Excavating, Timbering, and the Construction of the Brickwork of Tunnels, with the amount of Labour required for, and the Cost of, the various portions of the work. By F. W. SIMMS, M. Inst. C.E. Third Edition, Revised by D. K. CLARK, M.Inst.C.E. Imp. 8vo, 30s. cloth.

"The estimation in which Mr. Simms's book on tunnelling has been held for over thirty years cannot be more truly expressed than in the words of the late Professor Rankine:—'The best source of information on the subject of tunnels is Mr. F. W. Simms's work on Practical Tunnelling.'"—*Architect.*
"It has been regarded from the first as a text book of the subject . . . Mr. Clark has added immensely to the value of the book."—*Engineer.*

Tunnel Shafts.

THE CONSTRUCTION OF LARGE TUNNEL SHAFTS: A Practical and Theoretical Essay. By J. H. WATSON BUCK, M. Inst. C.E., Resident Engineer, London and North-Western Railway. Illustrated with Folding Plates, royal 8vo, 12s. cloth.

"Many of the methods given are of extreme practical value to the mason, and the observations on the form of arch, the rules for ordering the stone, and the construction of the templates, will be found of considerable use. We commend the book to the engineering profession."—*Building News.*

"Will be regarded by civil engineers as of the utmost value, and calculated to save much time and obviate many mistakes."—*Colliery Guardian.*

Cast and Wrought Iron Bridge Construction.

A COMPLETE AND PRACTICAL TREATISE ON CAST AND WROUGHT IRON BRIDGE CONSTRUCTION, including Iron Foundations. In Three Parts—Theoretical, Practical, and Descriptive. By WILLIAM HUMBER, A.-M. Inst. C.E., and M. Inst. M.E. Third Edition, revised and much improved, with 115 Double Plates (20 of which now first appear in this edition), and numerous Additions to the Text. In 2 vols., imp. 4to, £6 16s. 6d. half-bound in morocco.

"A very valuable contribution to the standard literature of civil engineering. In addition to elevations, plans, and sections, large scale details are given, which very much enhance the instructive worth of those illustrations."—*Civil Engineer and Architect's Journal.*

"Mr. Humber's stately volumes, lately issued—in which the most important bridges erected during the last five years, under the direction of the late Mr. Brunel, Sir W. Cubitt, Mr. Hawkshaw, Mr. Page, Mr. Fowler. Mr. Hemans, and others among our most eminent engineers, are drawn and specified in great detail."—*Engineer.*

Oblique Bridges.

A PRACTICAL AND THEORETICAL ESSAY ON OBLIQUE BRIDGES. With 13 large Plates. By the late GEORGE WATSON BUCK, M.I.C.E. Fourth Edition, revised by his Son, J. H. WATSON BUCK, M.I.C.E.; and with the addition of Description to Diagrams for Facilitating the Construction of Oblique Bridges, by W. H. BARLOW, M.I.C.E. Royal 8vo, 12s. cloth.

"The standard text-book for all engineers regarding skew arches is Mr. Buck's treatise, and it would be impossible to consult a better."—*Engineer.*

"Mr. Buck's treatise is recognised as a standard text-book, and his treatment has divested the subject of many of the intricacies supposed to belong to it. As a guide to the engineer and architect, on a confessedly difficult subject, Mr. Buck's work is unsurpassed."—*Building News.*

Oblique Arches.

A PRACTICAL TREATISE ON THE CONSTRUCTION OF OBLIQUE ARCHES. By JOHN HART. Third Edition, with Plates. Imperial 8vo, 8s. cloth.

Statics, Graphic and Analytic.

GRAPHIC AND ANALYTIC STATICS, in their Practical Application to the Treatment of Stresses in Roofs, Solid Girders, Lattice, Bowstring, and Suspension Bridges, Braced Iron Arches and Piers, and other Frameworks. By R. HUDSON GRAHAM, C.E. Containing Diagrams and Plates to Scale. With numerous Examples, many taken from existing Structures. Specially arranged for Class-work in Colleges and Universities. Second Edition, Revised and Enlarged. 8vo, 16s. cloth.

"Mr. Graham's book will find a place wherever graphic and analytic statics are used or studied."—*Engineer.*

"The work is excellent from a practical point of view, and has evidently been prepared with much care. The directions for working are ample, and are illustrated by an abundance of well-selected examples. It is an excellent text-book for the practical draughtsman."—*Athenæum.*

Girders, Strength of.

GRAPHIC TABLE for Facilitating the Computation of the Weights of Wrought Iron and Steel Girders, &c., for Parliamentary and other Estimates. By J. H. WATSON BUCK, M. Inst. C.E. On a Sheet, 2s. 6d.

Strains, Calculation of.

A HANDY BOOK FOR THE CALCULATION OF STRAINS IN GIRDERS AND SIMILAR STRUCTURES AND THEIR STRENGTH. Consisting of Formulæ and Corresponding Diagrams, with numerous details for Practical Application, &c. By WILLIAM HUMBER, A.-M. Inst. C.E., &c. Fifth Edition. Crown 8vo, with nearly 100 Woodcuts and 3 Plates, 7s. 6d. cloth.

"The formulæ are neatly expressed, and the diagrams good."—*Athenæum.*

"We heartily commend this really *handy* book to our engineer and architect readers."—*English Mechanic.*

Trusses.

TRUSSES OF WOOD AND IRON. Practical Applications of Science in Determining the Stresses, Breaking Weights, Safe Loads, Scantlings, and Details of Construction. With Complete Working Drawings. By WILLIAM GRIFFITHS, Surveyor, Assistant Master, Tranmere School of Science and Art. Oblong 8vo, 4s. 6d. cloth.

"This handy little book enters so minutely into every detail connected with the construction of roof trusses that no student need be ignorant of these matters."—*Practical Engineer.*

Strains in Ironwork.

THE STRAINS ON STRUCTURES OF IRONWORK; with Practical Remarks on Iron Construction. By F. W. SHEILDS, M.I.C.E. 8vo, 5s. cloth.

Barlow's Strength of Materials, Enlarged by Humber.

A TREATISE ON THE STRENGTH OF MATERIALS; with Rules for application in Architecture, the Construction of Suspension Bridges, Railways, &c. By PETER BARLOW, F.R.S. A New Edition, revised by his Sons, P. W. BARLOW, F.R.S., and W. H. BARLOW, F.R.S.; to which are added, Experiments by HODGKINSON, FAIRBAIRN, and KIRKALDY; and Formulæ for Calculating Girders, &c. Arranged and Edited by WM. HUMBER, A.-M. Inst. C.E. Demy 8vo, 400 pp., with 19 large Plates and numerous Woodcuts, 18s. cloth.

"Valuable alike to the student, tyro, and the experienced practitioner, it will always rank in future, as it has hitherto done, as the standard treatise on that particular subject."—*Engineer.*
"There is no greater authority than Barlow."—*Building News.*
"As a scientific work of the first class, it deserves a foremost place on the bookshelves of every civil engineer and practical mechanic."—*English Mechanic.*

Cast Iron and other Metals, Strength of.

A PRACTICAL ESSAY ON THE STRENGTH OF CAST IRON and other Metals. By THOMAS TREDGOLD, C.E. Fifth Edition, including HODGKINSON's Experimental Researches. 8vo, 12s. cloth.

Practical Mathematics.

MATHEMATICS FOR PRACTICAL MEN: Being a Common-place Book of Pure and Mixed Mathematics. Designed chiefly for the Use of Civil Engineers, Architects and Surveyors. By OLINTHUS GREGORY, LL.D., F.R.A.S., Enlarged by HENRY LAW, C.E. Fourth Ed., carefully revised by J. R. YOUNG, formerly Professor of Mathematics, Belfast College. With 13 Plates, 8vo, £1 1s. cloth.

"The engineer or architect will here find ready to his hand rules for solving nearly every mathematical difficulty that may arise in his practice. The rules are in all cases explained by means of examples, in which every step of the process is clearly worked out."—*Builder.*
"One of the most serviceable books for practical mechanics. . . . It is an instructive book for the student, and a Text-book for him who, having once mastered the subjects it treats of, needs occasionally to refresh his memory upon them."—*Building News.*

Railway Working.

SAFE RAILWAY WORKING: A Treatise on Railway Accidents, their Cause and Prevention; with a Description of Modern Appliances and Systems. By CLEMENT E. STRETTON, C.E., Vice-President and Consulting Engineer, Amalgamated Society of Railway Servants. With Illustrations and Coloured Plates. Third Edition, Enlarged. Crown 8vo, 3s. 6d. cloth.

"A book for the engineer, the directors, the managers; and, in short, all who wish for information on railway matters will find a perfect encyclopædia in 'Safe Railway Working.'"—*Railway Review.*
"We commend the remarks on railway signalling to all railway managers, especially where a uniform code and practice is advocated."—*Herepath's Railway Journal.*
"The author may be congratulated on having collected, in a very convenient form, much valuable information on the principal questions affecting the safe working of railways."—*Railway Engineer.*

Heat, Expansion by.

EXPANSION OF STRUCTURES BY HEAT. By JOHN KEILY, C.E., late of the Indian Public Works Department. Crown 8vo, 3s. 6d. cloth.

"The aim the author has set before him, viz., to show the effects of heat upon metallic and other structures, is a laudable one, for this is a branch of physics upon which the engineer or architect can find but little reliable and comprehensive data in books."—*Builder.*

Field Fortification.

A TREATISE ON FIELD FORTIFICATION, The Attack of Fortresses, Military Mining, and Reconnoitring. By Professor Colonel I. S. MACAULAY. Sixth Edition, crown 8vo, with separate Atlas of 12 Plates, 12s. cloth.

MR. HUMBER'S GREAT WORK ON MODERN ENGINEERING.

Complete in Four Volumes, imperial 4to, price £12 12s. half-morocco. Each volume sold separately as follows:—

A RECORD OF THE PROGRESS OF MODERN ENGINEERING. FIRST SERIES.
Comprising Civil, Mechanical, Marine, Hydraulic, Railway, Bridge, and other Engineering Works, &c. By WILLIAM HUMBER, A.-M. Inst. C.E., &c. Imp. 4to, with 36 Double Plates, drawn to a large scale, Photographic Portrait of John Hawkshaw, C.E., F.R.S., &c., and copious descriptive Letterpress, Specifications, &c., £3 3s. half-morocco.

LIST OF THE PLATES AND DIAGRAMS.
VICTORIA STATION AND ROOF, L. B. & S. C. R. (8 PLATES); SOUTHPORT PIER (2 PLATES); VICTORIA STATION AND ROOF, L. C. & D. AND G. W. R. (6 PLATES); ROOF OF CREMORNE MUSIC HALL; BRIDGE OVER G. N. RAILWAY; ROOF OF STATION, DUTCH RHENISH RAIL (2 PLATES); BRIDGE OVER THE THAMES, WEST LONDON EXTENSION RAILWAY (5 PLATES); ARMOUR PLATES; SUSPENSION BRIDGE, THAMES (4 PLATES); THE ALLEN ENGINE; SUSPENSION BRIDGE, AVON (3 PLATES); UNDERGROUND RAILWAY (3 PLATES).

"Handsomely lithographed and printed. It will find favour with many who desire to preserve in a permanent form copies of the plans and specifications prepared for the guidance of the contractors for many important engineering works."—*Engineer.*

HUMBER'S PROGRESS OF MODERN ENGINEERING. SECOND SERIES.
Imp. 4to, with 3 Double Plates, Photographic Portrait of Robert Stephenson, C.E., M.P., F.R.S., &c., and copious descriptive Letterpress, Specifications, &c., £3 3s. half-morocco.

LIST OF THE PLATES AND DIAGRAMS.
BIRKENHEAD DOCKS, LOW WATER BASIN (15 PLATES); CHARING CROSS STATION ROOF, C.C. RAILWAY (3 PLATES); DIGSWELL VIADUCT, GREAT NORTHERN RAILWAY; ROBBERY WOOD VIADUCT, GREAT NORTHERN RAILWAY; IRON PERMANENT WAY; CLYDACH VIADUCT; MERTHYR, TREDEGAR, AND ABERGAVENNY RAILWAY; EBBW VIADUCT, MERTHYR, TREDEGAR, AND ABERGAVENNY RAILWAY; COLLEGE WOOD VIADUCT, CORNWALL RAILWAY; DUBLIN WINTER PALACE ROOF (3 PLATES); BRIDGE OVER THE THAMES, L. C. and D. RAILWAY (6 PLATES); ALBERT HARBOUR, GREENOCK (4 PLATES).

"Mr. Humber has done the profession good and true service, by the fine collection of examples he has here brought before the profession and the public."—*Practical Mechanic's Journal.*

HUMBER'S PROGRESS OF MODERN ENGINEERING. THIRD SERIES.
Imp. 4to, with 40 Double Plates, Photographic Portrait of J. R. M'Clean, late Pres. Inst. C.E., and copious descriptive Letterpress, Specifications, &c., £3 3s. half-morocco.

LIST OF THE PLATES AND DIAGRAMS.
MAIN DRAINAGE, METROPOLIS.—*North Side.*—MAP SHOWING INTERCEPTION OF SEWERS; MIDDLE LEVEL SEWER (2 PLATES); OUTFALL SEWER, BRIDGE OVER RIVER LEA (3 PLATES); OUTFALL SEWER, BRIDGE OVER MARSH LANE, NORTH WOOLWICH RAILWAY, AND BOW AND BARKING RAILWAY JUNCTION; OUTFALL SEWER, BRIDGE OVER BOW AND BARKING RAILWAY (3 PLATES); OUTFALL SEWER, BRIDGE OVER EAST LONDON WATERWORKS' FEEDER (2 PLATES); OUTFALL SEWER RESERVOIR (2 PLATES); OUTFALL SEWER, TUMBLING BAY AND OUTLET; OUTFALL SEWER, PENSTOCKS. *South Side.*—OUTFALL SEWER, BERMONDSEY BRANCH (2 PLATES); OUTFALL SEWER, RESERVOIR AND OUTLET (4 PLATES); OUTFALL SEWER, FILTH HOIST; SECTIONS OF SEWERS (NORTH AND SOUTH SIDES).
THAMES EMBANKMENT.—SECTION OF RIVER WALL; STEAMBOAT PIER, WESTMINSTER (2 PLATES); LANDING STAIRS BETWEEN CHARING CROSS AND WATERLOO BRIDGES; YORK GATE (2 PLATES); OVERFLOW AND OUTLET AT SAVOY STREET SEWER (3 PLATES); STEAMBOAT PIER, WATERLOO BRIDGE (3 PLATES); JUNCTION OF SEWERS, PLANS AND SECTIONS; GULLIES, PLANS, AND SECTIONS; ROLLING STOCK; GRANITE AND IRON FORTS.

"The drawings have a constantly increasing value, and whoever desires to possess clear representations of the two great works carried out by our Metropolitan Board will obtain Mr. Humber's volume."—*Engineer.*

HUMBER'S PROGRESS OF MODERN ENGINEERING. FOURTH SERIES.
Imp. 4to, with 36 Double Plates, Photographic Portrait of John Fowler, late Pres. Inst. C.E., and copious descriptive Letterpress, Specifications, &c., £3 3s. half-morocco.

LIST OF THE PLATES AND DIAGRAMS.
ABBEY MILLS PUMPING STATION, MAIN DRAINAGE, METROPOLIS (4 PLATES); BARROW DOCKS (5 PLATES); MANQUIS VIADUCT, SANTIAGO AND VALPARAISO RAILWAY (2 PLATES); ADAM'S LOCOMOTIVE, ST. HELEN'S CANAL RAILWAY (2 PLATES); CANNON STREET STATION ROOF, CHARING CROSS RAILWAY (3 PLATES); ROAD BRIDGE OVER THE RIVER MOKA (2 PLATES); TELEGRAPHIC APPARATUS FOR MESOPOTAMIA; VIADUCT OVER THE RIVER WYE, MIDLAND RAILWAY (3 PLATES); ST. GERMANS VIADUCT, CORNWALL RAILWAY (2 PLATES); WROUGHT-IRON CYLINDER FOR DIVING BELL; MILLWALL DOCKS (6 PLATES); MILROY'S PATENT EXCAVATOR; METROPOLITAN DISTRICT RAILWAY (6 PLATES); HARBOURS, PORTS, AND BREAKWATERS (3 PLATES).

"We gladly welcome another year's issue of this valuable publication from the able pen of Mr. Humber. The accuracy and general excellence of this work are well known, while its usefulness in giving the measurements and details of some of the latest examples of engineering, as carried out by the most eminent men in the profession, cannot be too highly prized."—*Artizan.*

THE POPULAR WORKS OF MICHAEL REYNOLDS
("THE ENGINE DRIVER'S FRIEND").

Locomotive-Engine Driving.

LOCOMOTIVE-ENGINE DRIVING: A Practical Manual for Engineers in Charge of Locomotive Engines. By MICHAEL REYNOLDS, Member of the Society of Engineers, formerly Locomotive Inspector, L. B. and S. C. R. Ninth Edition. Including a KEY TO THE LOCOMOTIVE ENGINE. With Illustrations and Portrait of Author. Crown 8vo, 4s. 6d. cloth.

"Mr. Reynolds has supplied a want, and has supplied it well. We can confidently recommend the book not only to the practical driver, but to everyone who takes an interest in the performance of locomotive engines."—*The Engineer*.

"Mr. Reynolds has opened a new chapter in the literature of the day. This admirable practical treatise, of the practical utility of which we have to speak in terms of warm commendation."—*Athenæum*.

"Evidently the work of one who knows his subject thoroughly."—*Railway Service Gazette*.

"Were the cautions and rules given in the book to become part of the every-day working of our engine-drivers, we might have fewer distressing accidents to deplore."—*Scotsman*.

Stationary Engine Driving.

STATIONARY ENGINE DRIVING: A Practical Manual for Engineers in Charge of Stationary Engines. By MICHAEL REYNOLDS. Fifth Edition, Enlarged. With Plates and Woodcuts. Crown 8vo, 4s. 6d. cloth.

"The author is thoroughly acquainted with his subjects, and his advice on the various points treated is clear and practical. . . . He has produced a manual which is an exceedingly useful one for the class for whom it is specially intended."—*Engineering*.

"Our author leaves no stone unturned. He is determined that his readers shall not only know something about the stationary engine, but all about it."—*Engineer*.

"An engineman who has mastered the contents of Mr. Reynolds's book will require but little actual experience with boilers and engines befo e he can be trusted to look after them."—*English Mechanic*.

The Engineer, Fireman, and Engine-Boy.

THE MODEL LOCOMOTIVE ENGINEER, FIREMAN, AND ENGINE-BOY. Comprising a Historical Notice of the Pioneer Locomotive Engines and their Inventors. By MICHAEL REYNOLDS. Second Edition, with Revised Appendix. With numerous Illustrations, and Portrait of George Stephenson. Crown 8vo, 4s. 6d. cloth. [*Just published*

"From the technical knowledge of the author, it will appeal to the railway man of to-day more forcibly than anything written by Dr. Smiles. . . . The volume contains information of a technical kind, and facts that every driver should be familiar with."—*English Mechanic*.

"We should be glad to see this book in the possession of everyone in the kingdom who has ever laid, or is to lay, hands on a locomotive engine."—*Iron*.

Continuous Railway Brakes.

CONTINUOUS RAILWAY BRAKES: A Practical Treatise on the several Systems in Use in the United Kingdom; their Construction and Performance. With copious Illustrations and numerous Tables. By MICHAEL REYNOLDS. Large crown 8vo, 9s. cloth.

"A popular explanation of the different brakes. It will be of great assistance in forming public opinion, and will be studied with benefit by those who take an interest in the brake."—*English Mechanic*.

"Written with sufficient technical detail to enable the principal and relative connection of the various parts of each particular brake to be readily grasped."—*Mechanical World*.

Engine-Driving Life.

ENGINE-DRIVING LIFE: Stirring Adventures and Incidents in the Lives of Locomotive Engine-Drivers. By MICHAEL REYNOLDS. Third and Cheaper Edition. Crown 8vo, 1s. 6d. cloth.

"From first to last perfectly fascinating. Wilkie Collins's most thrilling conceptions are thrown into the shade by true incidents, endless in their variety, related in every page."—*North British Mail*.

"Anyone who wishes to get a real insight into railway life cannot do better than read 'Engine-Driving Life' for himself, and if he once takes it up he will find that the author's enthusiasm and real love of the engine-driving profession will carry him on till he has read every page."—*Saturday Review*.

Pocket Companion for Enginemen.

THE ENGINEMAN'S POCKET COMPANION and Practical Educator for Enginemen, Boiler Attendants, and Mechanics. By MICHAEL REYNOLDS. With Forty-five Illustrations and numerous Diagrams. Third Edition, Revised. Royal 18mo, 3s. 6d. strongly bound for pocket wear.

"This admirable work is well suited to accomplish its object, being the honest workmanship of a competent engineer."—*Glasgow Herald*.

"A most meritorious work, giving in a succinct and practical form all the information an engine-minder desirous of mastering the scientific principles of his daily calling would require."—*The Miller*.

"A boon to those who are striving to become efficient mechanics."—*Daily Chronicle*.

MARINE ENGINEERING, SHIPBUILDING, NAVIGATION, etc.

Pocket-Book for Naval Architects and Shipbuilders.

THE NAVAL ARCHITECT'S AND SHIPBUILDER'S POCKET-BOOK OF FORMULÆ, RULES, AND TABLES, AND MARINE ENGINEER'S AND SURVEYOR'S HANDY BOOK OF REFERENCE. By CLEMENT MACKROW, Member of the Institution of Naval Architects, Naval Draughtsman. Fifth Edition, Revised and Enlarged to 700 pages, with upwards of 300 Illustrations. Fcap., 12s. 6d. strongly bound in leather.

SUMMARY OF CONTENTS.

SIGNS AND SYMBOLS, DECIMAL FRACTIONS.—TRIGONOMETRY.—PRACTICAL GEOMETRY.—MENSURATION.—CENTRES AND MOMENTS OF FIGURES.—MOMENTS OF INERTIA AND RADII OF GYRATION.—ALGEBRAICAL EXPRESSIONS FOR SIMPSON'S RULES.—MECHANICAL PRINCIPLES.—CENTRE OF GRAVITY.—LAWS OF MOTION.—DISPLACEMENT, CENTRE OF BUOYANCY.—CENTRE OF GRAVITY OF SHIP'S HULL.—STABILITY CURVES AND METACENTRES.—SEA AND SHALLOW-WATER WAVES.—ROLLING OF SHIPS.—PROPULSION AND RESISTANCE OF VESSELS.—SPEED TRIALS.—SAILING, CENTRE OF EFFORT.— DISTANCES DOWN RIVERS, COAST LINES.— STEERING AND RUDDERS OF VESSELS.—LAUNCHING CALCULATIONS AND VELOCITIES.—WEIGHT OF MATERIAL AND GEAR.—GUN PARTICULARS AND WEIGHT.—STANDARD GAUGES.—RIVETED JOINTS AND RIVETING. — STRENGTH AND TESTS OF MATERIALS.—BINDING AND SHEARING STRESSES, ETC.—STRENGTH OF SHAFTING, PILLARS, WHEELS, ETC.—HYDRAULIC DATA, ETC.—CONIC SECTIONS, CATENARIAN CURVES.— MECHANICAL POWERS, WORK.—BOARD OF TRADE REGULATIONS FOR BOILERS AND ENGINES.—BOARD OF TRADE REGULATIONS FOR SHIPS.—LLOYD'S RULES FOR BOILERS.—LLOYD'S WEIGHT OF CHAINS.—LLOYD'S SCANTLINGS FOR SHIPS.—DATA OF ENGINES AND VESSELS.—SHIPS' FITTINGS AND TESTS.—SEASONING PRESERVING TIMBER—MEASUREMENT OF TIMBER.—ALLOYS, PAINTS, VARNISHES.—DATA FOR STOWAGE.—ADMIRALTY TRANSPORT REGULATIONS.—RULES FOR HORSE-POWER, SCREW PROPELLERS, ETC.—PERCENTAGES FOR BUTT STRAPS, ETC.—PARTICULARS OF YACHTS.—MASTING AND RIGGING VESSELS.—DISTANCES OF FOREIGN PORTS.—TONNAGE TABLES.—VOCABULARY OF FRENCH AND ENGLISH TERMS.—ENGLISH WEIGHTS AND MEASURES.—FOREIGN WEIGHTS AND MEASURES.—DECIMAL EQUIVALENTS.—FOREIGN MONEY.—DISCOUNT AND WAGE TABLES.—USEFUL NUMBERS AND READY RECKONERS.—TABLES OF CIRCULAR MEASURES.—TABLES OF AREAS OF AND CIRCUMFERENCES OF CIRCLES.—TABLES OF AREAS OF SEGMENTS OF CIRCLES.—TABLES OF SQUARES AND CUBES AND ROOTS OF NUMBERS.—TABLES OF LOGARITHMS OF NUMBERS.—TABLES OF HYPERBOLIC LOGARITHMS.—TABLES OF NATURAL SINES, TANGENTS, ETC.—TABLES OF LOGARITHMIC SINES, TANGENTS, ETC.

"In these days of advanced knowledge a work like this is of the greatest value. It contains a vast amount of information. We unhesitatingly say that it is the most valuable compilation for its specific purpose that has ever been printed. No naval architect, engineer, surveyor, or seaman, wood or iron shipbuilder, can afford to be without this work."—*Nautical Magazine.*

"The first edition of this book appeared in 1879, and its blue and yellow covers have become a familiar object in most shipyard drawing offices. We have now before us the fifth edition revised and enlarged, and we notice that it appeals to a wider circle of readers than its predecessors, since it claims to contain information valuable to marine engineers, as well as to naval architects and shipbuilders. We may say at once that both author and publishers are to be congratulated upon the production of a very useful, compendious and elegant book, well printed and well bound. The revision appears to have been done with care, and nearly all the errors in the earlier editions have been expunged. . . . The book is one of exceptional merit."—*Engineer,* Nov. 4, 1892.

"Should be used by all who are engaged in the construction or design of vessels. . . . Will be found to contain the most useful tables and formulæ required by shipbuilders, carefully collected from the best authorities, and put together in a popular and simple form."—*Engineer,* Nov. 7, 1879.

"The professional shipbuilder has now, in a convenient and accessible form, reliable data for solving many of the numerous problems that present themselves in the course of his work."—*Iron.*

"There is no doubt that a pocket-book of this description must be a necessity in the shipbuilding trade. . . . The volume contains a mass of useful information clearly expressed and presented in a handy form."—*Marine Engineer.*

Marine Engineering.

MARINE ENGINES AND STEAM VESSELS: A Treatise on. By ROBERT MURRAY, C.E. Eighth Edition, thoroughly Revised, with considerable Additions by the Author and by GEORGE CARLISLE, C.E., Senior Surveyor to the Board of Trade at Liverpool. 12mo, 5s. cloth boards.

"Well adapted to give the young steamship engineer or marine engine and boiler maker a general introduction into his practical work."—*Mechanical World.*

"We feel sure that this thoroughly revised edition will continue to be as popular in the future as it has been in the past, as, for its size, it contains more useful information than any similar treatise."—*Industries.*

"The information given is both sound and sensible, and well qualified to direct young sea-going hands on the straight road to the extra chief's certificate. . . . Most useful to surveyors, inspectors, draughtsmen, and all young engineers who take an interest in their profession."—*Glasgow Herald.*

English-French Dictionary of Sea Terms.

TECHNICAL DICTIONARY OF SEA TERMS, PHRASES AND WORDS USED IN THE ENGLISH AND FRENCH LANGUAGES. For the Use of Seamen, Engineers, Pilots, Ship-builders, Ship-owners and Ship-brokers. Compiled by W. PIRRIE, late of the African Steamship Company. Fcap. 8vo, 5s. cloth limp. [*Just published.*

B

Electric Lighting of Ships.

ELECTRIC SHIP LIGHTING: A Handbook on the Practical Fitting and Running of Ship's Electrical Plant, for the Use of Shipowners and Builders, Marine Electricians and Sea-going Engineers in Charge. By J. W. URQUHART, Author of "Electric Light," "Dynamo Construction," &c. With numerous Illustrations. Crown 8vo, 7s. 6d cloth.

Pocket-Book for Marine Engineers.

A POCKET-BOOK OF USEFUL TABLES AND FORMULÆ FOR MARINE ENGINEERS. By FRANK PROCTOR, A.I.N.A. Third Edition. Royal 32mo, leather, gilt edges, with strap, 4s.

"We recommend it to our readers as going far to supply a long-felt want."—*Naval Science.*
"A most useful companion to all marine engineers."—*United Service Gazette.*

Introduction to Marine Engineering.

ELEMENTARY ENGINEERING: A Manual for Young Marine Engineers and Apprentices. In the Form of Questions and Answers on Metals, Alloys, Strength of Materials, Construction and Management of Marine Engines and Boilers, Geometry, &c. &c. With an Appendix of Useful Tables. By JOHN SHERREN BREWER, Government Marine Surveyor, Hongkong. Second Edition, Revised, small crown 8vo, 2s. cloth.

"Contains much valuable information for the class for whom it is intended, especially in the chapters on the management of boilers and engines."—*Nautical Magazine.*
"A useful introduction to the more elaborate text books."—*Scotsman.*
"To a student who has the requisite desire and resolve to attain a thorough knowledge, Mr. Brewer offers decidedly useful help."—*Athenæum.*

Navigation.

PRACTICAL NAVIGATION. Consisting of THE SAILOR'S SEA-BOOK, by JAMES GREENWOOD and W. H. ROSSER; together with the requisite Mathematical and Nautical Tables for the Working of the Problems, by HENRY LAW, C.E., and Professor J. R. YOUNG. Illustrated. 12mo, 7s. strongly half-bound.

Drawing for Marine Engineers.

LOCKIE'S MARINE ENGINEER'S DRAWING-BOOK. Adapted to the Requirements of the Board of Trade Examinations. By JOHN LOCKIE, C.E. With 22 Plates, Drawn to Scale. Royal 8vo, 3s. 6d. cloth.

"The student who learns from these drawings will have nothing to unlearn."—*Engineer.*
"The examples chosen are essentially practical, and are such as should prove of service to engineers generally, while admirably fulfilling their specific purpose."—*Mechanical World.*

Sailmaking.

THE ART AND SCIENCE OF SAILMAKING. By SAMUEL B. SADLER, Practical Sailmaker, late in the employment of Messrs. Ratsey and Lapthorne, of Cowes and Gosport. With Plates and other Illustrations. Small 4to, 12s. 6d. cloth.

"This extremely practical work gives a complete education in all the branches of the manufacture, cutting out, roping, seaming, and goring. It is copiously illustrated, and will form a first-rate text-book and guide."—*Portsmouth Times.*
"The author of this work has rendered a distinct service to all interested in the art of sailmaking. The subject of which he treats is a congenial one. Mr. Sadler is a practical sailmaker, and has devoted years of careful observation and study to the subject; and the results of the experience thus gained he has set forth in the volume before us."—*Steamship.*

Chain Cables.

CHAIN CABLES AND CHAINS. Comprising Sizes and Curves of Links, Studs, &c., Iron for Cables and Chains, Chain Cable and Chain Making, Forming and Welding Links, Strength of Cables and Chains, Certificates for Cables, Marking Cables, Prices of Chain Cables and Chains, Historical Notes, Acts of Parliament, Statutory Tests, Charges for Testing, List of Manufacturers of Cables, &c. &c. By THOMAS W. TRAILL, F.E.R.N., M.Inst.C.E., Engineer-Surveyor-in-Chief, Board of Trade, Inspector of Chain Cable and Anchor Proving Establishments, and General Superintendent, Lloyd's Committee on Proving Establishments. With numerous Tables, Illustrations, and Lithographic Drawings. Folio, £2 2s. cloth, bevelled boards.

"It contains a vast amount of valuable information. Nothing seems to be wanting to make it a complete and standard work of reference on the subject."—*Nautical Magazine.*

MINING AND METALLURGY.

Mining Machinery.

MACHINERY FOR METALLIFEROUS MINES: A Practical Treatise for Mining Engineers, Metallurgists and Managers of Mines. By E. HENRY DAVIES, M.E., F.G.S. Crown 8vo, 580 pp., with upwards of 300 Illustrations. 12s. 6d. cloth. [*Just published.*

"Mr. Davies, in this handsome volume, has done the advanced student and the manager of mines good service. Almost every kind of machinery in actual use is carefully described, and the woodcuts and plates are good."—*Athenæum.*

"From cover to cover the work exhibits all the same characteristics which excite the confidence and attract the attention of the student as he peruses the first page. The work may safely be recommended. By its publication the literature connected with the industry will be enriched, and the reputation of its author enhanced."—*Mining Journal.*

"Mr. Davies has endeavoured to bring before his readers the best of everything in modern mining appliances. His work carries internal evidence of the author's impartiality, and this constitutes one of the great merits of the book. Throughout his work the criticisms are based on his own or other reliable experience."—*Iron and Steel Trades' Journal.*

"The work deals with nearly every class of machinery or apparatus likely to be met with or required in connection with metalliferous mining, and is one which we have every confidence in recommending."—*Practical Engineer.*

"Invaluable to mining engineers, metallurgists, and mine managers."—*The Mining Review., Denver, Colorado, U.S.A.*

Metalliferous Minerals and Mining.

A TREATISE ON METALLIFEROUS MINERALS AND MINING. By D. C. DAVIES, F.G.S., Mining Engineer, &c., Author of "A Treatise on Slate and Slate Quarrying." Fifth Edition, thoroughly Revised and much Enlarged by his Son, E. HENRY DAVIES, M.E., F.G.S. With about 150 Illustrations. Crown 8vo, 12s. 6d. cloth.

"Neither the practical miner nor the general reader, interested in mines, can have a better book for his companion and his guide."—*Mining Journal.*

"We are doing our readers a service in calling their attention to this valuable work."—*Mining World.*

"A book that will not only be useful to the geologist, the practical miner, and the metallurgist; but also very interesting to the general public."—*Iron.*

"As a history of the present state of mining throughout the world this book has a real value, and it supplies an actual want."—*Athenæum.*

Earthy Minerals and Mining.

A TREATISE ON EARTHY AND OTHER MINERALS AND MINING. By D. C. DAVIES, F.G.S., Author of "Metalliferous Minerals," &c. Third Edition, Revised and Enlarged, by his Son, E. HENRY DAVIES, M.E., F.G.S. With about 100 Illusts. Crown 8vo, 12s. 6d. cloth.

"We do not remember to have met with any English work on mining matters that contains the same amount of information packed in equally convenient form."—*Academy.*

"We should be inclined to rank it as among the very best of the handy technical and trades manuals which have recently appeared."—*British Quarterly Review.*

Metalliferous Mining in the United Kingdom.

BRITISH MINING: A Treatise on the History, Discovery, Practical Development, and Future Prospects of Metalliferous Mines in the United Kingdom. By ROBERT HUNT, F.R.S., Keeper of Mining Records; Editor of "Ure's Dictionary of Arts, Manufactures, and Mines," &c. Upwards of 950 pp., with 230 Illustrations. Second Edition, Revised. Super-royal 8vo, £2 2s. cloth.

"One of the most valuable works of reference of modern times. Mr. Hunt, as Keeper of Mining Records of the United Kingdom, has had opportunities for such a task not enjoyed by anyone else, and has evidently made the most of them. . . . The language and style adopted are good, and the treatment of the various subjects laborious, conscientious, and scientific."—*Engineering.*

"The book is, in fact, a treasure-house of statistical information on mining subjects, and we know of no other work embodying so great a mass of matter of this kind. Were this the only merit of Mr. Hunt's volume it would be sufficient to render it indispensable in the library of everyone interested in the development of the mining and metallurgical industries of this country."—*Athenæum.*

"A mass of information not elsewhere available, and of the greatest value to those who may be interested in our great mineral industries."—*Engineer.*

Underground Pumping Machinery.

MINE DRAINAGE: Being a Complete and Practical Treatise on Direct-Acting Underground Steam Pumping Machinery, with a Description of a large number of the best known Engines, their General Utility and the Special Sphere of their Action, the Mode of their Application, and their merits compared with other forms of Pumping Machinery. By STEPHEN MICHELL. 8vo, 15s. cloth.

"Will be highly esteemed by colliery owners and lessees, mining engineers, and students generally who require to be acquainted with the best means of securing the drainage of mines. It is a most valuable work, and stands almost alone in the literature of steam pumping machinery."—*Colliery Guardian.*

"Much valuable information is given, so that the book is thoroughly worthy of an extensive circulation amongst practical men and purchasers of machinery."—*Mining Journal.*

Prospecting for Gold and other Metals.

THE PROSPECTOR'S HANDBOOK : A Guide for the Prospector and Traveller in Search of Metal-Bearing and other Valuable Minerals. By J. W. ANDERSON, M.A. (Camb.), F.R.G.S., Author of "Fiji and New Caledonia." Sixth Edition, thoroughly Revised and much Enlarged. Small crown 8vo, 3s. 6d. cloth ; or, 4s. 6d. leather, pocket-book form, with tuck. [*Just published.*

"Will supply a much felt want, especially among Colonists, in whose way are so often thrown many mineralogical specimens the value of which it is difficult to determine."—*Engineer.*

"How to find commercial minerals, and how to identify them when they are found, are the leading points to which attention is directed. The author has managed to pack as much practical detail into his pages as would supply material for a book three times its size."—*Mining Journal.*

Mining Notes and Formulæ.

NOTES AND FORMULÆ FOR MINING STUDENTS. By JOHN HERMAN MERIVALE, M.A., Certificated Colliery Manager, Professor of Mining in the Durham College of Science, Newcastle-upon-Tyne. Third Edition, Revised and Enlarged. Small crown 8vo, 2s. 6d. cloth.

"Invaluable to anyone who is working up for an examination on mining subjects."—*Iron and Coal Trades' Review.*

"The author has done his work in an exceedingly creditable manner, and has produced a book that will be of service to students, and those who are practically engaged in mining operations."—*Engineer.*

Handybook for Miners.

THE MINER'S HANDBOOK : A Handy Book of Reference on the subjects of Mineral Deposits, Mining Operations, Ore Dressing, &c. For the Use of Students and others interested in Mining matters. Compiled by JOHN MILNE, F.R.S., Professor of Mining in the Imperial University of Japan. Revised Edition. Fcap. 8vo, 7s. 6d. leather. [*Just published.*

"Professor Milne's handbook is sure to be received with favour by all connected with mining, and will be extremely popular among students."—*Athenæum.*

Miners' and Metallurgists' Pocket-Book.

A POCKET-BOOK FOR MINERS AND METALLURGISTS. Comprising Rules, Formulæ, Tables, and Notes, for Use in Field and Office Work. By F. DANVERS POWER, F.G.S., M.E. Fcap. 8vo, 9s. leather.

"This excellent book is an admirable example of its kind, and ought to find a large sale amongst English-speaking prospectors and mining engineers."—*Engineering.*

"Miners and metallurgists will find in this work a useful *vade-mecum* containing a mass of rules, formulæ, tables, and various other information, the necessity for reference to which occurs in their daily duties."—*Iron.*

Mineral Surveying and Valuing.

THE MINERAL SURVEYOR AND VALUER'S COMPLETE GUIDE. Comprising a Treatise on Improved Mining Surveying and the Valuation of Mining Properties, with New Traverse Tables. By WM. LINTERN. Third Edition, Enlarged. 12mo, 4s. cloth.

"Mr. Lintern's book forms a valuable and thoroughly trustworthy guide."—*Iron and Coal Trades' Review.*

Asbestos and its Uses.

ASBESTOS : Its Properties, Occurrence, and Uses. With some Account of the Mines of Italy and Canada. By ROBERT H. JONES. With Eight Collotype Plates and other Illustrations. Crown 8vo, 12s. 6d. cloth.

"An interesting and invaluable work."—*Colliery Guardian.*

Explosives.

A HANDBOOK ON MODERN EXPLOSIVES. Being a Practical Treatise on the Manufacture and Application of Dynamite, Gun-Cotton, Nitro-Glycerine and other Explosive Compounds. Including the Manufacture of Collodion-Cotton. By M. EISSLER, Mining Engineer and Metallurgical Chemist, Author of "The Metallurgy of Gold," "The Metallurgy of Silver," &c. With about 100 Illustrations. Crown 8vo, 10s. 6d. cloth.

"Useful not only to the miner, but also to officers of both services to whom blasting and the use of explosives generally may at any time become a necessary auxiliary."—*Nature.*

"A veritable mine of information on the subject of explosives employed for military, mining and blasting purposes."—*Army and Navy Gazette.*

Colliery Management.

THE COLLIERY MANAGER'S HANDBOOK: A Comprehensive Treatise on the Laying-out and Working of Collieries, Designed as a Book of Reference for Colliery Managers, and for the Use of Coal-Mining Students preparing for First-class Certificates. By CALEB PAMELY, Mining Engineer and Surveyor; Member of the North of England Institute of Mining and Mechanical Engineers; and Member of the South Wales Institute of Mining Engineers. With nearly 500 Plans, Diagrams, and other Illustrations. Second Edition, Revised, with Additions, medium 8vo, about 700 pp. Price £1 5s. strongly bound.

SUMMARY OF CONTENTS.

GEOLOGY.	ON THE FRICTION OF AIR IN MINES.
SEARCH FOR COAL.	THE PRIESTMAN OIL ENGINE; PETROLEUM AND
MINERAL LEASES AND OTHER HOLDINGS.	NATURAL GAS.
SHAFT SINKING.	SURVEYING AND PLANNING.
FITTING UP THE SHAFT AND SURFACE AR-	SAFETY LAMPS AND FIRE-DAMP DETECTORS.
RANGEMENTS.	SUNDRY AND INCIDENTAL OPERATIONS AND AP-
STEAM BOILERS AND THEIR FITTINGS.	PLIANCES.
TIMBERING AND WALLING.	COLLIERY EXPLOSIONS.
NARROW WORK AND METHODS OF WORKING.	MISCELLANEOUS QUESTIONS AND ANSWERS.
UNDERGROUND CONVEYANCE.	*Appendix*: SUMMARY OF REPORT OF H.M. COM-
DRAINAGE.	MISSIONERS ON ACCIDENTS IN MINES.
THE GASES MET WITH IN MINES; VENTILATION.	

*** OPINIONS OF THE PRESS.

"Mr. Pamely has not only given us a comprehensive reference book of a very high order, suitable to the requirements of mining engineers and colliery managers, but at the same time has provided mining students with a class-book that is as interesting as it is instructive."—*Colliery Manager*.

"Mr. Pamely's work is eminently suited to the purpose for which it is intended—being clear, interesting, exhaustive, rich in detail, and up to date, giving descriptions of the very latest machines in every department. . . . A mining engineer could scarcely go wrong who followed this work."—*Colliery Guardian*.

"This is the most complete 'all-round' work on coal-mining published in the English language. . . . No library of coal-mining books is complete without it."—*Colliery Engineer* (Scranton, Pa., U.S.A.).

"Mr. Pamely's work is in all respects worthy of our admiration. No person in any responsible position connected with mines should be without a copy."—*Westminster Review*.

Coal and Iron.

THE COAL AND IRON INDUSTRIES OF THE UNITED KINGDOM. Comprising a Description of the Coal Fields, and of the Principal Seams of Coal, with Returns of their Produce and its Distribution, and Analyses of Special Varieties. Also, an Account of the occurrence of Iron Ores in Veins or Seams; Analyses of each Variety; and a History of the Rise and Progress of Pig Iron Manufacture. By RICHARD MEADE, Assistant Keeper of Mining Records. With Maps 8vo, £1 8s. cloth.

"The book is one which must find a place on the shelves of all interested in coal and iron production, and in the iron, steel, and other metallurgical industries."—*Engineer*.

"Of this book we may unreservedly say that it is the best of its class which we have ever met. . . . A book of reference which no one engaged in the iron or coal trades should omit from his library."—*Iron and Coal Trades' Review*.

Coal Mining.

COAL AND COAL MINING, A Rudimentary Treatise on. By the late Sir WARINGTON W. SMYTH, M.A., F.R.S., &c., Chief Inspector of the Mines of the Crown. Seventh Edition, Revised and Enlarged. With numerous Illustrations, 12mo, 4s. cloth boards.

"As an outline is given of every known coal-field in this and other countries, as well as of the principal methods of working, the book will doubtless interest a very large number of readers."—*Mining Journal*.

Subterraneous Surveying.

SUBTERRANEOUS SURVEYING, Elementary and Practical Treatise on; with and without the Magnetic Needle. By THOMAS FENWICK, Surveyor of Mines, and THOMAS BAKER, C.E. Illustrated. 12mo, 3s. cloth boards.

Granite Quarrying.

GRANITES AND OUR GRANITE INDUSTRIES. By GEORGE F. HARRIS, F.G.S., Membre de la Société Belge de Géologie, Lecturer on Economic Geology at the Birkbeck Institution, &c. With Illustrations. Crown 8vo, 2s. 6d. cloth.

"A clearly and well-written manual for persons engaged or interested in the granite industry."—*Scotsman*.

"An interesting work, which will be deservedly esteemed."—*Colliery Guardian*.

"An exceedingly interesting and valuable monograph on a subject which has hitherto received unaccountably little attention in the shape of systematic literary treatment."—*Scottish Leader*.

Gold, Metallurgy of.

THE METALLURGY OF GOLD: A Practical Treatise on the Metallurgical Treatment of Gold-bearing Ores. Including the Processes of Concentration and Chlorination, and the Assaying, Melting, and Refining of Gold. By M. EISSLER, Mining Engineer and Metallurgical Chemist, formerly Assistant Assayer of the U.S. Mint, San Francisco. Third Edition, Revised and greatly Enlarged. With 187 Illustrations. Crown 8vo, 12s. 6d. cloth.

"This book thoroughly deserves its title of a 'Practical Treatise.' The whole process of gold milling, from the breaking of the quartz to the as ay of the bullion, is described in clear and orderly narrative and with much, but not too much, fulness of detail."—*Saturday Review.*

"The work is a storehouse of information and valuable data, and we strongly recommend it to all professional men engaged in the gold-mining industry."—*Mining Journal.*

Gold Extraction.

THE CYANIDE PROCESS OF GOLD EXTRACTION; and its Practical Application on the Witwatersrand Gold Fields in South Africa. By M. EISSLER, M.E., Mem. Inst. Mining and Metallurgy, Author of "The Metallurgy of Gold," &c. With Diagrams and Working Drawings. Large crown 8vo, 7s. 6d. cloth. [*Just published.*

"This book is just what was needed to acquaint mining men with the actual working of a process which is not only the most popular, but is, as a general rule, the most successful for the extraction of gold from tailings."—*Mining Journal.*

"The work will prove invaluable to all interested in gold mining, whether metallurgists or as investors."—*Chemical News.*

Silver, Metallurgy of.

THE METALLURGY OF SILVER: A Practical Treatise on the Amalgamation, Roasting, and Lixiviation of Silver Ores. Including the Assaying, Melting, and Refining of Silver Bullion. By M. EISSLER, Author of "The Metallurgy of Gold," &c. Second Edition, Enlarged. Crown 8vo, 10s. 6d. cloth.

"A practical treatise, and a technical work which we are convinced will supply a long felt want amongst practical men, and at the same time be of value to students and others indirectly connected with the industries."—*Mining Journal.*

"From first to last the book is thoroughly sound and reliable."—*Colliery Guardian.*

"For chemists, practical miners, assayers, and investors alike, we do not know of any work on the subject so handy and yet so comprehensive."—*Glasgow Herald.*

Lead, Metallurgy of.

THE METALLURGY OF ARGENTIFEROUS LEAD: A Practical Treatise on the Smelting of Silver-Lead Ores and the Refining of Lead Bullion. Including Reports on various Smelting Establishments and Descriptions of Modern Smelting Furnaces and Plants in Europe and America. By M. EISSLER, M.E., Author of "The Metallurgy of Gold," &c. Crown 8vo, 400 pp., with 183 Illustrations, 12s. 6d. cloth.

"The numerous metallurgical processes, which are fully and extensively treated of, embrace all the stages experienced in the passage of the lead from the various natural states to its issue from the refinery as an article of commerce."—*Practical Engineer.*

"The present volume fully maintains the reputation of the author. Those who wish to obtain a thorough insight into the present state of this industry cannot do better than read this volume, and all mining engineers cannot fail to find many useful hints and suggestions in it."—*Industries.*

"This is the work of an expert for experts, by whom it will be prized as an indispensable textbook."—*Bristol Mercury.*

Iron, Metallurgy of.

METALLURGY OF IRON. Containing History of Iron Manufacture, Methods of Assay, and Analyses of Iron Ores, Processes of Manufacture of Iron and Steel, &c. By H. BAUERMAN, F.G.S., A.R.S.M. With numerous Illustrations. Sixth Edition, Revised and Enlarged. 12mo, 5s. 6d. cloth.

Iron Mining.

THE IRON ORES OF GREAT BRITAIN AND IRELAND: Their Mode of Occurrence, Age and Origin, and the Methods of Searching for and Working Them. With a Notice of some of the Iron Ores of Spain. By J. D. KENDALL, F.G.S., Mining Engineer. Crown 8vo, 16s. cloth.

"The author has a thorough practical knowledge of his subject, and has supplemented a careful study of the available literature by unpublished information derived from his own observations The result is a very useful volume which cannot fail to be of value to all interested in the iron industry of the country."—*Industries.*

"Mr. Kendall is a great authority on this subject and writes from personal observation."—*Colliery Guardian.*

ELECTRICITY, ELECTRICAL ENGINEERING, ETC.

Dynamo Management.

THE MANAGEMENT OF DYNAMOS: A Handybook of Theory and Practice for the Use of Mechanics, Engineers, Students and others in Charge of Dynamos. By G. W. LUMMIS PATERSON. With numerous Illustrations. Crown 8vo, 3s. 6d. cloth. [*Just published.*

Electrical Engineering.

THE ELECTRICAL ENGINEER'S POCKET-BOOK OF MODERN RULES, FORMULÆ, TABLES, AND DATA. By H. R. KEMPE, M. Inst. E.E., A. M. Inst. C.E., Technical Officer, Postal Telegraphs, Author of "A Handbook of Electrical Testing," &c. Second Edition, Thoroughly Revised, with Additions. With numerous Illustrations. Royal 32mo, oblong, 5s. leather.

"There is very little in the shape of formulæ or data which the electrician is likely to want in a hurry which cannot be found in its pages."—*Practical Engineer.*
"A very useful book of reference for daily use in practical electrical engineering and its various applications to the industries of the present day."—*Iron.*
"It is the best book of its kind."—*Electrical Engineer.*
"The Electrical Engineer's Pocket-Book is a good one."—*Electrician.*
"Strongly recommended to those engaged in the electrical industries."—*Electrical Review.*

Electric Lighting.

ELECTRIC LIGHT FITTING: A Handbook for Working Electrical Engineers, embodying Practical Notes on Installation Management. By J. W. URQUHART, Electrician, Author of "Electric Light," &c. With numerous Illusts. Second Edition, Revised, with Additional Chapters. Crown 8vo, 5s. cloth.

"This volume deals with what may be termed the mechanics of electric lighting, and is addressed to men who are already engaged in the work, or are training for it. The work traverses a great deal of ground, and may be read as a sequel to the same author's useful work on 'Electric Light,'"—*Electrician.*
"This is an attempt to state in the simplest language the precautions which should be adopted in installing the electric light, and to give information for the guidance of those who have to run the plant when installed. The book is well worth the perusal of the workman, for whom it is written."—*Electrical Review.*
"Eminently practical and useful. . . . . Ought to be in the hands of everyone in charge of an electric light plant."—*Electrical Engineer.*
"Mr. Urquhart has succeeded in producing a really capital book, which we have no hesitation in recommending to working electricians and electrical engineers."—*Mechanical World.*

Electric Light.

ELECTRIC LIGHT: Its Production and Use, Embodying Plain Directions for the Treatment of Dynamo-Electric Machines, Batteries, Accumulators, and Electric Lamps. By J. W. URQUHART, C.E., Author of "Electric Light Fitting," "Electroplating," &c. Fifth Edition, carefully Revised, with Large Additions and 145 Illustrations. Crown 8vo, 7s. 6d. cloth.

"The whole ground of electric lighting is more or less covered and explained in a very clear and concise manner."—*Electrical Review.*
"Contains a good deal of very interesting information, especially in the parts where the author gives dimensions and working costs."—*Electrical Engineer.*
"A *vade-mecum* of the salient facts connected with the science of electric lighting."—*Electrician.*
"You cannot for your purpose have a better book than 'Electric Light,' by Urquhart."—*Engineer.*
"The book is by far the best that we have yet met with on the subject."—*Athenæum.*

Construction of Dynamos.

DYNAMO CONSTRUCTION: A Practical Handbook for the Use of Engineer Constructors and Electricians-in-Charge. Embracing Framework Building, Field Magnet and Armature Winding and Grouping, Compounding, &c. With Examples of leading English, American, and Continental Dynamos and Motors. By J. W. URQUHART, Author of "Electric Light," &c. Second Edition, Enlarged. With 114 Illustrations. Crown 8vo, 7s. 6d. cloth.

"Mr. Urquhart's book s the first one which deals with these matters in such a way that the engineering student can understand them. The book is very readable, and the author leads his readers up to difficult subjects by reasonably simple tests."—*Engineering Review.*
"The author deals with his subject in a style so popular as to make his volume a handbook of great practical value to engineer contractors and electricians in charge of lighting installations."—*Scotsman.*
"'Dynamo Construction' more than sustains the high character of the author's previous publications. It is sure to be widely read by the large and rapidly-increasing number of practical electricians."—*Glasgow Herald.*
"A book for which a demand has long existed."—*Mechanical World.*

A New Dictionary of Electricity.

THE STANDARD ELECTRICAL DICTIONARY. A Popular Dictionary of Words and Terms Used in the Practice of Electrical Engineering. Containing upwards of 3,000 Definitions. By T. O'CONNOR SLOANE, A.M., Ph.D., Author of "The Arithmetic of Electricity," &c. &c. Crown 8vo, 630 pp., 350 Illustrations, 7s. 6d. cloth. [*Just published.*

"The work has many attractive features in it, and is, beyond doubt, a well put together and useful publication. The amount of ground covered may be gathered from the fact that in the index about 5,000 references will be found. The inclusion of such comparatively modern words as 'impedence,' 'reluctance,' &c., shows that the author has desired to be up to date, and indeed there are other indications of carefulness of compilation. The work is one which does the author great credit and it should prove of great value, especially to students."—*Electrical Review.*

"Very complete and contains a large amount of useful information."—*Industries.*

"An encyclopædia of electrical science in the compass of a dictionary. The information given is sound and clear. The book is well printed, well illustrated, and well up to date, and may be confidently recommended."—*Builder.*

"The volume is excellently printed and illustrated, and should form part of the library of every one who is directly or indirectly connected with electrical matters."—*Hardware Trade Journal.*

Electric Lighting of Ships.

ELECTRIC SHIP-LIGHTING: A Handbook on the Practical Fitting and Running of Ship's Electrical Plant. For the Use of Shipowners and Builders, Marine Electricians, and Sea-going Engineers in Charge. By J. W. URQUHART, C.E., Author of "Electric Light," &c. With 88 Illusts., crown 8vo, 7s. 6d. cloth.

"The subject of ship electric lighting is one of vast importance, and Mr. Urquhart is to be highly complimented for placing such a valuable work at the service of marine electricians."—*The Steamship.*

"Distinctly a book which of its kind stands almost alone, and for which there should be a demand."—*Electrical Review.*

Country House Electric Lighting.

ELECTRIC LIGHT FOR COUNTRY HOUSES: A Practical Handbook on the Erection and Running of Small Installations, with Particulars of the Cost of Plant and Working. By J. H. KNIGHT. Crown 8vo, 1s. wrapper.
[*Just published.*

"The book contains excellent advice and many practical hints for the help of those who wish to light their own houses."—*Building News.*

Electric Lighting.

THE ELEMENTARY PRINCIPLES OF ELECTRIC LIGHTING. By ALAN A. CAMPBELL SWINTON, Associate I.E.E. Third Edition, Enlarged and Revised. With Sixteen Illustrations. Crown 8vo, 1s. 6d. cloth.

"Anyone who desires a short and thoroughly clear exposition of the elementary principles of electric-lighting cannot do better than read this little work."—*Bradford Observer.*

Dynamic Electricity.

THE ELEMENTS OF DYNAMIC ELECTRICITY AND MAGNETISM. By PHILIP ATKINSON, A.M., Ph.D., Author of "Elements of Static Electricity," &c. Crown 8vo, 417 pp., with 120 Illustrations, 10s. 6d. cloth.

Electric Motors, &c.

THE ELECTRIC TRANSFORMATION OF POWER and its Application by the Electric Motor, including Electric Railway Construction. By P. ATKINSON, A.M., Ph.D. With 96 Illustrations. Crown 8vo, 7s. 6d. cloth.

Dynamo Construction.

HOW TO MAKE A DYNAMO: A Practical Treatise for Amateurs. Containing numerous Illustrations and Detailed Instructions for Constructing a Small Dynamo to Produce the Electric Light. By ALFRED CROFTS. Fourth Edition, Revised and Enlarged. Crown 8vo, 2s. cloth.

"The instructions given in this unpretentious little book are sufficiently clear and explicit to enable any amateur mechanic possessed of average skill and the usual tools to be found in an amateur's workshop, to build a practical dynamo machine."—*Electrician.*

Text-Book of Electricity.

THE STUDENT'S TEXT-BOOK OF ELECTRICITY. By H. M. NOAD, F.R.S. Cheaper Edition. 650 pp., with 470 Illustrations. Crown 8vo, 9s. cloth.

ARCHITECTURE, BUILDING, etc.

Building Construction.

PRACTICAL BUILDING CONSTRUCTION: A Handbook for Students Preparing for Examinations, and a Book of Reference for Persons Engaged in Building. By JOHN PARNELL ALLEN, Surveyor, Lecturer on Building Construction at the Durham College of Science, Newcastle-on-Tyne. Medium 8vo, 450 pages, with 1,000 Illustrations. 12s. 6d. cloth. [*Just published.*

"The most complete exposition of building construction we have seen. It contains all that is necessary to prepare students for the various examinations in building construction."—*Building News.*

"The author depends nearly as much on his diagrams as on his type. The pages suggest the hand of a man of experience in building operations—and the volume must be a blessing to many teachers as well as to students."—*The Architect.*

"The work is sure to prove a formidable rival to great and small competitors alike, and bids fair to take a permanent place as a favourite students' text-book. The large number of illustrations deserve particular mention for the great merit they possess for purposes of reference, in exactly corresponding to convenient scales."—*Jour. Inst. Brit. Archts.*

The New London Building Act, 1894.

THE LONDON BUILDING ACT, 1894. With the By-Laws and Regulations of the London County Council, and Introduction, Notes, Cases and Index. By ALEX. J. DAVID, B.A., LL.M., of the Inner Temple, Barrister-at-Law. Crown 8vo, 3s. 6d. cloth. [*Just published.*

"To all architects and district surveyors and builders, Mr. David's manual will be welcome."—*Building News.*

"The volume will doubtless be eagerly consulted by the building fraternity."—*Illustrated Carpenter and Builder.*

Concrete.

CONCRETE: ITS NATURE AND USES. A Book for Architects, Builders, Contractors, and Clerks of Works. By GEORGE L. SUTCLIFFE, A.R.I.B.A. 350 pages, with numerous Illustrations. Crown 8vo, 7s. 6d. cloth.

"The author treats a difficult subject in a lucid manner. The manual fills a long felt gap. It is careful and exhaustive; equally useful as a student's guide and an architect's book of reference."—*Journal of Royal Institution of British Architects.*

"There is room for this new book, which will probably be for some time the standard work on the subject for a builder's purpose."—*Glasgow Herald.*

Mechanics for Architects.

THE MECHANICS OF ARCHITECTURE: A Treatise on Applied Mechanics, especially Adapted to the Use of Architects. By E. W. TARN, M.A., Author of "The Science of Building," &c. Second Edition, Enlarged. Illustrated with 125 Diagrams. Crown 8vo, 7s. 6d. cloth.

"The book is a very useful and helpful manual of architectural mechanics, and really contains sufficient to enable a careful and painstaking student to grasp the principles bearing upon the majority of building problems. . . . Mr. Tarn has added, by this volume, to the debt of gratitude which is owing to him by architectural students for the many valuable works which he has produced for their use."—*The Builder.*

"The mechanics in the volume are really mechanics, and are harmoniously wrought in with the distinctive professional matter proper to the subject. The diagrams and type are commendably clear."—*The Schoolmaster.*

The New Builder's Price Book, 1896.

LOCKWOOD'S BUILDER'S PRICE BOOK FOR 1896. A Comprehensive Handbook of the Latest Prices and Data for Builders, Architects, Engineers, and Contractors. Re-constructed, Re-written, and Greatly Enlarged. By FRANCIS T. W. MILLER. 800 closely-printed pages, crown 8vo, 4s. cloth.

"This book is a very useful one, and should find a place in every English office connected with the building and engineering professions."—*Industries.*

"An excellent book of reference."—*Architect.*

"In its new and revised form this Price Book is what a work of this kind should be—comprehensive, reliable, well arranged, legible, and well bound."—*British Architect.*

Designing Buildings.

THE DESIGN OF BUILDINGS: Being Elementary Notes on the Planning, Sanitation and Ornamentive Formation of Structures, based on Modern Practice. Illustrated with Nine Folding Plates. By W. WOODLEY. 8vo, 6s. cloth.

Sir William Chambers's Treatise on Civil Architecture.

THE DECORATIVE PART OF CIVIL ARCHITECTURE. By Sir WILLIAM CHAMBERS, F.R.S. With Portrait, Illustrations, Notes, and an EXAMINATION OF GRECIAN ARCHITECTURE, by JOSEPH GWILT, F.S.A. Revised and Edited by W. H. LEEDS. 66 Plates, 4to, 21s. cloth.

Villa Architecture.

A HANDY BOOK OF VILLA ARCHITECTURE : Being a Series of Designs for Villa Residences in various Styles. With Outline Specifications and Estimates. By C. WICKES, Architect, Author of "The Spires and Towers of England," &c. 61 Plates, 4to, £1 11s. 6d. half-morocco, gilt edges.
"The whole of the designs bear evidence of their being the work of an artistic architect, and they will prove very valuable and suggestive."—*Building News.*

Text-Book for Architects.

THE ARCHITECT'S GUIDE : Being a Text-book of Useful Information for Architects, Engineers, Surveyors, Contractors, Clerks of Works, &c. &c. By FREDERICK ROGERS, Architect. Third Edition. Cr. 8vo, 3s. 6d. cloth.
"As a text-book of useful information for architects, engineers, surveyors, &c., it would be hard to find a handier or more complete little volume."—*Standard.*

Taylor and Cresy's Rome.

THE ARCHITECTURAL ANTIQUITIES OF ROME. By the late G. L. TAYLOR, Esq., F.R.I.B.A., and EDWARD CRESY, Esq. New Edition, thoroughly Revised by the Rev. ALEXANDER TAYLOR, M.A. (son of the late G. L. Taylor, Esq.), Fellow of Queen's College, Oxford, and Chaplain of Gray's Inn. Large folio, with 130 Plates, £3 3s. half-bound.
"Taylor and Cresy's work has from its first publication been ranked among those professional books which cannot be bettered."—*Architect.*

Linear Perspective.

ARCHITECTURAL PERSPECTIVE. The whole Course and Operations of the Draughtsman in Drawing a Large House in Linear Perspective. Illustrated by 43 Folding Plates. By F. O. FERGUSON. Second Edition, Enlarged. 8vo, 3s. 6d. boards. [*Just published.*
"It is the most intelligible of the treatises on this ill-treated subject that I have met with."— E. INGRESS BELL, ESQ., *in the R.I.B.A. Journal.*

Architectural Drawing.

PRACTICAL RULES ON DRAWING, for the Operative Builder and Young Student in Architecture. By GEORGE PYNE. With 14 Plates, 4to, 7s. 6d. boards.

Vitruvius' Architecture.

THE ARCHITECTURE OF MARCUS VITRUVIUS POLLIO. Translated by JOSEPH GWILT, F.S.A., F.R.A.S. New Edition, Revised by the Translator. With 23 Plates, fcap. 8vo, 5s. cloth.

Designing, Measuring, and Valuing.

THE STUDENT'S GUIDE TO THE PRACTICE OF MEASURING AND VALUING ARTIFICERS' WORK. Containing Directions for taking Dimensions, Abstracting the same, and bringing the Quantities into Bill, with Tables of Constants for Valuation of Labour, and for the Calculation of Areas and Solidities. Originally edited by EDWARD DOBSON, Architect. With Additions by E. WYNDHAM TARN, M.A. Sixth Edition. With 8 Plates and 63 Woodcuts. Crown 8vo, 7s. 6d. cloth.
"This edition will be found the most complete treatise on the principles of measuring and valuing artificers' work that has yet been published."—*Building News.*

Pocket Estimator and Technical Guide.

THE POCKET TECHNICAL GUIDE, MEASURER, AND ESTIMATOR FOR BUILDERS AND SURVEYORS. Containing Technical Directions for Measuring Work in all the Building Trades, Complete Specifications for Houses, Roads, and Drains, and an Easy Method of Estimating the parts of a Building collectively. By A. C. BEATON. Seventh Edition. Waistcoat-pocket size, 1s. 6d. gilt edges.
"No builder, architect, surveyor, or valuer should be without his 'Beaton.'"—*Building News.*

Donaldson on Specifications.

THE HANDBOOK OF SPECIFICATIONS; or, Practical Guide to the Architect, Engineer, Surveyor, and Builder, in drawing up Specifications and Contracts for Works and Constructions. Illustrated by Precedents of Buildings actually executed by eminent Architects and Engineers. By Professor T. L. DONALDSON, P.R.I.B.A., &c. New Edition, in One large Vol., 8vo, with upwards of 1,000 pages of Text, and 33 Plates, £1 11s. 6d. cloth.
". . . Valuable as a record, and more valuable still as a book of precedents. . . . Suffice it to say that Donaldson's 'Handbook of Specifications' must be bought by all architects."—*Builder.*

Bartholomew and Rogers' Specifications.

SPECIFICATIONS FOR PRACTICAL ARCHITECTURE. A Guide to the Architect, Engineer, Surveyor, and Builder. With an Essay on the Structure and Science of Modern Buildings. Upon the Basis of the Work by ALFRED BARTHOLOMEW, thoroughly Revised, Corrected, and greatly added to by FREDERICK ROGERS, Architect. Third Edition, Revised, with Additions. 8vo, 15s. cloth.

"The collection of specifications prepared by Mr. Rogers on the basis of Bartholomew's work is too well known to need any recommendation from us. It is one of the books with which every young architect must be equipped."—*Architect.*

Construction.

THE SCIENCE OF BUILDING: An Elementary Treatise on the Principles of Construction. By E. WYNDHAM TARN, M.A., Architect. Third Edition, Revised and Enlarged, with 59 Engravings. Fcap. 8vo, 4s. cloth.

"A very valuable book, which we strongly recommend to all students."—*Builder.*

House Building and Repairing.

THE HOUSE-OWNER'S ESTIMATOR; or, What will it Cost to Build, Alter, or Repair? A Price Book for Unprofessional People, as well as the Architectural Surveyor and Builder. By J. D. SIMON. Edited by F. T. W. MILLER, A.R.I.B.A. Fourth Edition. Crown 8vo, 3s. 6d. cloth.

"In two years it will repay its cost a hundred times over."—*Field.*

Cottages and Villas.

COUNTRY AND SUBURBAN COTTAGES AND VILLAS: How to Plan and Build Them. Containing 33 Plates, with Introduction, General Explanations, and Description of each Plate. By JAMES W. BOGUE, Architect, Author of "Domestic Architecture," &c. 4to, 10s. 6d. cloth.

Building ; Civil and Ecclesiastical.

A BOOK ON BUILDING, Civil and Ecclesiastical, including Church Restoration; with the Theory of Domes and the Great Pyramid, &c. By Sir EDMUND BECKETT, Bart., LL.D., F.R.A.S. Second Edition. Fcap. 8vo, 5s. cloth.

"A book which is always amusing and nearly always instructive."—*The Times.*

Sanitary Houses, etc.

THE SANITARY ARRANGEMENT OF DWELLING-HOUSES: A Handbook for Householders and Owners of Houses. By A. J. WALLIS-TAYLER, A.M.Inst. C.E. With numerous Illustrations. Crown 8vo, 2s. 6d. cloth.

[*Just published.*

"This book will be largely read; it will be of considerable service to the public. It is well arranged, easily read, and for the most part devoid of technical terms."—*Lancet.*

Ventilation of Buildings.

VENTILATION. A Text-Book to the Practice of the Art of Ventilating Buildings. By W. P. BUCHAN, R.P. 12mo, 4s cloth.

"Contains a great amount of useful practical information, as thoroughly interesting as it is technically reliable.'"—*British Architect.*

The Art of Plumbing.

PLUMBING. A Text-Book to the Practice of the Art or Craft of the Plumber. By W. P. BUCHAN, R.P. Sixth Edition, Enlarged. 12mo, 4s. cloth.

"A text book which may be safely put in the hands of every young plumber."—*Builder.*

Geometry for the Architect, Engineer, &c.

PRACTICAL GEOMETRY, for the Architect, Engineer, and Mechanic. Giving Rules for the Delineation and Application of various Geometrical Lines, Figures, and Curves. By E. W. TARN, M.A., Architect. 8vo, 9s. cloth.

"No book with the same objects in view has ever been published in which the clearness of the rules laid down and the illustrative diagrams have been so satisfactory."—*Scotsman.*

The Science of Geometry.

THE GEOMETRY OF COMPASSES; or, Problems Resolved by the mere Description of Circles, and the use of Coloured Diagrams and Symbols. By OLIVER BYRNE. Coloured Plates. Crown 8vo, 3s. 6d. cloth.

CARPENTRY, TIMBER, etc.

Tredgold's Carpentry, Revised and Enlarged by Tarn.

THE ELEMENTARY PRINCIPLES OF CARPENTRY: A Treatise on the Pressure and Equilibrium of Timber Framing, the Resistance of Timber, and the Construction of Floors, Arches, Bridges, Roofs, Uniting Iron and Stone with Timber, &c. To which is added an Essay on the Nature and Properties of Timber, &c., with Descriptions of the kinds of Wood used in Building; also numerous Tables of the Scantlings of Timber for different purposes, the Specific Gravities of Materials, &c. By THOMAS TREDGOLD, C.E. With an Appendix of Specimens of Various Roofs of Iron and Stone, Illustrated. Seventh Edition, thoroughly Revised and considerably Enlarged by E. WYNDHAM TARN, M.A., Author of "The Science of Building," &c. With 61 Plates, Portrait of the Author, and several Woodcuts. In One large Vol., 4to, 25s. cloth.

"Ought to be in every architect's and every builder's library."—*Builder.*
"A work whose monumental excellence must commend it wherever skilful carpentry is concerned. The author's principles are rather confirmed than impaired by time. The additional plates are of great intrinsic value."—*Building News.*

Woodworking Machinery.

WOODWORKING MACHINERY: Its Rise, Progress, and Construction. With Hints on the Management of Saw Mills and the Economical Conversion of Timber. Illustrated with Examples of Recent Designs by leading English, French, and American Engineers. By M. POWIS BALE, A.M.Inst.C.E., M.I.M.E. Second Edition, Revised, with large Additions, large crown 8vo, 440 pages, 9s. cloth. [*Just published.*

"Mr. Bale is evidently an expert on the subject, and he has collected so much information that his book is all-sufficient for builders and others engaged in the conversion of timber."—*Architect.*
"The most comprehensive compendium of wood-working machinery we have seen. The author is a thorough master of his subject."—*Building News.*

Saw Mills.

SAW MILLS: Their Arrangement and Management, and the Economical Conversion of Timber. (A Companion Volume to "Woodworking Machinery.") By M. POWIS BALE. Crown 8vo, 10s. 6d. cloth.

"The *administration* of a large sawing establishment is discussed, and the subject examined from a financial standpoint. Hence the size, shape, order, and disposition of saw-mills and the like are gone into in detail, and the course of the timber is traced from its reception to its delivery in its converted state. We could not desire a more complete or practical treatise."—*Builder.*

Nicholson's Carpentry.

THE CARPENTER'S NEW GUIDE; or, Book of Lines for Carpenters; comprising all the Elementary Principles essential for acquiring a knowledge of Carpentry. Founded on the late PETER NICHOLSON's standard work. A New Edition, Revised by ARTHUR ASHPITEL, F.S.A. Together with Practical Rules on Drawing, by GEORGE PYNE. With 74 Plates, 4to, £1 1s. cloth.

Handrailing and Stairbuilding.

A PRACTICAL TREATISE ON HANDRAILING: Showing New and Simple Methods for Finding the Pitch of the Plank, Drawing the Moulds, Bevelling, Jointing-up, and Squaring the Wreath. By GEORGE COLLINGS. Second Edition, Revised and Enlarged, to which is added A TREATISE ON STAIRBUILDING. With Plates and Diagrams. 12mo, 2s. 6d. cloth limp.

"Will be found of practical utility in the execution of this difficult branch of joinery."—*Builder.*
"Almost every difficult phase of this somewhat intricate branch of joinery is elucidated by the aid of plates and explanatory letterpress."—*Furniture Gazette.*

Circular Work.

CIRCULAR WORK IN CARPENTRY AND JOINERY: A Practical Treatise on Circular Work of Single and Double Curvature. By GEORGE COLLINGS. With Diagrams. Second Edition, 12mo, 2s. 6d. cloth limp.

"An excellent example of what a book of this kind should be. Cheap in price, clear in definition, and practical in the examples selected."—*Builder.*

Handrailing.

HANDRAILING COMPLETE IN EIGHT LESSONS. On the Square-Cut System. By J. S. GOLDTHORP, Teacher of Geometry and Building Construction at the Halifax Mechanic's Institute. With Eight Plates and over 150 Practical Exercises. 4to, 3s. 6d. cloth.

"Likely to be of considerable value to joiners and others who take a pride in good work. The arrangement of the book is excellent. We heartily commend it to teachers and students."—*Timber Trades Journal.*

Timber Merchant's Companion.

THE TIMBER MERCHANT'S AND BUILDER'S COMPANION. Containing New and Copious Tables of the Reduced Weight and Measurement of Deals and Battens, of all sizes, from One to a Thousand Pieces, and the relative Price that each size bears per Lineal Foot to any given Price per Petersburgh Standard Hundred ; the Price per Cube Foot of Square Timber to any given Price per Load of 50 Feet ; the proportionate Value of Deals and Battens by the Standard, to Square Timber by the Load of 50 Feet ; the readiest mode of ascertaining the Price of Scantling per Lineal Foot of any size, to any given Figure per Cube Foot, &c. &c. By WILLIAM DOWSING. Fourth Edition, Revised and Corrected. Cr. 8vo, 3s. cloth.

"Everything is as concise and clear as it can possibly be made. There can be no doubt that every timber merchant and builder ought to possess it."—*Hull Advertiser.*

"We are glad to see a fourth edition of these admirable tables, which for correctness and simplicity of arrangement leave nothing to be desired."—*Timber Trades' Journal.*

Practical Timber Merchant.

THE PRACTICAL TIMBER MERCHANT : Being a Guide for the use of Building Contractors, Surveyors, Builders, &c., comprising useful Tables for all purposes connected with the Timber Trade, Marks of Wood, Essay on the Strength of Timber, Remarks on the Growth of Timber, &c. By W. RICHARDSON. Second Edition. Fcap. 8vo, 3s. 6d. cloth. [*Just published.*

"This handy manual contains much valuable information for the use of timber merchants, builders, foresters, and all others connected with the growth, sale, and manufacture of timber."—*Journal of Forestry.*

Packing-Case Makers, Tables for.

PACKING-CASE TABLES ; showing the number of Superficial Feet in Boxes or Packing-Cases, from six inches square and upwards. By W. RICHARDSON, Timber Broker. Third Edition. Oblong 4to, 3s. 6d. cloth.

"Invaluable labour-saving tables."—*Ironmonger.* "Will save much labour and calculation."—*Grocer.*

Superficial Measurement.

THE TRADESMAN'S GUIDE TO SUPERFICIAL MEASUREMENT. Tables calculated from 1 to 200 inches in length, by 1 to 108 inches in breadth. For the use of Architects, Surveyors, Engineers, Timber Merchants, Builders, &c. By JAMES HAWKINGS. Fourth Edition. Fcap., 3s. 6d. cloth.

"A useful collection of tables to facilitate rapid calculation of surfaces. The exact area of any surface of which the limits have been ascertained can be instantly determined. The book will be found of the greatest utility to all engaged in building operations."—*Scotsman.*

"These tables will be found of great assistance to all who require to make calculations in superficial measurement."—*English Mechanic.*

Forestry.

THE ELEMENTS OF FORESTRY. Designed to afford Information concerning the Planting and Care of Forest Trees for Ornament or Profit, with suggestions upon the Creation and Care of Woodlands. By F. B. HOUGH. Large crown 8vo, 10s. cloth.

Timber Importer's Guide.

THE TIMBER IMPORTER'S, TIMBER MERCHANT'S, AND BUILDER'S STANDARD GUIDE. By RICHARD E. GRANDY. Comprising :—An Analysis of Deal Standards, Home and Foreign, with Comparative Values and Tabular Arrangements for fixing Net Landed Cost on Baltic and North American Deals, including all intermediate Expenses, Freight, Insurance, &c. &c. ; together with copious Information for the Retailer and Builder. Third Edition, Revised. 12mo, 2s. cloth limp.

"Everything it pretends to be: built up gradually, it leads one from a forest to a treenail, and throws in, as a makeweight, a host of material concerning bricks, columns, cisterns, &c."—*English Mechanic.*

DECORATIVE ARTS, etc.

Woods and Marbles, Imitation of.

SCHOOL OF PAINTING FOR THE IMITATION OF WOODS AND MARBLES, as Taught and Practised by A. R. VAN DER BURG and P. VAN DER BURG, Directors of the Rotterdam Painting Institution. Royal folio, 18½ by 12½ in., Illustrated with 24 full-size Coloured Plates; also 12 plain Plates, comprising 154 Figures. Second and Cheaper Edition. Price £1 11s. 6d.

LIST OF PLATES.

1. VARIOUS TOOLS REQUIRED FOR WOOD PAINTING.—2, 3. WALNUT; PRELIMINARY STAGES OF GRAINING AND FINISHED SPECIMEN.—4. TOOLS USED FOR MARBLE PAINTING AND METHOD OF MANIPULATION.—5, 6. ST. REMI MARBLE; EARLIER OPERATIONS AND FINISHED SPECIMEN.—7. METHODS OF SKETCHING DIFFERENT GRAINS, KNOTS, &c.—8, 9. ASH: PRELIMINARY STAGES AND FINISHED SPECIMEN.—10. METHODS OF SKETCHING MARBLE GRAINS.—11, 12. BRECHE MARBLE; PRELIMINARY STAGES OF WORKING AND FINISHED SPECIMEN.—13. MAPLE; METHODS OF PRODUCING THE DIFFERENT GRAINS.—14, 15. BIRD'S-EYE MAPLE; PRELIMINARY STAGES AND FINISHED SPECIMEN.—16. METHODS OF SKETCHING THE DIFFERENT SPECIES OF WHITE MARBLE.—17, 18. WHITE MARBLE; PRELIMINARY STAGES OF PROCESS AND FINISHED SPECIMEN.—19. MAHOGANY; SPECIMENS OF VARIOUS GRAINS AND METHODS OF MANIPULATION,—20, 21. MAHOGANY; EARLIER STAGES AND FINISHED SPECIMEN.—22, 23, 24. SIENNA MARBLE; VARIETIES OF GRAIN, PRELIMINARY STAGES AND FINISHED SPECIMEN.—25, 26, 27. JUNIPER WOOD; METHODS OF PRODUCING GRAIN, &c.; PRELIMINARY STAGES AND FINISHED SPECIMEN.—28, 29, 30. VERT DE MER MARBLE; VARIETIES OF GRAIN AND METHODS OF WORKING, UNFINISHED AND FINISHED SPECIMENS.—31, 32, 33. OAK; VARIETIES OF GRAIN, TOOLS EMPLOYED AND METHODS OF MANIPULATION, PRELIMINARY STAGES AND FINISHED SPECIMEN.—34, 35, 36, WAULSORT MARBLE; VARIETIES OF GRAIN, UNFINISHED AND FINISHED SPECIMENS.

"Those who desire to attain skill in the art of painting woods and marbles will find advantage in consulting this book. . . . Some of the Working Men's Clubs should give their young men the opportunity to study it."—*Builder.*

"A comprehensive guide to the art. The explanations of the processes, the manipulation and management of the colours, and the beautifully executed plates will not be the least valuable to the student who aims at making his work a faithful transcript of nature."—*Building News.*

"Students and novices are fortunate who are able to become the possessors of so noble a work."—*The Architect.*

House Decoration.

ELEMENTARY DECORATION: A Guide to the Simpler Forms of Everyday Art. Together with **PRACTICAL HOUSE DECORATION**. By JAMES W. FACEY. With numerous Illustrations. In One Vol., 5s. strongly half-bound.

House-Painting, Graining, etc.

HOUSE-PAINTING, GRAINING, MARBLING, AND SIGN WRITING, A Practical Manual of. By ELLIS A. DAVIDSON. Sixth Edition. With Coloured Plates and Wood Engravings. 12mo, 6s. cloth boards.

"A mass of information, of use to the amateur and of value to the practical man."—*English Mechanic.*

Decorators, Receipts for.

THE DECORATOR'S ASSISTANT: A Modern Guide to Decorative Artists and Amateurs, Painters, Writers, Gilders, &c. Containing upwards of 600 Receipts, Rules and Instructions; with a variety of Information for General Work connected with every Class of Interior and Exterior Decorations, &c., Sixth Edition. 152 pp., crown 8vo, 1s. in wrapper.

"Full of receipts of value to decorators, painters, gilders, &c. The book contains the gist of larger treatises on colour and technical processes. It would be difficult to meet with a work so full of varied information on the painter's art."—*Building News.*

Moyr Smith on Interior Decoration.

ORNAMENTAL INTERIORS, ANCIENT AND MODERN. By J. MOYR SMITH. Super-royal 8vo, with Thirty-two full-page Plates and numerous smaller Illustrations, handsomely bound in cloth, gilt top, 18s.

"The book is well illustrated and handsomely got up, and contains some true criticism and a good many good examples of decorative treatment."—*The Builder.*

British and Foreign Marbles.

MARBLE DECORATION and the Terminology of British and Foreign Marbles. A Handbook for Students. By GEORGE H. BLAGROVE, Author of "Shoring and its Application," &c. With 28 Illustrations. Cr. 8vo, 3s. 6d. cloth.

"This most useful and much wanted handbook should be in the hands of every architect and builder."—*Building World.*

"A carefully and usefully written treatise; the work is essentially practical."—*Scotsman.*

Marble Working, etc.

MARBLE AND MARBLE WORKERS: A Handbook for Architects, Artists, Masons, and Students. By ARTHUR LEE, Author of "A Visit to Carrara," "The Working of Marble," &c. Small crown 8vo, 2s. cloth.

"A really valuable addition to the technical literature of architects and masons."—*Building News.*

DELAMOTTE'S WORKS ON ILLUMINATION AND ALPHABETS.

A PRIMER OF THE ART OF ILLUMINATION, for the Use of Beginners; with a Rudimentary Treatise on the Art, Practical Directions for its Exercise, and Examples taken from Illuminated MSS., printed in Gold and Colours. By F. DELAMOTTE. New and Cheaper Edition. Small 4to, 6s. ornamental boards.

"The examples of ancient MSS. recommended to the student, which, with much good sense, the author chooses from collections accessible to all, are selected with judgment and knowledge, as well as taste."—*Athenæum.*

ORNAMENTAL ALPHABETS, Ancient and Mediæval, from the Eighth Century, with Numerals; including Gothic, Church-Text, large and small, German, Italian, Arabesque, Initials for Illumination, Monograms, Crosses, &c., for the use of Architectural and Engineering Draughtsmen, Missal Painters, Masons, Decorative Painters, Lithographers, Engravers, Carvers, &c. &c. Collected and Engraved by F. DELAMOTTE, and printed in Colours. New and Cheaper Edition. Royal 8vo, oblong, 2s. 6d. ornamental boards.

"For those who insert enamelled sentences round gilded chalices, who blazon shop legends over shop-doors, who letter church walls with pithy sentences from the Decalogue, this book will be useful."—*Athenæum.*

EXAMPLES OF MODERN ALPHABETS, Plain and Ornamental, including German, Old English, Saxon, Italic, Perspective, Greek, Hebrew, Court Hand, Engrossing, Tuscan, Riband, Gothic, Rustic, and Arabesque; with several Original Designs, and an Analysis of the Roman and Old English Alphabets, large and small, and Numerals, for the use of Draughtsmen, Surveyors, Masons, Decorative Painters, Lithographers, Engravers, Carvers, &c. Collected and Engraved by F. DELAMOTTE, and printed in Colours. New and Cheaper Edition. Royal 8vo, oblong, 2s. 6d. ornamental boards.

"There is comprised in it every possible shape into which the letters of the alphabet and numerals can be formed, and the talent which has been expended in the conception of the various plain and ornamental letters is wonderful."—*Standard.*

MEDIÆVAL ALPHABETS AND INITIALS FOR ILLUMINATORS. By F. G. DELAMOTTE. Containing 21 Plates and Illuminated Title, printed in Gold and Colours. With an Introduction by J. WILLIS BROOKS. Fourth and Cheaper Edition. Small 4to, 4s. ornamental boards.

"A volume in which the letters of the alphabet come forth glorified in gilding and all the colours of the prism interwoven and intertwined and intermingled."—*Sun.*

THE EMBROIDERER'S BOOK OF DESIGN. Containing Initials, Emblems, Cyphers, Monograms, Ornamental Borders, Ecclesiastical Devices, Mediæval and Modern Alphabets, and National Emblems. Collected by F. DELAMOTTE, and printed in Colours. Oblong royal 8vo, 1s. 6d. ornamental wrapper.

"The book will be of great assistance to ladies and young children who are endowed with the art of plying the needle in this most ornamental and useful pretty work."—*East Anglian Times.*

Wood Carving.

INSTRUCTIONS IN WOOD-CARVING FOR AMATEURS; with Hints on Design. By A LADY. With Ten Plates. New and Cheaper Edition. Crown 8vo, 2s. in emblematic wrapper.

"The handicraft of the wood-carver, so well as a book can impart it, may be learnt from 'A Lady's publication."—*Athenæum.*

NATURAL SCIENCE, etc.

The Heavens and their Origin.

THE VISIBLE UNIVERSE: Chapters on the Origin and Construction of the Heavens. By J. E. GORE, F.R.A.S., Author of "Star Groups," &c. Illustrated by 6 Stellar Photographs and 12 Plates. Demy 8vo, 16s. cloth.

"A valuable and lucid summary of recent astronomical theory, rendered more valuable and attractive by a series of stellar photographs and other illustrations."—*The Times.*

"In presenting a clear and concise account of the present state of our knowledge, Mr. Gore has made a valuable addition to the literature of the subject."—*Nature.*

"Mr. Gore's 'Visible Universe' is one of the finest works on astronomical science that has recently appeared in our language. In spirit and in method it is scientific from cover to cover, but the style is so clear and attractive that it will be as acceptable and as readable to those who make no scientific pretensions as to those who devote themselves specially to matters astronomical."—*Leeds Mercury.*

"As interesting as a novel, and instructive withal; the text being made still more luminous by stellar photographs and other illustrations. . . . A most valuable book."—*Manchester Examiner.*

The Constellations.

STAR GROUPS: A Student's Guide to the Constellations. By J. ELLARD GORE, F.R.A.S., M.R.I.A., &c., Author of "The Visible Universe," "The Scenery of the Heavens." With 30 Maps. Small 4to, 5s. cloth, silvered.

"A knowledge of the principal constellations visible in our latitudes may be easily acquired from the thirty maps and accompanying text contained in this work."—*Nature.*

"The volume contains thirty maps showing stars of the sixth magnitude—the usual naked-eye limit—and each is accompanied by a brief commentary, adapted to facilitate recognition and bring to notice objects of special interest. For the purpose of a preliminary survey of the 'midnight pomp' of the heavens, nothing could be better than a set of delineations averaging scarcely twenty square inches in area, and including nothing that cannot at once be identified."—*Saturday Review.*

"A very compact and handy guide to the constellations."—*Athenæum.*

Astronomical Terms.

AN ASTRONOMICAL GLOSSARY; or, Dictionary of Terms used in Astronomy. With Tables of Data and Lists of Remarkable and Interesting Celestial Objects. By J. ELLARD GORE, F.R.A.S., Author of "The Visible Universe," &c. Small crown 8vo, 2s. 6d. cloth.

"A very useful little work for beginners in astronomy, and not to be despised by more advanced students."—*The Times.*

"A very handy book . . . the utility of which is much increased by its valuable tables of astronomical data."—*The Athenæum.*

"Astronomers of all kinds will be glad to have it for reference."—*Guardian.*

The Microscope.

THE MICROSCOPE: Its Construction and Management. Including Technique, Photo-micrography, and the Past and Future of the Microscope. By Dr. HENRI VAN HEURCK. Re-Edited and Augmented from the Fourth French Edition, and Translated by WYNNE E. BAXTER, F.G.S. 400 pages, with upwards of 250 Woodcuts, imp. 8vo, 18s., cloth.

"A translation of a well-known work, at once popular and comprehensive."—*Times.*
"The translation is as felicitous as it is accurate."—*Nature.*

The Microscope.

PHOTO-MICROGRAPHY. By Dr. H. VAN HEURCK. Extracted from the above Work. Royal 8vo, with Illustrations, 1s. sewed.

Astronomy.

ASTRONOMY. By the late Rev. ROBERT MAIN, M.A., F.R.S. Third Edition, Revised by WILLIAM THYNNE LYNN, B.A., F.R.A.S., formerly of the Royal Observatory, Greenwich. 12mo, 2s. cloth limp.

"A sound and simple treatise, very carefully edited, and a capital book for beginners."—*Knowledge.*
"Accurately brought down to the requirements of the present time by Mr. Lynn."—*Educational Times.*

Recent and Fossil Shells.

A MANUAL OF THE MOLLUSCA: Being a Treatise on Recent and Fossil Shells. By S. P. WOODWARD, A.L.S., F.G.S. With an Appendix on RECENT AND FOSSIL CONCHOLOGICAL DISCOVERIES by RALPH TATE, A.L.S., F.G.S. With 23 Plates and upwards of 300 Woodcuts. Reprint of Fourth Edition (1880). Crown 8vo, 7s. 6d. cloth.

"A most valuable storehouse of conchological and geological information."—*Science Gossip.*

Geology and Genesis.

THE TWIN RECORDS OF CREATION; or Geology and Genesis, their Perfect Harmony and Wonderful Concord. By G. W. V. LE VAUX. 8vo, 5s. cl.

"A valuable contribution to the evidences of Revelation, and disposes very conclusively of the arguments of those who would set God's Works against God's Word. No real difficulty is shirked, and no sophistry is left unexposed."—*The Rock.*

DR. LARDNER'S COURSE OF NATURAL PHILOSOPHY.

THE HANDBOOK OF MECHANICS. Enlarged and almost re-written by BENJAMIN LOEWY, F.R.A.S. With 378 Illustrations. Post 8vo, 6s. cloth.

"The perspicuity of the original has been retained, and chapters which had become obsolete have been replaced by others of more modern character. The explanations throughout are studiously popular, and care has been taken to show the application of the various branches of physics to the industrial arts, and to the practical business of life."—*Mining Journal.*

"Mr. Loewy has carefully revised the book, and brought it up to modern requirements."—*Nature.*

"Natural philosophy has had few exponents more able or better skilled in the art of popularising the subject than Dr. Lardner: and Mr. Loewy is doing good service in fitting this treatise, and the others of the series, for use at the present time."—*Scotsman.*

THE HANDBOOK OF HYDROSTATICS AND PNEUMATICS. New Edition, Revised and Enlarged by BENJAMIN LOEWY, F.R.A.S. With 236 Illustrations. Post 8vo, 5s. cloth.

"For those 'who desire to attain an accurate knowledge of physical science without the profound methods of mathematical investigation,' this work is not merely intended, but well adapted."—*Chemical News.*

"The volume before us has been carefully edited, augmented to nearly twice the bulk of the former edition, and all the most recent matter has been added. . . . It is a valuable text-book."—*Nature.*

"Candidates for pass examinations will find it, we think, specially suited to their requirements."—*English Mechanic.*

THE HANDBOOK OF HEAT. Edited and almost entirely re-written by BENJAMIN LOEWY, F.R.A.S., &c. 117 Illustrations. Post 8vo, 6s. cloth.

"The style is always clear and precise, and conveys instruction without leaving any cloudiness or lurking doubts behind."—*Engineering.*

"A most exhaustive book on the subject on which it treats, and is so arranged that it can be understood by all who desire to attain an accurate knowledge of physical science. . . . Mr. Loewy has included all the latest discoveries in the varied laws and effects of heat."—*Standard.*

"A complete and handy text-book for the use of students and general readers."—*English Mechanic.*

THE HANDBOOK OF OPTICS. By DIONYSIUS LARDNER, D.C.L., formerly Professor of Natural Philosophy and Astronomy in University College, London. New Edition. Edited by T. OLVER HARDING, B.A. Lond., of University College, London. With 298 Illustrations. Small 8vo, 448 pages, 5s. cloth.

"Written by one of the ablest English scientific writers, beautifully and elaborately illustrated."—*Mechanic's Magazine.*

THE HANDBOOK OF ELECTRICITY, MAGNETISM, AND ACOUSTICS. By Dr. LARDNER. Ninth Thousand. Edited by GEO. CAREY FOSTER, B.A., F.C.S. With 400 Illustrations. Small 8vo, 5s. cloth.

"The book could not have been entrusted to anyone better calculated to preserve the terse and lucid style of Lardner, while correcting his errors and bringing up his work to the present state of scientific knowledge."—*Popular Science Review.*

THE HANDBOOK OF ASTRONOMY. Forming a Companion to the "Handbook of Natural Philosophy." By DIONYSIUS LARDNER, D.C.L., formerly Professor of Natural Philosophy and Astronomy in University College, London. Fourth Edition. Revised and Edited by EDWIN DUNKIN, F.R.A.S., Royal Observatory, Greenwich. With 38 Plates and upwards of 100 Woodcuts. In One Vol., small 8vo, 550 pages, 9s. 6d. cloth.

"Probably no other book contains the same amount of information in so compendious and well-arranged a form—certainly none at the price at which this is offered to the public."—*Athenæum.*

"We can do no other than pronounce this work a most valuable manual of astronomy, and we strongly recommend it to all who wish to acquire a general—but at the same time correct—acquaintance with this sublime science."—*Quarterly Journal of Science.*

"One of the most deservedly popular books on the subject . . . We would recommend not only the student of the elementary principles of the science, but him who aims at mastering the higher and mathematical branches of astronomy, not to be without this work beside him."—*Practical Magazine.*

Geology.

RUDIMENTARY TREATISE ON GEOLOGY, PHYSICAL AND HISTORICAL. Consisting of "PHYSICAL GEOLOGY," which sets forth the Leading Principles of the Science; and "HISTORICAL GEOLOGY," which treats of the Mineral and Organic Conditions of the Earth at each successive epoch, especial reference being made to the British Series of Rocks. By RALPH TATE, A.L.S., F.G.S., &c. &c. With 250 Illustrations. 12mo, 5s. cloth boards.

"The fulness of the matter has elevated the book into a manual. Its information is exhaustive and well arranged."—*School Board Chronicle.*

C

DR. LARDNER'S MUSEUM OF SCIENCE AND ART.

THE MUSEUM OF SCIENCE AND ART. Edited by DIONYSIUS LARDNER, D.C.L., formerly Professor of Natural Philosophy and Astronomy in University College, London. With upwards of 1,200 Engravings on Wood. In 6 double volumes, £1 1s., in a new and elegant cloth binding; or handsomely bound in half morocco, 31s. 6d.

⁎ OPINIONS OF THE PRESS.

"This series, besides affording popular but sound instruction on scientific subjects, with which the humblest man in the country ought to be acquainted, also undertakes that teaching of 'Common Things' which every well wisher of his kind is anxious to promote. Many thousand copies of this serviceable publication have been printed, in the belief and hope that the desire for instruction and improvement widely prevails; and we have no fear that such enlightened faith will meet with disappointment."—*The Times.*

"A cheap and interesting publication, alike informing and attractive. The papers combine subjects of importance and great scientific knowledge, considerable inductive powers, and a popular style of treatment."—*Spectator.*

"The 'Museum of Science and Art' is the most valuable contribution that has ever been made to the scientific instruction of every class of society."—Sir DAVID BREWSTER, in the *North British Review.*

"Whether we consider the liberality and beauty of the illustrations, the charm of the writing, or the durable interest of the matter, we must express our belief that there is hardly to be found among the new books one that would be welcomed by people of so many ages and classes as a valuable present."—*Examiner.*

⁎ *Separate books formed from the above, suitable for Workmen's Libraries, Science Classes, &c.*

Common Things Explained. Containing Air, Earth, Fire, Water, Time, Man, the Eye, Locomotion, Colour, Clocks and Watches, &c. 233 Illustrations, cloth gilt, 5s.

The Microscope. Containing Optical Images, Magnifying Glasses, Origin and Description of the Microscope, Microscopic Objects, the Solar Microscope, Microscopic Drawing and Engraving, &c. 147 Illustrations, cloth gilt, 2s.

Popular Geology. Containing Earthquakes and Volcanoes, the Crust of the Earth, &c. 201 Illustrations, cloth gilt, 2s. 6d.

Popular Physics. Containing Magnitude and Minuteness, the Atmosphere, Meteoric Stones, Popular Fallacies, Weather Prognostics, the Thermometer, the Barometer, Sound, &c. 85 Illustrations, cloth gilt, 2s. 6d.

Steam and its Uses. Including the Steam Engine, the Locomotive, and Steam Navigation. 89 Illustrations, cloth gilt, 2s.

Popular Astronomy. Containing How to observe the Heavens. The Earth, Sun, Moon, Planets. Light, Comets, Eclipses, Astronomical Influences, &c. 182 Illustrations, cloth gilt, 4s. 6d.

The Bee and White Ants: Their Manners and Habits. With Illustrations of Animal Instinct and Intelligence. 135 Illustrations, cloth gilt, 2s.

The Electric Telegraph Popularized. To render intelligible to all who can Read, irrespective of any previous Scientific Acquirements, the various forms of Telegraphy in Actual Operation. 100 Illustrations, cloth gilt, 1s. 6d.

Dr. Lardner's School Handbooks.

NATURAL PHILOSOPHY FOR SCHOOLS. By Dr. LARDNER. 328 Illustrations. Sixth Edition. One Vol., 3s. 6d. cloth.

"A very convenient class-book for junior students in private schools. It is intended to convey, in clear and precise terms, general notions of all the principal divisions of Physical Science."—*British Quarterly Review.*

ANIMAL PHYSIOLOGY FOR SCHOOLS. By Dr. LARDNER. With 190 Illustrations. Second Edition. One Vol., 3s. 6d. cloth.

"Clearly written, well arranged, and excellently illustrated."—*Gardener's Chronicle.*

Lardner and Bright on the Electric Telegraph.

THE ELECTRIC TELEGRAPH. By Dr. LARDNER. Revised and Re-written by E. B. BRIGHT, F.R.A.S. 140 Illustrations. Small 8vo, 2s. 6d. cloth.

"One of the most readable books extant on the Electric Telegraph."—*English Mechanic.*

CHEMICAL MANUFACTURES, CHEMISTRY, etc.

Chemistry for Engineers, etc.

ENGINEERING CHEMISTRY: A Practical Treatise for the Use of Analytical Chemists, Engineers, Iron Masters, Iron Founders, Students and others. Comprising Methods of Analysis and Valuation of the Principal Materials used in Engineering Work, with numerous Analyses, Examples and Suggestions. By H. JOSHUA PHILLIPS, F.I.C., F.C.S., Formerly Analytical and Consulting Chemist to the Great Eastern Railway. Second Edition, Revised and Enlarged. Crown 8vo, 400 pp., with Illustrations, 10s. 6d. cloth. [*Just published.*

"In this work the author has rendered no small service to a numerous body of practical men. . . . The analytical methods may be pronounced most satisfactory, being as accurate as the despatch required of engineering chemists permits."—*Chemical News.*

"Those in search of a handy treatise on the subject of analytical chemistry as applied to the every-day requirements of workshop practice will find this volume of great assistance."—*Iron.*

"The book will be very useful to those who require a handy and concise résumé of approved methods of analysing and valuing metals, oils, fuels, &c. It is, in fact, a work for chemists, a guide to the routine of the engineering laboratory. . . . The book is full of good things. As a handbook of technical analysis, it is very welcome."—*Builder.*

"The analytical methods given are, as a whole, such as are likely to give rapid and trustworthy results in experienced hands. . . . There is much excellent descriptive matter in the work, the chapter on 'Oils and Lubrication' being specially noticeable in this respect."—*Engineer.*

Explosives and Dangerous Goods.

DANGEROUS GOODS: Their Sources and Properties, Modes of Storage and Transport. With Notes and Comments on Accidents arising therefrom, together with the Government and Railway Classifications, Acts of Parliament, &c. A Guide for the Use of Government and Railway Officials, Steamship Owners, Insurance Companies and Manufacturers and Users of Explosives and Dangerous Goods. By H. JOSHUA PHILLIPS, F.I.C., F.C.S., Author of "Engineering, Chemistry," &c. Crown 8vo, 350 pages, 9s. cloth. [*Just ready.*

The Alkali Trade, Manufacture of Sulphuric Acid, &c.

A MANUAL OF THE ALKALI TRADE, including the Manufacture of Sulphuric Acid, Sulphate of Soda, and Bleaching Powder. By JOHN LOMAS, Alkali Manufacturer, Newcastle-upon-Tyne and London. With 232 Illustrations and Working Drawings, and containing 390 pages of Text. Second Edition, with Additions. Super-royal 8vo, £1 10s. cloth.

"This book is written by a manufacturer for manufacturers. The working details of the most approved forms of apparatus are given, and these are accompanied by no less than 232 wood engravings, all of which may be used for the purposes of construction. Every step in the manufacture is very fully described in this manual, and each improvement explained."—*Athenæum.*

"We find not merely a sound and luminous explanation of the chemical principles of the trade, but a notice of numerous matters which have a most important bearing on the successful conduct of alkali works, but which are generally overlooked by even experienced technological authors."—*Chemical Review.*

The Blowpipe.

THE BLOWPIPE IN CHEMISTRY, MINERALOGY, AND GEOLOGY. Containing all known Methods of Anhydrous Analysis, many Working Examples, and Instructions for Making Apparatus. By Lieut.-Colonel W. A. ROSS, R.A., F.G.S. With 120 Illustrations. Second Edition, Enlarged. Crown 8vo, 5s. cloth.

"The student who goes conscientiously through the course of experimentation here laid down will gain a better insight into inorganic chemistry and mineralogy than if he had 'got up' any of the best text-books of the day, and passed any number of examinations in their contents."—*Chemical News.*

Commercial Chemical Analysis.

THE COMMERCIAL HANDBOOK OF CHEMICAL ANALYSIS; or, Practical Instructions for the determination of the Intrinsic or Commercial Value of Substances used in Manufactures, in Trades, and in the Arts. By A. NORMANDY. New Edition by H. M. NOAD, Ph.D., F.R.S. Crown 8vo, 12s. 6d. cloth.

"We strongly recommend this book to our readers as a guide, alike indispensable to the housewife as to the pharmaceutical practitioner."—*Medical Times.*

Dye-Wares and Colours.

THE MANUAL OF COLOURS AND DYE-WARES: Their Properties, Applications, Valuations, Impurities, and Sophistications. For the use of Dyers, Printers, Drysalters, Brokers, &c. By J. W. SLATER. Second Edition, Revised and greatly Enlarged, crown 8vo, 7s. 6d. cloth.

"A complete encyclopædia of the *materia tinctoria*. The information given respecting each article is full and precise, and the methods of determining the value of articles such as these, so liable to sophistication, are given with clearness, and are practical as well as valuable."—*Chemist and Druggist.*

"There is no other work which covers precisely the same ground. To students preparing for examinations in dyeing and printing it will prove exceedingly useful."—*Chemical News.*

Modern Brewing and Malting.

A HANDYBOOK FOR BREWERS: Being a Practical Guide to the Art of Brewing and Malting. Embracing the Conclusions of Modern Research which bear upon the Practice of Brewing. By HERBERT EDWARDS WRIGHT. M.A. Crown 8vo, 530 pp., 12s. 6d. cloth.

"May be consulted with advantage by the student who is preparing himself for examinational tests, while the scientific brewer will find in it a résumé of all the most important discoveries of modern times. The work is written throughout in a clear and concise manner, and the author takes great care to discriminate between vague theories and well-established facts."—*Brewers' Journal.*

"We have great pleasure in recommending this handybook, and have no hesitation in saying that it is one of the best—if not the best—which has yet been written on the subject of beer-brewing in this country, it should have a place on the shelves of every brewer's library."—*Brewer's Guardian.*

"Although the requirements of the student are primarily considered, an acquaintance of half-an-hour's duration cannot fail to impress the practical brewer with the sense of having found a trustworthy guide and practical counsellor in brewery matters."—*Chemical Trade Journal.*

Analysis and Valuation of Fuels.

FUELS: SOLID, LIQUID, AND GASEOUS: Their Analysis and Valuation. For the Use of Chemists and Engineers. By H. J. PHILLIPS, F.C.S., Formerly Analytical and Consulting Chemist to the Great Eastern Railway. Second Edition, Revised and Enlarged. Crown 8vo, 5s. cloth.

"Ought to have its place in the laboratory of every metallurgical establishment, and wherever fuel is used on a large scale."—*Chemical News.*

Pigments.

THE ARTISTS' MANUAL OF PIGMENTS. Showing their Composition, Conditions of Permanency, Non-Permanency, and Adulterations; Effects in Combination with Each Other and with Vehicles; and the most Reliable Tests of Purity. By H. C. STANDAGE. Second Edition, crown 8vo, 2s. 6d. cloth.

"This work is indeed *multum-in-parvo*, and we can, with good conscience, recommend it to all who come in contact with pigments, whether as makers, dealers, or users."—*Chemical Review.*

Gauging. Tables and Rules for Revenue Officers, Brewers, &c.

A POCKET BOOK OF MENSURATION AND GAUGING: Containing Tables, Rules, and Memoranda for Revenue Officers, Brewers, Spirit Merchants, &c. By J. B. MANT (Inland Revenue). Second Edition, Revised. 18mo, 4s. leather.

"This handy and useful book is adapted to the requirements of the Inland Revenue Department, and will be a favourite book of reference. The range of subjects is comprehensive, and the arrangement simple and clear."—*Civilian.* "Should be in the hands of every practical brewer."—*Brewers' Journal.*

INDUSTRIAL ARTS, TRADES AND MANUFACTURES.

Cotton Spinning.

COTTON MANUFACTURE: A Manual of Practical Instruction in the Processes of Opening, Carding, Combing, Drawing, Doubling and Spinning of Cotton, the Methods of Dyeing &c. For the Use of Operatives, Overlookers and Manufacturers. By JOHN LISTER, Technical Instructor, Pendleton. 8vo, 7s. 6d. cloth. [*Just published.*

"This invaluable volume is a distinct advance in the literature of cotton manufacture."—*Machinery.*
"It is thoroughly reliable, fulfilling nearly all the requirements desired."—*Glasgow Herald.*

Flour Manufacture, Milling, etc.

FLOUR MANUFACTURE: A Treatise on Milling Science and Practice. By FRIEDRICH KICK, Imperial Regierungsrath, Professor of Mechanical Technology in the Imperial German Polytechnic Institute, Prague. Translated from the Second Enlarged and Revised Edition with Supplement. By H. H. P. POWLES, Assoc. Memb. Institution of Civil Engineers. Nearly 400 pp. Illustrated with 28 Folding Plates, and 167 Woodcuts. Roy. 8vo, 25s. cloth.

"This valuable work is, and will remain, the standard authority on the science of milling. . . . The miller who has read and digested this work will have laid the foundation, so to speak, of a successful career; he will have acquired a number of general principles which he can proceed to apply. In this handsome volume we at last have the accepted text-book of modern milling in good, sound English, which has little, if any, trace of the German idiom."—*The Miller.*

"The appearance of this celebrated work in English is very opportune, and British millers will, we are sure, not be slow in availing themselves of its pages."—*Millers' Gazette.*

Agglutinants.

CEMENTS, PASTES, GLUES AND GUMS: A Practical Guide to the Manufacture and Application of the various Agglutinants required in the Building, Metal-Working, Wood-Working, and Leather-Working Trades, and for Workshop, Laboratory or Office Use. With upwards of 900 Recipes and Formulæ. By H. C. STANDAGE, Chemist. Crown 8vo, 2s. 6d. cloth.

"We have pleasure in speaking favourably of this volume. So far as we have had experience, which is not inconsiderable, this manual is trustworthy."—*Athenæum*.

"As a revelation of what are considered trade secrets, this book will arouse an amount of curiosity among the large number of industries it touches."—*Daily Chronicle*.

"In this goodly collection of recipes it would be strange if a cement for any purpose cannot be found."—*Oil and Colourman's Journal*.

Soap-making.

THE ART OF SOAP-MAKING: A Practical Handbook of the Manufacture of Hard and Soft Soaps, Toilet Soaps, &c. Including many New Processes, and a Chapter on the Recovery of Glycerine from Waste Leys. By ALEXANDER WATT. Fifth Edition, Revised, with an Appendix on Modern Candlemaking. Crown 8vo, 7s. 6d. cloth. [*Just published*.

"The work will prove very useful, not merely to the technological student, but to the practical soap-boiler who wishes to understand the theory of his art."—*Chemical News*.

"A thoroughly practical treatise on an art which has almost no literature in our language. We congratulate the author on the success of his endeavour to fill a void in English technical literature."—*Nature*.

Paper Making.

PRACTICAL PAPER-MAKING: A Manual for Paper-makers and Owners and Managers of Paper-Mills. With Tables, Calculations, &c. By G. CLAPPERTON, Paper-maker With Illustrations of Fibres from Micro-photographs. Crown 8vo, 5s. cloth. [*Just published*.

"The author caters for the requirements of responsible mill hands, apprentices, &c., whilst his manual will be found of great service to students of technology, as well as to veteran paper-makers and mill owners. The illustrations form an excellent feature."—*The World's Paper Trade Review*.

"We recommend everybody interested in the trade to get a copy of this thoroughly practical book."—*Paper Making*.

Paper Making

THE ART OF PAPER MAKING: A Practical Handbook of the Manufacture of Paper from Rags, Esparto, Straw, and other Fibrous Materials. Including the Manufacture of Pulp from Wood Fibre, with a Description of the Machinery and Appliances used. To which are added Details of Processes for Recovering Soda from Waste Liquors. By ALEXANDER WATT, Author of "The Art of Soap-Making." With Illustrations. Crown 8vo, 7s. 6d. cloth.

"It may be regarded as the standard work on the subject. The book is full of valuable information. The 'Art of Paper-making,' is in every respect a model of a text-book, either for a technical class, or for the private student."—*Paper and Printing Trades Journal*.

Leather Manufacture.

THE ART OF LEATHER MANUFACTURE: Being a Practical Handbook, in which the Operations of Tanning, Currying, and Leather Dressing are fully Described, and the Principles of Tanning Explained, and many Recent Processes Introduced; as also Methods for the Estimation of Tannin, and a Description of the Arts of Glue Boiling, Gut Dressing, &c. By ALEXANDER WATT, Author of "Soap-Making," &c. Second Edition. Crown 8vo, 9s. cloth.

"A sound, comprehensive treatise on tanning and its accessories. The book is an eminently valuable production, which redounds to the credit of both author and publishers."—*Chemical Review*.

Boot and Shoe Making.

THE ART OF BOOT AND SHOE-MAKING: A Practical Handbook, including Measurement, Last-Fitting, Cutting-Out, Closing and Making, with a Description of the most approved Machinery Employed. By JOHN B. LENO, late Editor of *St. Crispin*, and *The Boot and Shoe-Maker*. 12mo, 2s. cloth.

"This excellent treatise is by far the best work ever written. The chapter on clicking, which shows how waste may be prevented, will save fifty times the price of the book."—*Scottish Leather Trader*.

Dentistry Construction.

MECHANICAL DENTISTRY: A Practical Treatise on the Construction of the various kinds of Artificial Dentures. Comprising also Useful Formulæ, Tables, and Receipts for Gold Plate, Clasps, Solders, &c. &c. By C. HUNTER. Third Edition. With 100 Wood Engravings. Crown 8vo, 3s. 6d. cloth.

"We can strongly recommend Mr. Hunter's treatise to all students preparing for the profession of dentistry, as well as to every mechanical dentist."—*Dublin Journal of Medical Science*.

Wood Engraving.

WOOD ENGRAVING: A Practical and Easy Introduction to the Study of the Art. By W. N. BROWN. 12mo, 1s. 6d. cloth.

"The book is clear and complete, and will be useful to anyone wanting to understand the first elements of the beautiful art of wood engraving."—*Graphic.*

Horology.

A TREATISE ON MODERN HOROLOGY, in Theory and Practice. Translated from the French of CLAUDIUS SAUNIER, ex-Director of the School of Horology at Macon, by JULIEN TRIPPLIN, F.R.A.S., Besancon Watch Manufacturer, and EDWARD RIGG, M.A., Assayer in the Royal Mint. With Seventy-eight Woodcuts and Twenty-two Coloured Copper Plates. Second Edition. Super-royal 8vo, £2 2s. cloth; £2 10s. half-calf.

"There is no horological work in the English language at all to be compared to this production of M. Saunier's for clearness and completeness. It is alike good as a guide for the student and as a reference for the experienced horologist and skilled workman."—*Horological Journal.*

"The latest, the most complete, and the most reliable of those literary productions to which continental watchmakers are indebted for the mechanical superiority over their English brethren—in fact, the Book of Books, is M. Saunier's 'Treatise.'"—*Watchmaker, Jeweller, and Silversmith.*

Watch Adjusting.

THE WATCH ADJUSTER'S MANUAL: A Practical Guide for the Watch and Chronometer Adjuster in Making, Springing, Timing and Adjusting for Isochronism, Positions and Temperatures. By C. E. FRITTS. 370 pages, with Illustrations, 8vo, 16s. cloth. [*Just published.*

Watchmaking.

THE WATCHMAKER'S HANDBOOK. Intended as a Workshop Companion for those engaged in Watchmaking and the Allied Mechanical Arts. Translated from the French of CLAUDIUS SAUNIER, and enlarged by JULIEN TRIPPLIN, F.R.A.S., and EDWARD RIGG, M.A., Assayer in the Royal Mint. Third Edition. Crown 8vo, 9s. cloth.

"Each part is truly a treatise in itself. The arrangement is good and the language is clear and concise. It is an admirable guide for the young watchmaker."—*Engineering.*

"It is impossible to speak too highly of its excellence. It fulfils every requirement in a handbook intended for the use of a workman. Should be found in every workshop."—*Watch and Clockmaker.*

Watches and Timekeepers.

A HISTORY OF WATCHES AND OTHER TIMEKEEPERS. By JAMES F. KENDAL, M.B.H. Inst., 1s. 6d. boards; or 2s. 6d. cloth, gilt.

"Mr. Kendal's book, for its size, is the best which has yet appeared on this subject in the English language."—*Industries.*

"Open the book where you may, there is interesting matter in it concerning the ingenious devices of the ancient or modern horologer."—*Saturday Review.*

Electrolysis of Gold, Silver, Copper, &c.

ELECTRO-DEPOSITION: A Practical Treatise on the Electrolysis of Gold, Silver, Copper, Nickel, and other Metals and Alloys. With descriptions of Voltaic Batteries, Magneto and Dynamo-Electric Machines, Thermopiles, and of the Materials and Processes used in every Department of the Art, and several Chapters on ELECTRO-METALLURGY. By ALEXANDER WATT, Author of "Electro-Metallurgy," &c. Third Edition, Revised. Crown 8vo, 9s., cloth.

"Eminently a book for the practical worker in electro-deposition. It contains practical descriptions of methods, processes and materials, as actually pursued and used in the workshop."—*Engineer.*

Electro-Metallurgy.

ELECTRO-METALLURGY: Practically Treated. By ALEXANDER WATT. Tenth Edition, including the most recent Processes. 12mo, 4s. cloth.

"From this book both amateur and artisan may learn everything necessary for the successful prosecution of electroplating."—*Iron.*

Working in Gold.

THE JEWELLER'S ASSISTANT IN THE ART OF WORKING IN GOLD: A Practical Treatise for Masters and Workmen, Compiled from the Experience of Thirty Years' Workshop Practice. By GEORGE E. GEE, Author of "The Goldsmith's Handbook," &c. Crown 8vo, 7s. 6d. cloth.

"This manual of technical education is apparently destined to be a valuable auxiliary to a handicraft which is certainly capable of great improvement."—*The Times.*

"Very useful in the workshop, the knowledge is practical, having been acquired by long experience, and all the recipes and directions are guaranteed to be successful."—*Jeweller and Metalworker.*

Electroplating.

ELECTROPLATING: A Practical Handbook on the Deposition of Copper, Silver, Nickel, Gold, Aluminium, Brass, Platinum, &c. &c. By J. W. URQUHART, C.E., Author of "Electric Light," &c. Third Edition, Revised, with Additions. Numerous Illustrations. Crown 8vo, 5s. cloth.
"An excellent practical manual."—*Engineering.*
"An excellent work, giving the newest information."—*Horological Journal.*

Electrotyping.

ELECTROTYPING: The Reproduction and Multiplication of Printing Surfaces and Works of Art by the Electro-deposition of Metals. By J. W. URQUHART, C.E. Crown 8vo, 5s. cloth.
"The book is thoroughly practical; the reader is, therefore, conducted through the leading laws of electricity, then through the metals used by electrotypers, the apparatus, and the depositing processes, up to the final preparation of the work."—*Art Journal.*

Goldsmiths' Work.

THE GOLDSMITH'S HANDBOOK. By GEORGE E. GEE, Jeweller, &c. Fourth Edition. 12mo, 3s. 6d. cloth boards.
"A good, sound educator, and will be generally accepted as an authority."—*Horological Journal.*

Silversmiths' Work.

THE SILVERSMITH'S HANDBOOK. By GEORGE E. GEE, Jeweller, &c. Third Edition, with numerous Illustrations. 12mo, 3s. 6d. cloth boards.
"The chief merit of the work is its practical character. . . . The workers in the trade will speedily discover its merits when they sit down to study it."—*English Mechanic.*
*** *The above two works together, strongly half-bound, price 7s.*

Sheet Metal Working.

THE SHEET METAL WORKER'S INSTRUCTOR: For Zinc, Sheet Iron, Copper, and Tin Plate Workers. Containing Rules for describing the Patterns required in the Different Branches of the Trade. By R. H. WARN, Tin Plate Worker. With Thirty-two Plates. 8vo, 7s. 6d. cloth.

Bread and Biscuit Baking.

THE BREAD AND BISCUIT BAKER'S AND SUGAR-BOILER'S ASSISTANT. Including a large variety of Modern Recipes. With Remarks on the Art of Bread-making. By ROBERT WELLS. Second Edition. Cr. 8vo, 2s. cloth.
"A large number of wrinkles for the ordinary cook, as well as the baker."—*Saturday Review.*

Confectionery for Hotels and Restaurants.

THE PASTRYCOOK AND CONFECTIONER'S GUIDE. For Hotels, Restaurants, and the Trade in general, adapted also for Family Use. By R. WELLS, Author of "The Bread and Biscuit Baker." Crown 8vo, 2s. cloth.
"We cannot speak too highly of this really excellent work. In these days of keen competition our readers cannot do better than purchase this book."—*Baker's Times.*

Ornamental Confectionery.

ORNAMENTAL CONFECTIONERY: A Guide for Bakers, Confectioners and Pastrycooks; including a Variety of Modern Recipes, and Remarks on Decorative and Coloured Work. With 129 Original Designs. By ROBERT WELLS. Crown 8vo, 5s. cloth gilt.
"A valuable work, practical, and should be in the hands of every baker and confectioner. The llustrative designs are alone worth treble the amount charged for the whole work."—*Baker's Times.*

Flour Confectionery.

THE MODERN FLOUR CONFECTIONER, Wholesale and Retail. Containing a large Collection of Recipes for Cheap Cakes, Biscuits, &c. With Remarks on the Ingredients Used in their Manufacture. By ROBERT WELLS, Author of "The Bread and Biscuit Baker," &c. Crown 8vo, 2s. cloth.
"The work is of a decidedly practical character, and in every recipe regard is had to economical working."—*North British Daily Mail.*

Laundry Work.

LAUNDRY MANAGEMENT. A Handbook for Use in Private and Public Laundries. Including Descriptive Accounts of Modern Machinery and Appliances for Laundry Work. By the EDITOR of "The Laundry Journal." With numerous Illustrations. Second Edition. Crown 8vo, 2s. 6d. cloth.
"This book should certainly occupy an honoured place on the shelves of all housekeepers who wish to keep themselves *au courant* of the newest appliances and methods."—*The Queen.*

HANDYBOOKS FOR HANDICRAFTS.

BY PAUL N. HASLUCK,
Editor of "Work" (New Series), Author of "Lathe Work," "Milling Machines," &c.
Crown 8vo, 144 pages, cloth, price 1s. each.

☞ *These* HANDYBOOKS *have been written to supply information for* WORKMEN, STUDENTS, *and* AMATEURS *in the several Handicrafts, on the actual* PRACTICE *of the* WORKSHOP, *and are intended to convey in plain language* TECHNICAL KNOWLEDGE *of the several* CRAFTS. *In describing the processes employed, and the manipulation of material, workshop terms are used; workshop practice is fully explained; and the text is freely illustrated with drawings of modern tools, appliances, and processes.*

THE METAL TURNER'S HANDYBOOK. A Practical Manual for Workers at the Foot-Lathe. With over 100 Illustrations. Price 1s.
"The book will be of service alike to the amateur and the artisan turner. It displays thorough knowledge of the subject."—*Scotsman.*

THE WOOD TURNER'S HANDYBOOK. A Practical Manual for Workers at the Lathe. With over 100 Illustrations. Price 1s.
"We recommend the book to young turners and amateurs. A multitude of workmen have hitherto sought in vain for a manual of this special industry."—*Mechanical World.*

THE WATCH JOBBER'S HANDYBOOK. A Practical Manual on Cleaning, Repairing, and Adjusting. With upwards of 100 Illustrations. Price 1s.
"We strongly advise all young persons connected with the watch trade to acquire and study this inexpensive work."—*Clerkenwell Chronicle.*

THE PATTERN MAKER'S HANDYBOOK. A Practical Manual on the Construction of Patterns for Founders. With upwards of 100 Illustrations. 1s.
"A most valuable, if not indispensable, manual for the pattern maker."—*Knowledge.*

THE MECHANIC'S WORKSHOP HANDYBOOK. A Practical Manual on Mechanical Manipulation, embracing Information on various Handicraft Processes. With Useful Notes and Miscellaneous Memoranda. Comprising about 200 Subjects. Price 1s.
"A very clever and useful book, which should be found in every workshop; and it should certainly find a place in all technical schools."—*Saturday Review.*

THE MODEL ENGINEER'S HANDYBOOK. A Practical Manual on the Construction of Model Steam Engines. With upwards of 100 Illustrations. 1s.
"Mr. Hasluck has produced a very good little book."—*Builder.*

THE CLOCK JOBBER'S HANDYBOOK. A Practical Manual on Cleaning, Repairing, and Adjusting. With upwards of 100 Illustrations. Price 1s.
"It is of inestimable service to those commencing the trade."—*Coventry Standard.*

THE CABINET WORKER'S HANDYBOOK. A Practical Manual on the Tools, Materials, Appliances, and Processes employed in Cabinet Work. With upwards of 100 Illustrations. Price 1s.
"Mr. Hasluck's thoroughgoing little Handybook is amongst the most practical guides we have seen for beginners in cabinet-work."—*Saturday Review.*

THE WOODWORKER'S HANDYBOOK OF MANUAL INSTRUCTION. Embracing Information on the Tools, Materials, Appliances and Processes Employed in Woodworking. With 104 Illustrations. Price 1s. [*Just published.*

THE METALWORKER'S HANDYBOOK. With upwards of 100 Illustrations. [*In preparation.*

OPINIONS OF THE PRESS.

"Written by a man who knows, not only how work ought to be done, but how to do it, and how to convey his knowledge to others."—*Engineering.*

"Mr. Hasluck writes admirably, and gives complete instructions."—*Engineer.*

"Mr. Hasluck combines the experience of a practical teacher with the manipulative skill and scientific knowledge of processes of the trained mechanician, and the manuals are marvels of what can be produced at a popular price."—*Schoolmaster.*

"Helpful to workmen of all ages and degrees of experience."—*Daily Chronicle.*

"Practical, sensible, and remarkably cheap."—*Journal of Education.*

"Concise, clear, and practical."—*Saturday Review.*

COMMERCE, COUNTING-HOUSE WORK, TABLES, etc.

Commercial Education.

LESSONS IN COMMERCE. By Professor R. GAMBARO, of the Royal High Commercial School at Genoa. Edited and Revised by JAMES GAULT, Professor of Commerce and Commercial Law in King's College, London. Crown 8vo, 3s. 6d. cloth.

"The publishers of this work have rendered considerable service to the cause of commercial education by the opportune production of this volume. . . . The work is peculiarly acceptable to English readers and an admirable addition to existing class books. In a phrase, we think the work attains its object in furnishing a brief account of those laws and customs of British trade with which the commercial man interested therein should be familiar."—*Chamber of Commerce Journal.*

"An invaluable guide in the hands of those who are preparing for a commercial career, and, in fact the information it contains on matters of business should be impressed on every one."—*Counting House.*

Foreign Commercial Correspondence.

THE FOREIGN COMMERCIAL CORRESPONDENT: Being Aids to Commercial Correspondence in Five Languages—English, French, German, Italian, and Spanish. By CONRAD E. BAKER. Second Edition. Cr. 8vo, 3s. 6d. cl.

"Whoever wishes to correspond in all the languages mentioned by Mr. Baker cannot do better than study this work, the materials of which are excellent and conveniently arranged. They consist not of entire specimen letters, but—what are far more useful—short passages, sentences, or phrases expressing the same general idea in various forms."—*Athenæum.*

"A careful examination has convinced us that it is unusually complete, well arranged and reliable. The book is a thoroughly good one."—*Schoolmaster.*

Commercial French.

A NEW BOOK OF COMMERCIAL FRENCH: Grammar—Vocabulary—Correspondence—Commercial Documents—Geography—Arithmetic—Lexicon. By P. CARROUÉ, Professor in the City High School J.—B. Say (Paris). Crown 8vo, 4s. 6d. cloth. [*Just published.*

Accounts for Manufacturers.

FACTORY ACCOUNTS: Their Principles and Practice. A Handbook for Accountants and Manufacturers, with Appendices on the Nomenclature of Machine Details; the Income Tax Acts; the Rating of Factories; Fire and Boiler Insurance; the Factory and Workshop Acts, &c., including also a Glossary of Terms and a large number of Specimen Rulings. By EMILE GARCKE and J. M. FELLS. Fourth Edition, Revised and Enlarged. Demy 8vo, 250 pages. 6s. strongly bound.

"A very interesting description of the requirements of Factory Accounts. . . . The principle of assimilating the Factory Accounts to the general commercial books is one which we thoroughly agree with."—*Accountants' Journal.*

"Characterised by extreme thoroughness. There are few owners of factories who would not derive great benefit from the perusal of this most admirable work."—*Local Government Chronicle.*

Modern Metrical Units and Systems.

MODERN METROLOGY: A Manual of the Metrical Units and Systems of the present Century. With an Appendix containing a proposed English System. By LEWIS D'A. JACKSON, A.-M. Inst. C.E., Author of "Aid to Survey Practice," &c. Large crown 8vo, 12s. 6d. cloth.

"We recommend the work to all interested in the practical reform of our weights and measures."—*Nature*

The Metric System and the British Standards.

A SERIES OF METRIC TABLES, in which the British Standard Measures and Weights are compared with those of the Metric System at present in Use on the Continent. By C. H. DOWLING, C.E. 8vo, 10s. 6d. strongly bound.

"Mr. Dowling's Tables are well put together as a ready reckoner for the conversion of one system into the other."—*Athenæum.*

Iron and Metal Trades' Calculator.

THE IRON AND METAL TRADES' COMPANION: For expeditiously ascertaining the Value of any Goods bought or sold by Weight, from 1s. per cwt. to 112s. per cwt., and from one farthing per pound to one shilling per pound. By THOMAS DOWNIE. Strongly bound in leather, 396 pp., 9s.

"A most useful set of tables, nothing like them before existed."—*Building News.*

"Although specially adapted to the iron and metal trades, the tables will be found useful in every other business in which merchandise is bought and sold by weight."—*Railway News.*

Chadwick's Calculator for Numbers and Weights Combined.

THE NUMBER, WEIGHT, AND FRACTIONAL CALCULATOR. Containing upwards of 250,000 Separate Calculations, showing at a glance the value at 422 different rates, ranging from $\frac{1}{16}$th of a Penny to 20s. each, or per cwt., and £20 per ton, of any number of articles consecutively, from 1 to 470.—Any number of cwts., qrs., and lbs., from 1 cwt. to 470 cwts.—Any number of tons, cwts., qrs., and lbs., from 1 to 1,000 tons. By WILLIAM CHADWICK, Public Accountant. Third Edition, Revised and Improved. 8vo, 18s. strongly bound.

☞ *Is adapted for the use of Accountants and Auditors, Railway Companies, Canal Companies, Shippers, Shipping Agents, General Carriers, &c. Ironfounders, Brassfounders, Metal Merchants, Iron Manufacturers, Ironmongers, Engineers, Machinists, Boiler Makers, Millwrights, Roofing, Bridge and Girder Makers, Colliery Proprietors, &c. Timber Merchants, Builders, Contractors, Architects, Surveyors, Auctioneers, Valuers, Brokers, Mill Owners and Manufacturers, Mill Furnishers, Merchants, and General Wholesale Tradesmen. Also for the Apportionment of Mileage Charges for Railway Traffic.*

"It is as easy of reference for any answer or any number of answers as a dictionary, and the references are even more quickly made. For making up accounts or estimates the book must prove invaluable to all who have any considerable quantity of calculations involving price and measure in any combination to do."—*Engineer*.

"The most perfect work of the kind yet prepared."—*Glasgow Herald*.

Harben's Comprehensive Weight Calculator.

THE WEIGHT CALCULATOR: Being a Series of Tables upon a New and Comprehensive Plan, exhibiting at one Reference the exact Value of any Weight from 1 lb. to 15 tons, at 300 Progressive Rates, from 1d. to 168s. per cwt., and containing 186,000 Direct Answers, which, with their Combinations, consisting of a single addition (mostly to be performed at sight), will afford an aggregate of 10,266,000 Answers; the whole being calculated and designed to ensure correctness and promote despatch. By HENRY HARBEN, Accountant. Fourth Edition, carefully corrected. Royal 8vo, strongly half-bound, £1 5s.

"A practical and useful work of reference for men of business generally."—*Ironmonger*.

"Of priceless value to business men. It is a necessary book in all mercantile offices."—*Sheffield Independent*.

Harben's Comprehensive Discount Guide.

THE DISCOUNT GUIDE. Comprising several Series of Tables for the use of Merchants, Manufacturers, Ironmongers, and others, by which may be ascertained the exact Profit arising from any mode of using Discounts, either in the Purchase or Sale of Goods, and the method of either Altering a Rate of Discount, or Advancing a Price, so as to produce, by one operation, a sum that will realise any required profit after allowing one or more Discounts: to which are added Tables of Profit or Advance from 1¼ to 90 per cent., Tables of Discount from 1¼ to 98¾ per cent., and Tables of Commission, &c., from ⅛ to 10 per cent. By HENRY HARBEN, Accountant. New Edition, Corrected. Demy 8vo, £1 5s. half-bound.

"A book such as this can only be appreciated by business men, to whom the saving of time means saving of money. We have the high authority of Professor J. R. Young that the tables throughout the work are constructed upon strictly accurate principles. The work is a model of typographical clearness, and must prove of great value to merchants, manufacturers, and general traders."—*British Trade Journal*.

A New Series of Calculators.

DIRECT CALCULATORS: A Series of Tables and Calculations varied in arrangement to suit the needs of Particular Trades. By M. B. COTSWORTH. The Series comprises 13 distinct books, at prices ranging from 2s. 6d. to 10s. 6d. (Detailed prospectus on application).

New Wages Calculator.

TABLES OF WAGES at 54, 52, 50 and 48 Hours per Week. Showing the Amounts of Wages from One-quarter-of-an-hour to Sixty-four hours in each case at Rates of Wages advancing by One Shilling from 4s. to 55s. per week. By THOS. GARBUTT, Accountant. Square crown 8vo, 6s. half-bound.

[*Just published.*]

Iron Shipbuilders' and Merchants' Weight Tables.

IRON-PLATE WEIGHT TABLES: For Iron Shipbuilders, Engineers, and Iron Merchants. Containing the Calculated Weights of upwards of 150,000 different sizes of Iron Plates from 1 foot by 6 in. by ¼ in. to 10 feet by 5 feet by 1 in. Worked out on the basis of 40 lbs. to the square foot of Iron of 1 inch in thickness. By H. BURLINSON and W. H. SIMPSON. 4to, 25s. half-bound.

AGRICULTURE, FARMING, GARDENING, etc.

Dr. Fream's New Edition of "The Standard Treatise on Agriculture."
THE COMPLETE GRAZIER AND FARMER'S AND CATTLE BREEDER'S ASSISTANT: A Compendium of Husbandry. Originally Written by WILLIAM YOUATT. Thirteenth Edition, entirely Re-written, considerably Enlarged, and brought up to the Present Requirements of Agricultural Practice, by WILLIAM FREAM, LL.D., Steven Lecturer in the University of Edinburgh, Author of "The Elements of Agriculture," &c. Royal 8vo, 1,100 pp., with over 450 Illustrations. Price £1 11s. 6d. strongly and handsomely bound.

EXTRACT FROM PUBLISHERS' ADVERTISEMENT.

"A treatise that made its original appearance in the first decade of the century, and that enters upon its Thirteenth edition before the century has run its course, has undoubtedly established its position as a work of permanent value. . . . The phenomenal progress of the last dozen years in the Practice and Science of Farming has rendered it necessary, however, that the volume should be re-written, . . . and for this undertaking the Publishers were fortunate enough to secure the services of Dr. FREAM, whose high attainments in all matters pertaining to agriculture have been so emphatically recognised by the highest professional and official authorities. In carrying out his editorial duties, Dr. FREAM has been favoured with valuable contributions by Prof. J. WORTLEY AXE, Mr. E. BROWN, Dr. BERNARD DYER, Mr. W. J. MALDEN, Mr. R. H. REW, Prof. SHELDON, Mr. J. SINCLAIR, Mr. SANDERS SPENCER, and others.

"As regards the illustrations of the work, no pains have been spared to make them as representative and characteristic as possible, so as to be practically useful to the Farmer and Grazier."

SUMMARY OF CONTENTS.

BOOK I. ON THE VARIETIES, BREEDING, REARING, FATTENING AND MANAGEMENT OF CATTLE.
BOOK II. ON THE ECONOMY AND MANAGEMENT OF THE DAIRY.
BOOK III. ON THE BREEDING, REARING, AND MANAGEMENT OF HORSES.
BOOK IV. ON THE BREEDING, REARING, AND FATTENING OF SHEEP.
BOOK V. ON THE BREEDING, REARING, AND FATTENING OF SWINE.
BOOK VI. ON THE DISEASES OF LIVE STOCK.
BOOK VII. ON THE BREEDING, REARING, AND MANAGEMENT OF POULTRY.
BOOK VIII. ON FARM OFFICES AND IMPLEMENTS OF HUSBANDRY.
BOOK IX. ON THE CULTURE AND MANAGEMENT OF GRASS LANDS.
BOOK X. ON THE CULTIVATION AND APPLICATION OF GRASSES, PULSE AND ROOTS.
BOOK XI. ON MANURES AND THEIR APPLICATION TO GRASS LAND AND CROPS.
BOOK XII. MONTHLY CALENDARS OF FARMWORK.

*** OPINIONS OF THE PRESS ON THE NEW EDITION.

"Dr. Fream is to be congratulated on the successful attempt he has made to give us a work which will at once become the standard classic of the farm practice of the country. We believe that it will be found that it has no compeer among the many works at present in existence. . . . The illustrations are admirable, while the frontispiece, which represents the well-known bull, New Year's Gift, bred by the Queen, is a work of art."—*The Times*.

"The book must be recognised as occupying the proud position of the most exhaustive work of reference in the English language on the subject with which it deals."—*Athenæum*.

"The most comprehensive guide to modern farm practice that exists in the English language to-day. . . . The book is one that ought to be on every farm and in the library of every land owner."—*Mark Lane Express*.

"In point of exhaustiveness and accuracy the work will certainly hold a pre-eminent and unique position among books dealing with scientific agricultural practice. It is, in fact, an agricultural library of itself."—*North British Agriculturist*.

"A compendium of authoritative and well-ordered knowledge on every conceivable branch of the work of the live stock farmer; probably without an equal in this or any other country."—*Yorkshire Post*.

"The best and brightest guide to the practice of husbandry: one that has no superior—no equal we might truly say—among the agricultural literature now before the public. . . . In every section in which we have tested it, the work has been found thoroughly up to date."—*Bell's Weekly Messenger*.

British Farm Live Stock.
FARM LIVE STOCK OF GREAT BRITAIN. By ROBERT WALLACE, F.L.S., F.R.S.E., &c., Professor of Agriculture and Rural Economy in the University of Edinburgh. Third Edition, thoroughly Revised and considerably Enlarged. With over 120 Phototypes of Prize Stock. Demy 8vo, 384 pp., with 79 Plates and Maps. Price 12s. 6d., cloth.

"A really complete work on the history, breeds, and management of the farm stock of Great Britain, and one which is likely to find its way to the shelves of every country gentleman's library."—*The Times*.

"The latest edition of 'Farm Live Stock of Great Britain' is a production to be proud of, and its issue not the least of the services which its author has rendered to agricultural science."—*Scottish Farmer*.

"The book is very attractive, . . . and we can scarcely imagine the existence of a farmer who would not like to have a copy of this beautiful and useful work."—*Mark Lane Express*.

"A work which will long be regarded as a standard authority whenever a concise history and description of the breeds of live stock in the British Isles is required."—*Bell's Weekly Messenger*.

Dairy Farming.

BRITISH DAIRYING: A Handy Volume on the Work of the Dairy-Farm. For the Use of Technical Instruction Classes, Students in Agricultural Colleges and the Working Dairy-Farmer. By Prof. J. P. SHELDON, late Special Commissioner of the Canadian Government, Author of "Dairy Farming," "The Farm and the Dairy," &c. With numerous Illustrations. Crown 8vo, 2s. 6d. cloth.

"Confidently recommended as a useful text-book on dairy farming."—*Agricultural Gazette.*
"Probably the best half-crown manual on dairy work that has yet been produced."—*North British Agriculturist.*
"It is the soundest little work we have yet seen on the subject."—*The Times.*

Dairy Manual.

MILK, CHEESE AND BUTTER: A Practical Handbook on their Properties and the Processes of their Production. Including a Chapter on Cream and the Methods of its Separation from Milk. By JOHN OLIVER, late Principal of the Western Dairy Institute, Berkeley. With Coloured Plates and 200 Illustrations. Crown 8vo, 7s. 6d. cloth. [*Just published.*

"An exhaustive and masterly production. It may be cordially recommended to all students and practitioners of dairy science."—*N.B. Agriculturist.*
"We strongly recommend this very comprehensive and carefully-written book to dairy-farmers and students of dairying. It is a distinct acquisition to the library of the agriculturist."—*Agricultural Gazette.*

Agricultural Facts and Figures.

NOTE-BOOK OF AGRICULTURAL FACTS AND FIGURES FOR FARMERS AND FARM STUDENTS. By PRIMROSE McCONNELL, B.Sc. Fifth Edition. Royal 32mo, roan, gilt edges, with band, 4s.

"Literally teems with information and we can cordially recommend it to all connected with agriculture."—*North British Agriculturist.*

Small Farming.

SYSTEMATIC SMALL FARMING; or, The Lessons of my Farm. Being an Introduction to Modern Farm Practice for Small Farmers. By R. SCOTT BURN, Author of "Outlines of Modern Farming," &c. Crown 8vo, 6s. cloth.

"This is the completest book of its class we have seen, and one which every amateur farmer will read with pleasure, and accept as a guide."—*Field.*

Modern Farming.

OUTLINES OF MODERN FARMING. By R. SCOTT BURN. Soils, Manures, and Crops—Farming and Farming Economy—Cattle, Sheep, and Horses—Management of Dairy, Pigs, and Poultry—Utilization of Town-Sewage, Irrigation, &c. Sixth Edition. In one vol., 1,250 pp., half-bound, profusely Illustrated, 12s.

"The aim of the author has been to make his work at once comprehensive and trustworthy, and in this aim he has succeeded to a degree which entitles him to much credit."—*Morning Advertiser.*

Agricultural Engineering.

FARM ENGINEERING, THE COMPLETE TEXT-BOOK OF. Comprising Draining and Embanking; Irrigation and Water Supply; Farm Roads, Fences and Gates; Farm Buildings; Barn Implements and Machines; Field Implements and Machines; Agricultural Surveying, &c. By Professor JOHN SCOTT. In one vol., 1,150 pages, half-bound, with over 600 Illustrations, 12s.

"Written with great care, as well as with knowledge and ability. The author has done his work well; we have found him a very trustworthy guide wherever we have tested his statements. The volume will be of great value to agricultural students."—*Mark Lane Express.*

Agricultural Text-Book.

THE FIELDS OF GREAT BRITAIN: A Text-Book of Agriculture. Adapted to the Syllabus of the Science and Art Department. For Elementary and Advanced Students. By HUGH CLEMENTS (Board of Trade). Second Edition, Revised, with Additions. 18mo, 2s. 6d. cloth.

"It is a long time since we have seen a book which has pleased us more, or which contains such a vast and useful fund of knowledge."—*Educational Times.*

Tables for Farmers, &c.

TABLES, MEMORANDA, AND CALCULATED RESULTS for Farmers, Graziers, Agricultural Students, Surveyors, Land Agents, Auctioneers, &c. With a New System of Farm Book-keeping. Selected and Arranged by SIDNEY FRANCIS. Third Edition, Revised. 272 pp., waistcoat-pocket size, limp leather, 1s. 6d.

"Weighing less than 1 oz., and occupying no more space than a match box, it contains a mass of facts and calculations which has never before, in such handy form, been obtainable. Every operation on the farm is dealt with. The work may be taken as thoroughly accurate, the whole of the tables having been revised by Dr. Fream. We cordially recommend it."—*Bell's Weekly Messenger.*

Artificial Manures and Foods.
FERTILISERS AND FEEDING STUFFS: Their Properties and Uses. A Handbook for the Practical Farmer. By BERNARD DYER, D.Sc. (Lond.). With the Text of the Fertilisers and Feeding Stuffs Act of 1893, the Regulations and Forms of the Board of Agriculture, and Notes on the Act by A. J. DAVID, B.A., LL.M., of the Inner Temple, Barrister-at-Law. Crown 8vo, 120 pp., 1s. cloth. [*Just published.*
"An excellent shillingsworth. Dr. Dyer has done farmers good service in placing at their disposal so much useful information in so intelligible a form."—*The Times.*

The Management of Bees.
BEES FOR PLEASURE AND PROFIT: A Guide to the Manipulation of Bees, the Production of Honey, and the General Management of the Apiary. By G. GORDON SAMSON. With numerous Illustrations. Crown 8vo, 1s. cloth.
"The intending bee-keeper will find exactly the kind of information required to enable him to make a successful start with his hives. The author is a thoroughly competent teacher."—*Morning Post.*

Farm and Estate Book-keeping.
BOOK-KEEPING FOR FARMERS AND ESTATE OWNERS. A Practical Treatise, presenting, in Three Plans, a System adapted for all Classes of Farms. By JOHNSON M. WOODMAN, Chartered Accountant. Second Edition, Revised. Crown 8vo, 3s. 6d. cloth boards; or, 2s. 6d. cloth limp.
"The volume is a capital study of a most important subject."—*Agricultural Gazette.*
"Farmers and land agents will find the book more than repay its cost and study."—*Building News.*

Farm Account Book.
WOODMAN'S YEARLY FARM ACCOUNT BOOK. Giving a Weekly Labour Account and Diary, and showing the Income and Expenditure under each Department of Crops, Live Stock, Dairy, &c. &c. With Valuation, Profit and Loss Account, and Balance Sheet at the end of the Year. By JOHNSON M. WOODMAN, Chartered Accountant. Folio, 7s. 6d. half-bound.
"Contains every requisite form for keeping farm accounts readily and accurately."—*Agriculture.*

Early Fruits, Flowers and Vegetables.
THE FORCING-GARDEN; or, How to Grow Early Fruits, Flowers, and Vegetables. With Plans and Estimates for Building Glasshouses, Pits and Frames. With Illustrations. By SAMUEL WOOD. Crown 8vo, 3s. 6d. cloth.
"A good book, and fairly fills a place that was in some degree vacant. The book is written with great care, and contains a great deal of valuable teaching."—*Gardeners' Magazine.*

Good Gardening.
A PLAIN GUIDE TO GOOD GARDENING; or, How to Grow Vegetables, Fruits, and Flowers. By S. WOOD. Fourth Edition, with considerable Additions, &c., and numerous Illustrations. Crown 8vo, 3s. 6d. cloth.
"A very good book, and one to be highly recommended as a practical guide. The practical directions are excellent."—*Athenæum.*

Gainful Gardening.
MULTUM-IN-PARVO GARDENING; or, How to make One Acre of Land produce £620 a-year, by the Cultivation of Fruits and Vegetables; also, How to Grow Flowers in Three Glass Houses, so as to realise £176 per annum clear Profit. By SAMUEL WOOD, Author of "Good Gardening," &c. Fifth and Cheaper Edition, revised, with Additions. Crown 8vo, 1s. sewed.
"We are bound to recommend it as not only suited to the case of the amateur and gentleman's gardener, but to the market grower."—*Gardeners' Magazine.*

Gardening for Ladies.
THE LADIES' MULTUM-IN-PARVO FLOWER GARDEN, AND AMATEURS' COMPLETE GUIDE. With Illusts. By S. WOOD. Cr. 8vo, 3s. 6d. cloth.
"This volume contains a good deal of sound, common-sense instruction."—*Florist.*
"Full of shrewd hints and useful instructions, based on a lifetime of experience."—*Scotsman.*

Cultivation of the Potato.
POTATOES: How to Grow and Show Them. A Practical Guide to the Cultivation and General Treatment of the Potato. By JAMES PINK. Second Edition. Crown 8vo, 2s. cloth.

Market Gardening.
MARKET AND KITCHEN GARDENING. By Contributors to "The Garden." By C. W. SHAW, late Editor of "Gardening Illustrated." 3s. 6d. cloth.
"The most valuable compendium of kitchen and market-garden work published."—*Farmer.*

AUCTIONEERING, VALUING, LAND SURVEYING, ESTATE AGENCY, etc.

Auctioneer's Assistant.

THE APPRAISER, AUCTIONEER, BROKER, HOUSE AND ESTATE AGENT AND VALUER'S POCKET ASSISTANT, for the Valuation for Purchase, Sale, or Renewal of Leases, Annuities and Reversions, and of property generally; with Prices for Inventories, &c. By JOHN WHEELER, Valuer, &c. Sixth Edition, Re-written and greatly Extended by C. NORRIS, Surveyor, Valuer, &c. Royal 32mo, 5s. cloth.

"A neat and concise book of reference, containing an admirable and clearly-arranged list of prices for inventories, and a very practical guide to determine the value of furniture, &c."—*Standard.*

"Contains a large quantity of varied and useful information as to the valuation for purchase, sale, or renewal of leases, annuities and reversions, and of property generally, with prices for inventories, and a guide to determine the value of interior fittings and other effects."—*Builder.*

Auctioneering.

AUCTIONEERS: THEIR DUTIES AND LIABILITIES. A Manual of Instruction and Counsel for the Young Auctioneer. By ROBERT SQUIBBS, Auctioneer. Second Edition, Revised and partly Re-written. Demy 8vo, 12s. 6d. cloth.

*** OPINIONS OF THE PRESS.

"The standard text-book on the topics of which it treats."—*Athenæum.*

"The work is one of general excellent character, and gives much information in a compendious and satisfactory form."—*Builder.*

"May be recommended as giving a great deal of information on the law relating to auctioneers, in a very readable form."—*Law Journal.*

"Auctioneers may be congratulated on having so pleasing a writer to minister to their special needs."—*Solicitors' Journal.*

"Every auctioneer ought to possess a copy of this excellent work."—*Ironmonger.*

"Of great value to the profession. . . . We readily welcome this book from the fact that it treats the subject in a manner somewhat new to the profession."—*Estates Gazette.*

Inwood's Estate Tables.

TABLES FOR THE PURCHASING OF ESTATES, FREEHOLD, COPYHOLD, OR LEASEHOLD; ANNUITIES, ADVOWSONS, &C., and for the Renewing of Leases held under Cathedral Churches, Colleges, or other Corporate bodies, for Terms of Years certain, and for Lives; also for Valuing Reversionary Estates, Deferred Annuities, Next Presentations, &c.; together with SMART'S Five Tables of Compound Interest, and an Extension of the same to Lower and Intermediate Rates. By W. INWOOD. 24th Edition, with considerable Additions, and new and valuable Tables of Logarithms for the more Difficult Computations of the Interest of Money, Discount, Annuities, &c., by M. FÉDOR THOMAN, of the Société Crédit Mobilier of Paris. Crown 8vo, 8s. cloth.

"Those interested in the purchase and sale of estates, and in the adjustment of compensation cases, as well as in transactions in annuities, life insurances, &c., will find the present edition of eminent service."—*Engineering.*

"'Inwood's Tables' still maintain a most enviable reputation. The new issue has been enriched by large additional contributions by M. Fédor Thoman, whose carefully-arranged Tables cannot fail to be of the utmost utility."—*Mining Journal.*

Agricultural Valuer's Assistant.

THE AGRICULTURAL VALUER'S ASSISTANT. A Practical Handbook on the Valuation of Landed Estates; including Rules and Data for Measuring and Estimating the Contents, Weights and Values of Agricultural Produce and Timber, and the Values of Feeding Stuffs, Manures, and Labour; with Forms of Tenant-Right Valuations, Lists of Local Agricultural Customs, Scales of Compensation under the Agricultural Holdings Act, &c. &c. By TOM BRIGHT, Agricultural Surveyor. Second Edition, much Enlarged. Crown 8vo, 5s. cloth.

"Full of tables and examples in connection with the valuation of tenant-right, estates, labour, contents and weights of timber, and farm produce of all kinds."—*Agricultural Gazette.*

"An eminently practical handbook, full of practical tables and data of undoubted interest and value to surveyors and auctioneers in preparing valuations of all kinds."—*Farmer.*

Plantations and Underwoods.

POLE PLANTATIONS AND UNDERWOODS: A Practical Handbook on Estimating the Cost of Forming, Renovating, Improving, and Grubbing Plantations and Underwoods, their Valuation for Purposes of Transfer, Rental, Sale or Assessment. By TOM BRIGHT, Author of "The Agricultural Valuer's Assistant," &c. Crown 8vo, 3s. 6d. cloth.

"To valuers, foresters and agents it will be a welcome aid."—*North British Agriculturist.*

"Well calculated to assist the valuer in the discharge of his duties, and of undoubted interest and use both to surveyors and auctioneers in preparing valuations of all kinds."—*Kent Herald.*

Hudson's Land Valuer's Pocket-Book.

THE LAND VALUER'S BEST ASSISTANT: Being Tables on a very much improved Plan, for Calculating the Value of Estates. With Tables for reducing Scotch, Irish, and Provincial Customary Acres to Statute Measure, &c. By R. HUDSON, C.E. New Edition. Royal 32mo, leather, elastic band, 4s.
"Of incalculable value to the country gentleman and professional man."—*Farmers' Journal.*

Ewart's Land Improver's Pocket-Book.

THE LAND IMPROVER'S POCKET-BOOK OF FORMULÆ, TABLES, AND MEMORANDA required in any Computation relating to the Permanent Improvement of Landed Property. By JOHN EWART, Surveyor. Second Edition, Revised. Royal 32mo, oblong, 4s. leather.
"A compendious and handy little volume."—*Spectator.*

Complete Agricultural Surveyor's Pocket-Book.

THE LAND VALUER'S AND LAND IMPROVER'S COMPLETE POCKET-BOOK. Being the above Two Works bound together. 7s. 6d. leather.

House Property.

HANDBOOK OF HOUSE PROPERTY: A Popular and Practical Guide to the Purchase, Mortgage, Tenancy, and Compulsory Sale of Houses and Land, including the Law of Dilapidations and Fixtures: with Examples of all kinds of Valuations, Useful Information on Building and Suggestive Elucidations of Fine Art. By E. L. TARBUCK, Architect and Surveyor. Fifth Edition, Enlarged. 12mo, 5s. cloth.
"The advice is thoroughly practical."—*Law Journal.*
"For all who have dealings with house property, this is an indispensable guide."—*Decoration.*
"Carefully brought up to date, and much improved by the addition of a division on Fine Art. . . . A well-written and thoughtful work."—*Land Agent's Record.*

LAW AND MISCELLANEOUS.

Journalism.

MODERN JOURNALISM: A Handbook of Instruction and Counsel for the Young Journalist. By JOHN B. MACKIE, Fellow of the Institute of Journalists. Crown 8vo, 2s. cloth. [*Just published.*
"This invaluable guide to journalism is a work which all aspirants to a journalistic career will read with advantage."—*Journalist.*

Private Bill Legislation and Provisional Orders.

HANDBOOK FOR THE USE OF SOLICITORS AND ENGINEERS Engaged in Promoting Private Acts of Parliament and Provisional Orders, for the authorization of Railways, Tramways, Gas and Water Works, &c. By L. LIVINGSTON MACASSEY, of the Middle Temple, Barrister-at-Law, M. Inst. C.E. 8vo, 25s. cloth.

Law of Patents.

PATENTS FOR INVENTIONS, AND HOW TO PROCURE THEM. Compiled for the Use of Inventors, Patentees and others. By G. G. M. HARDINGHAM, Assoc. Mem. Inst. C.E., &c. Demy 8vo, 1s. 6d. cloth.

Labour Disputes.

CONCILIATION AND ARBITRATION IN LABOUR DISPUTES: A Historical Sketch and Brief Statement of the Present Position of the Question at Home and Abroad. By J. S. JEANS, Author of "England's Supremacy," &c. Crown 8vo, 200 pp., 2s. 6d. cloth. [*Just published.*
"Mr. Jeans is well qualified to write on this subject, both by his previous books and by his practical experience as an arbitrator."—*The Times.*

Pocket-Book for Sanitary Officials.

THE HEALTH OFFICER'S POCKET-BOOK: A Guide to Sanitary Practice and Law. For Medical Officers of Health, Sanitary Inspectors, Members of Sanitary Authorities, &c. By EDWARD F. WILLOUGHBY, M.D. (Lond.), &c. Fcap. 8vo, 7s. 6d., cloth. [*Just published.*
"A mine of condensed information of a pertinent and useful kind on the various subjects of which it treats. The matter seems to have been carefully compiled and arranged for facility of reference, and it is well illustrated by diagrams and woodcuts. The different subjects are succinctly but fully and scientifically dealt with."—*The Lancet.*
"Ought to be welcome to those for whose use it is designed, since it practically boils down a reference library into a pocket volume. . . . It combines, with an uncommon degree of efficiency, the qualities of accuracy, conciseness and comprehensiveness."—*Scotsman.*
"An excellent publication, dealing with the scientific, technical and legal matters connected with the duties of medical officers of health and sanitary inspectors."—*Local Government Journal.*

A Complete Epitome of the Laws of this Country.

EVERY MAN'S OWN LAWYER: A Handy-Book of the Principles of Law and Equity. With a Concise Dictionary of Legal Terms. By A BARRISTER. Thirty-second Edition, carefully Revised, and including New Acts of Parliament of 1894. Comprising the *Local Government Act*, 1894 (establishing District and Parish Councils); *Finance Act*, 1894 (imposing the New Death Duties); *Merchant Shipping Act*, 1894; *Prevention of Cruelty to Children Act*, 1894; *Building Societies Act*, 1894; *Notice of Accidents Act*, 1894; *Sale of Goods Act*, 1893; *Voluntary Conveyances Act*, 1893; *Married Women's Property Act*, 1893; *Trustee Act*, 1893; *Fertiliser and Feeding Stuffs Act*, 1893; *Betting and Loans (Infants) Act*, 1892; *Shop Hours Act*, 1892; *Small Holdings Act*, 1892; and many other important new Acts. Crown 8vo, 750 pp., price 6s. 8d. (saved at every consultation!), strongly bound in cloth. [*Just published.*

*** *The Book will be found to comprise (amongst other matter)*—

THE RIGHTS AND WRONGS OF INDIVIDUALS—LANDLORD AND TENANT—VENDORS AND PURCHASERS—LEASES AND MORTGAGES—PRINCIPAL AND AGENT—PARTNERSHIP AND COMPANIES—MASTERS, SERVANTS AND WORKMEN—CONTRACTS AND AGREEMENTS—BORROWERS, LENDERS AND SURETIES—SALE AND PURCHASE OF GOODS—CHEQUES, BILLS AND NOTES—BILLS OF SALE—BANKRUPTCY—RAILWAY AND SHIPPING LAW—LIFE, FIRE, AND MARINE INSURANCE—ACCIDENT AND FIDELITY INSURANCE—CRIMINAL LAW—PARLIAMENTARY ELECTIONS—COUNTY COUNCILS—DISTRICT COUNCILS—PARISH COUNCILS—MUNICIPAL CORPORATIONS—LIBEL AND SLANDER—PUBLIC HEALTH AND NUISANCES—COPYRIGHT, PATENTS, TRADE MARKS—HUSBAND AND WIFE—DIVORCE—INFANCY—CUSTODY OF CHILDREN—TRUSTEES AND EXECUTORS—CLERGY, CHURCHWARDENS, ETC.—GAME LAWS AND SPORTING—INNKEEPERS—HORSES AND DOGS—TAXES AND DEATH DUTIES—FORMS OF AGREEMENTS, WILLS, CODICILS, NOTICES, ETC.

☞ *The object of this work is to enable those who consult it to help themselves to the law; and thereby to dispense, as far as possible, with professional assistance and advice. There are many wrongs and grievances which persons submit to from time to time through not knowing how or where to apply for redress; and many persons have as great a dread of a lawyer's office as of a lion's den. With this book at hand it is believed that many a* SIX-AND-EIGHTPENCE *may be saved; many a wrong redressed; many a right reclaimed; many a law suit avoided; and many an evil abated. The work has established itself as the standard legal adviser of all classes, and has also made a reputation for itself as a useful book of reference for lawyers residing at a distance from law libraries, who are glad to have at hand a work embodying recent decisions and enactments.*

*** OPINIONS OF THE PRESS.

" It is a complete code of English Law written in plain language, which all can understand. . . Should be in the hands of every business man, and all who wish to abolish lawyers' bills."—*Weekly Times*

" A useful and concise epitome of the law, compiled with considerable care."—*Law Magazine.*

" A complete digest of the most useful facts which constitute English law."—*Globe.*

" This excellent handbook. . . . Admirably done, admirably arranged, and admirably cheap."—*Leeds Mercury.*

" A concise, cheap, and complete epitome of the English law. So plainly written that he who runs may read, and he who reads may understand."—*Figaro.*

" A dictionary of legal facts well put together. The book is a very useful one."—*Spectator.*

" A work which has long been wanted, which is thoroughly well done, and which we most cordially recommend."—*Sunday Times.*

" The latest edition of this popular book ought to be in every business establishment, and on every library table."—*Sheffield Post.*

" A complete epitome of the law; thoroughly intelligible to non-professional readers."—*Bell's Life.*

Legal Guide for Pawnbrokers.

THE PAWNBROKERS', FACTORS' AND MERCHANTS' GUIDE TO THE LAW OF LOANS AND PLEDGES. With the Statutes and a Digest of Cases. By H. C. FOLKARD, Esq., Barrister-at-Law. Fcap. 8vo, 3s. 6d. cloth.

The Law of Contracts.

LABOUR CONTRACTS: A Popular Handbook on the Law of Contracts for Works and Services. By DAVID GIBBONS. Fourth Edition, with Appendix of Statutes by T. F. UTTLEY, Solicitor. Fcap. 8vo, 3s. 6d. cloth.

The Factory Acts.

SUMMARY OF THE FACTORY AND WORKSHOP ACTS (1878-1891). For the Use of Manufacturers and Managers. By EMILE GARCKE and J. M. FELLS. (Reprinted from "FACTORY ACCOUNTS.") Crown 8vo, 6d. sewed.

Weale's Rudimentary Series.

London, 1862,
THE PRIZE MEDAL
Was awarded to the Publishers of
"WEALE'S SERIES."

A NEW LIST OF
WEALE'S SERIES
OF
RUDIMENTARY SCIENTIFIC WORKS.

☞ "WEALE'S SERIES includes Text-Books on almost every branch of Science and Industry, comprising such subjects as Agriculture, Architecture and Building, Civil Engineering, Fine Arts, Mechanics and Mechanical Engineering, Physical and Chemical Science, and many miscellaneous Treatises. The whole are constantly undergoing revision, and new editions, brought up to the latest discoveries in scientific research, are constantly issued. The prices at which they are sold are as low as their excellence is assured."—*American Literary Gazette.*

"Amongst the literature of technical education, WEALE'S SERIES has ever enjoyed a high reputation, and the additions being made by Messrs. CROSBY LOCKWOOD & SON render the series even more complete, and bring the information upon the several subjects down to the present time."—*Mining Journal.*

"Any persons wishing to acquire knowledge cannot do better than look through WEALE'S SERIES and get all the books they require. The Series is indeed an inexhaustible mine of literary wealth."—*The Metropolitan.*

"WEALE'S SERIES has become a standard as well as an unrivalled collection of treatises in all branches of art and science."—*Public Opinion.*

"The excellence of WEALE'S SERIES is now so well appreciated that it would be wasting our space to enlarge upon their general usefulness and value."—*Builder.*

"It is not too much to say that no books have ever proved more popular with or more useful to young engineers and others than the excellent treatises comprised in WEALE'S SERIES."—*Engineer.*

"The volumes of WEALE'S SERIES form one of the best collections of elementary technical books in any language."—*Architect.*

"A collection of technical manuals which is unrivalled."—*Weekly Dispatch.*

Philadelphia, 1876,
THE PRIZE MEDAL
Was awarded to the Publishers for
Books : Rudimentary Scientific.
"WEALE'S SERIES," &c.

CROSBY LOCKWOOD & SON,
7, STATIONERS' HALL COURT, LUDGATE HILL, LONDON, E.C.

D

WEALE'S RUDIMENTARY SCIENTIFIC SERIES.

⁂ The volumes of this Series are freely Illustrated with Woodcuts, or otherwise, where requisite. Throughout the following List it must be understood that the books are bound in limp cloth, unless otherwise stated; *but the volumes marked with a ‡ may also be had strongly bound in cloth boards for 6d. extra.*

N.B.—*In ordering from this List it is recommended, as a means of facilitating business and obviating error, to quote the numbers affixed to the volumes, as well as the titles and prices.*

CIVIL ENGINEERING, etc.

31. **WELLS AND WELL-SINKING.** By JOHN GEO. SWINDELL, A.R.I.B.A., and G. R. BURNELL, C.E. Revised Edition. With a New Appendix on the Qualities of Water. Illustrated 2/0
"Solid practical information, written in a concise and lucid style. The work can be recommended as a text-book for all surveyors, architects, &c."—*Iron and Coal Trades Review.*

35. **THE BLASTING AND QUARRYING OF STONE,** for Building and other Purposes. With Remarks on the Blowing up of Bridges. By Gen. Sir J. BURGOYNE, K.C.B. 1/6

43. **TUBULAR AND OTHER IRON GIRDER BRIDGES,** describing the Britannia and Conway Tubular Bridges. With a Sketch of Iron Bridges, &c. By G. DRYSDALE DEMPSEY, C.E. Fourth Edition . 2/0

44. **FOUNDATIONS AND CONCRETE WORKS.** With Practical Remarks on Footings, Planking, Sand, Concrete, Béton, Pile-driving, Caissons, and Cofferdams. By E. DOBSON, M.R.I.B.A. Seventh Edition . 1/6

60. **LAND AND ENGINEERING SURVEYING.** For Students and Practical Use. By T. BAKER, C.E. Fifteenth Edition, revised and corrected by J. R. YOUNG, formerly Professor of Mathematics, Belfast College. Illustrated with Plates and Diagrams . . . . 2/0‡

80*. **EMBANKING LANDS FROM THE SEA.** With Examples and Particulars of actual Embankments, &c. By JOHN WIGGINS, F.G.S. . 2/0

81. **WATER WORKS,** for the Supply of Cities and Towns. With a Description of the Principal Geological Formations of England as influencing Supplies of Water; and Details of Engines and Pumping Machinery for raising Water. By SAMUEL HUGHES, F.G.S., C.E. Enlarged Edition . . 4/0‡
"Every one who is debating how his village, town, or city shall be plentifully supplied with pure water should read this book."—*Newcastle Courant.*

117. **SUBTERRANEOUS SURVEYING.** By THOMAS FENWICK. Also the Method of Conducting Subterraneous Surveys without the use of the Magnetic Needle, and other modern Improvements. By T. BAKER, C.E. 2/6‡

118. **CIVIL ENGINEERING IN NORTH AMERICA,** A Sketch of. By DAVID STEVENSON, F.R.S.E., &c. Plates and Diagrams. . 3/0

167. **A TREATISE ON THE APPLICATION OF IRON TO THE CONSTRUCTION OF BRIDGES, ROOFS, AND OTHER WORKS.** By FRANCIS CAMPIN, C.E. Fourth Edition . . . . 2/6‡
"For numbers of young engineers the book is just the cheap, handy, first guide they want."—*Middlesbrough Weekly News.* "Remarkably accurate and well written."—*Artizan.*

197. **ROADS AND STREETS (THE CONSTRUCTION OF),** in Two Parts: I. THE ART OF CONSTRUCTING COMMON ROADS, by H. LAW, C.E., Revised by D. KINNEAR CLARK, C.E.; II. RECENT PRACTICE: Including Pavements of Stone, Wood, and Asphalte. By D. K. CLARK, C.E. 4/6‡
"A book which every borough surveyor and engineer must possess, and which will be of considerable service to architects, builders, and property owners generally."—*Building News.*

203. **SANITARY WORK IN THE SMALLER TOWNS AND IN VILLAGES.** By CHARLES SLAGG, Assoc. M. Inst. C.E. Second Edition, enlarged . . . . 3/0‡
"This is a very useful book. There is a great deal of work required to be done in the smaller towns and villages, and this little volume will help those who are willing to do it."—*Builder.*

☞ *The ‡ indicates that these vols. may be had strongly bound at 6d. extra.*

Civil Engineering, etc., *continued.*

212. *THE CONSTRUCTION OF GAS WORKS*, and the Manufacture and Distribution of Coal Gas. By S. HUGHES, C.E. Re-written by WILLIAM RICHARDS, C.E. Eighth Edition, with important Additions . 5/6‡
"Will be of infinite service alike to manufacturers, distributors, and consumers."—*Foreman Engineer.*

213. *PIONEER ENGINEERING:* A Treatise on the Engineering Operations connected with the Settlement of Waste Lands in New Countries. By EDWARD DOBSON, A.I. C.E. With numerous Plates. Second Edition . 4/6‡
"Mr. Dobson is familiar with the difficulties which have to be overcome in this class of work, and much of his advice will be valuable to young engineers proceeding to our colonies."—*Engineering.*

216. *MATERIALS AND CONSTRUCTION:* A Theoretical and Practical Treatise on the Strains, Designing, and Erection of Works of Construction. By FRANCIS CAMPIN, C.E. Second Edition, carefully revised. 3/0‡
"No better exposition of the practical application of the principles of construction has yet been published to our knowledge in such a cheap comprehensive form."—*Building News.*

219. *CIVIL ENGINEERING.* By HENRY LAW, M. Inst. C.E. Including a Treatise on HYDRAULIC ENGINEERING by G. R. BURNELL, M.I.C.E. Seventh Edition, revised, WITH LARGE ADDITIONS ON RECENT PRACTICE by D. KINNEAR CLARK, M. Inst. C.E. 6s. 6d., cloth boards . 7/6
"An admirable volume, which we warmly recommend to young engineers."—*Builder.*

260. *IRON BRIDGES OF MODERATE SPAN:* Their Construction and Erection. By HAMILTON WELDON PENDRED, late Inspector of Ironwork to the Salford Corporation. With 40 Illustrations . . . 2/0
"Students and engineers should obtain this book for constant and practical use."—*Colliery Guardian*

268. *THE DRAINAGE OF LANDS, TOWNS, AND BUILDINGS.* By G. D. DEMPSEY, C.E. Revised, with large Additions on Recent Practice in Drainage Engineering, by D. KINNEAR CLARK, M.I.C.E. Second Edition, corrected . . . . 4/6‡

MECHANICAL ENGINEERING, etc.

33. *CRANES*, the Construction of, and other Machinery for Raising Heavy Bodies for the Erection of Buildings, &c. By JOSEPH GLYNN, F.R.S. 1/6

34. *THE STEAM ENGINE.* By Dr. LARDNER. Illustrated . 1/6

59. *STEAM BOILERS:* Their Construction and Management. By R. ARMSTRONG, C.E. Illustrated 1/6
"A mass of information suitable for beginners."—*Design and Work.*

82. *THE POWER OF WATER*, as applied to drive Flour Mills, and to give motion to Turbines and other Hydrostatic Engines. By JOSEPH GLYNN, F.R.S., &c. New Edition, Illustrated 2/0

98. *PRACTICAL MECHANISM*, and Machine Tools. By T. BAKER, C.E. With Remarks on Tools and Machinery, by J. NASMYTH, C.E. 2/6

139. *THE STEAM ENGINE*, a Treatise on the Mathematical Theory of, with Rules and Examples for Practical Men. By T. BAKER, C.E. 1/6
"Teems with scientific information in reference to the steam-engine.—*Design and Work.*

164. *MODERN WORKSHOP PRACTICE*, as applied to Marine, Land, and Locomotive Engines, Floating Docks, Dredging Machines, Bridges, Ship-building, &c. By J. G. WINTON. Fourth Edition, Illustrated . . 3/6‡
"Whether for the apprentice determined to master his profession, or for the artisan bent upon raising himself to a higher position, this clearly written and practical treatise will be a great help."—*Scotsman.*

165. *IRON AND HEAT*, exhibiting the Principles concerned in the Construction of Iron Beams, Pillars, and Girders. By J. ARMOUR, C.E. . 2/6
"A very useful and thoroughly practical little volume."—*Mining Journal.*

166. *POWER IN MOTION:* Horse-power Motion, Toothed-Wheel Gearing, Long and Short Driving Bands, Angular Forces, &c. By JAMES ARMOUR, C.E. With 73 Diagrams. Third Edition . . . . 2/0‡
"The value of the knowledge imparted cannot well be over-estimated."—*Newcastle Weekly Chron.*

171. *THE WORKMAN'S MANUAL OF ENGINEERING DRAWING.* By JOHN MAXTON, Instructor in Engineering Drawing, Royal Naval College, Greenwich. Seventh Edition. 300 Plates and Diagrams 3 6‡
"A copy of it should be kept for reference in every drawing office."—*Engineering.*

☞ The ‡ indicates that these vols. may be had strongly bound at 6d. extra.

Mechanical Engineering, etc., *continued*.

190. **STEAM AND THE STEAM ENGINE**, Stationary and Portable, An Elementary Treatise on. Being an Extension of the Treatise on the Steam Engine of Mr. J. SEWELL. By D. K. CLARK, C.E. Third Edition 3/6‡
"Every essential part of the subject is treated of competently, and in a popular style."—*Iron*.

200. **FUEL, ITS COMBUSTION AND ECONOMY**. Consisting of an Abridgment of "A Treatise on the Combustion of Coal and the Prevention of Smoke." By C. W. WILLIAMS, A.I.C.E. With extensive Additions by D. KINNEAR CLARK, M. Inst. C.E. Third Edition, corrected 3/6‡
"Students should buy the book and read it, as one of the most complete and satisfactory treatises on the combustion and economy of fuel to be had."—*Engineer*.

202. **LOCOMOTIVE ENGINES**, A Rudimentary Treatise on. By G. D. DEMPSEY, C.E. With large Additions treating of the Modern Locomotive, by D. K. CLARK, M. Inst. C.E. With numerous Illustrations . 3/0‡
"A model of what an elementary technical book should be."—*Academy*.

211. **THE BOILERMAKER'S ASSISTANT** in Drawing, Templating, and Calculating Boiler Work, &c. By J. COURTNEY, Practical Boilermaker. Edited by D. K. CLARK, C.E. Third Edition, revised . . 2/0
"With very great care we have gone through the 'Boilermaker's Assistant,' and have to say that it has our unqualified approval. Scarcely a point has been omitted."—*Foreman Engineer*.

217. **SEWING MACHINERY**: Its Construction, History, &c. With full Technical Directions for Adjusting, &c. By J. W. URQUHART, C.E. 2/0
"A full description of the principles and construction of the leading machines, and minute instructions as to their management."—*Scotsman*.

223. **MECHANICAL ENGINEERING.** Comprising Metallurgy, Moulding, Casting, Forging, Tools, Workshop Machinery, Mechanical Manipulation, Manufacture of the Steam Engine, &c. By FRANCIS CAMPIN, C.E. Third Edition, Re-written and Enlarged . . . [*Just published*. 2/6‡
"A sound and serviceable text-book, quite up to date."—*Building News*.

236. **DETAILS OF MACHINERY.** Comprising Instructions for the Execution of various Works in Iron in the Fitting-Shop, Foundry, and Boiler-Yard. By FRANCIS CAMPIN, C.E. . . . . . 3/0‡
"A sound and practical handbook for all engaged in the engineering trades."—*Building World*.

237. **THE SMITHY AND FORGE**, including the Farrier's Art and Coach Smithing. By W. J. E. CRANE. Second Edition, revised . . 2/6‡
"The first modern English book on the subject. Great pains have been bestowed by the author upon the book; shoeing smiths will find it both useful and interesting."—*Builder*.

238. **THE SHEET-METAL WORKER'S GUIDE**: A Practical Handbook for Tinsmiths, Coppersmiths, Zincworkers, &c., with 46 Diagrams and Working Patterns. By W. J. E. CRANE. Second Edition, revised. . 1/6
"The author has acquitted himself with considerable tact in choosing his examples, and with no less ability in treating them."—*Plumber*.

251. **STEAM AND MACHINERY MANAGEMENT**: A Guide to the Arrangement and Economical Management of Machinery, with Hints on Construction and Selection. By M. POWIS BALE, M.Inst.M.E. . . 2/6‡
"Of high practical value."—*Colliery Guardian*.
"Gives the results of wide experience."—*Lloyd's Newspaper*.

254. **THE BOILER-MAKER'S READY RECKONER**, with Examples of Practical Geometry and Templating for the Use of Platers, Smiths, and Riveters. By JOHN COURTNEY. Edited by D. K. CLARK, M.I.C.E. Second Edition, revised, with Additions 4/0

* Nos. 211 *and* 254 *in One Vol., half-bound, entitled* "THE BOILERMAKER'S READY-RECKONER AND ASSISTANT." By J. COURTNEY and D. K. CLARK. *Price* 7s.
"A most useful work. No workman or apprentice should be without it."—*Iron Trade Circular*.

255. **LOCOMOTIVE ENGINE-DRIVING.** A Practical Manual for Engineers in charge of Locomotive Engines. By MICHAEL REYNOLDS, M.S.E. Eighth Edition. 3s. 6d. limp; cloth boards 4/6
"We can confidently recommend the book, not only to the practical driver, but to everyone who takes an interest in the performance of locomotive engines.—*The Engineer*.

256. **STATIONARY ENGINE-DRIVING.** A Practical Manual for Engineers in charge of Stationary Engines. By MICHAEL REYNOLDS, M.S.E. Fourth Edition. 3s. 6d. limp; cloth boards 4/6
"The author is thoroughly acquainted with his subjects, and has produced a manual which is exceedingly useful one for the class for whom it is specially intended."—*Engineering*.

☞ *The* ‡ *indicates that these vols. may be had strongly bound at* 6d. *extra.*

MINING, METALLURGY, etc.

4. *MINERALOGY*, Rudiments of. By A. RAMSAY, F.G.S. Third Edition, revised and enlarged. Woodcuts and Plates . . . 3/6‡

"The author throughout has displayed an intimate knowledge of his subject, and great facility in imparting that knowledge to others. The book is of great utility."—*Mining Journal.*

117. *SUBTERRANEOUS SURVEYING*, with and without the Magnetic Needle. By T. FENWICK and T. BAKER, C.E. Illustrated . . 2/6‡

135. *ELECTRO-METALLURGY*, Practically Treated. By ALEXANDER WATT. Ninth Edition, enlarged and revised. With Additional Illustrations, and including the most Recent Processes . . . . 3/6‡

"From this book both amateur and artisan may learn everything necessary."—*Iron.*

172. *MINING TOOLS*, Manual of. By WILLIAM MORGANS, Lecturer on Practical Mining at the Bristol School of Mines . . . 2/6

172*. *MINING TOOLS, ATLAS* of Engravings to Illustrate the above, containing 235 Illustrations of Mining Tools, drawn to Scale. 4to. . 4/6

"Students, Overmen, Captains, Managers, and Viewers may gain practical knowledge and useful hints by the study of Mr. Morgans' Manual."—*Colliery Guardian.*

176. *METALLURGY OF IRON*. Containing History of Iron Manufacture, Methods of Assay, and Analyses of Iron Ores, Processes of Manufacture of Iron and Steel, &c. By H. BAUERMAN, F.G.S., A.R.S.M. With numerous Illustrations. Sixth Edition, revised and enlarged . . . 5/0‡

"Carefully written, it has the merit of brevity and conciseness, as to less important points; while all material matters are very fully and thoroughly entered into."—*Standard.*

180. *COAL AND COAL MINING*, A Rudimentary Treatise on. By the late Sir WARINGTON W. SMYTH, M.A., F.R.S., &c., Chief Inspector of the Mines of the Crown. Seventh Edition, revised and enlarged . . 3/6‡

"Every portion of the volume appears to have been prepared with much care, and as an outline is given of every known coal-field in this and other countries, as well as of the two principal methods of working, the book will doubtless interest a very large number of readers."—*Mining Journal.*

195. *THE MINERAL SURVEYOR AND VALUER'S COMPLETE GUIDE*. Comprising a Treatise on Improved Mining Surveying and the Valuation of Mining Properties, with New Traverse Tables. By W. LINTERN, Mining and Civil Engineer. Third Edition, with an Appendix on Magnetic and Angular Surveying, with Records of the Peculiarities of Needle Disturbances. With Four Plates of Diagrams, Plans, &c. . . . 3/6‡

"Contains much valuable information, and is thoroughly trustworthy."—*Iron & Coal Trades Review.*

214. *SLATE AND SLATE QUARRYING*, Scientific, Practical, and Commercial. By D. C. DAVIES, F.G.S., Mining Engineer, &c. With numerous Illustrations and Folding Plates. Third Edition . . . 3/0‡

"One of the best and best-balanced treatises on a special subject that we have met with."—*Engineer.*

264. *A FIRST BOOK OF MINING AND QUARRYING*, with the Sciences connected therewith, for Primary Schools and Self Instruction. By J. H. COLLINS, F.G.S., Lecturer to the Miners' Association of Cornwall and Devon. Second Edition, with additions 1/6

"For those concerned in schools in the mining districts, this work is the very thing that should be in the hands of their schoolmasters."—*Iron.*

ARCHITECTURE, BUILDING, etc.

16. *ARCHITECTURE—ORDERS*—The Orders and their Æsthetic Principles. By W. H. LEEDS. Illustrated 1/6

17. *ARCHITECTURE—STYLES*—The History and Description of the Styles of Architecture of Various Countries, from the Earliest to the Present Period. By T. TALBOT BURY, F.R.I.B.A., &c. Illustrated . . 2/0

*** ORDERS AND STYLES OF ARCHITECTURE, *in One Vol., 3s. 6d.*

18. *ARCHITECTURE—DESIGN*—The Principles of Design in Architecture, as deducible from Nature and exemplified in the Works of the Greek and Gothic Architects. By EDW. LACY GARBETT, Architect. Illustrated 2/6

"We know no work that we would sooner recommend to an attentive reader desirous to obtain clear views of the nature of architectural art. The book is a valuable one."—*Builder.*

*** *The three preceding Works in One handsome Vol., half bound, entitled* "MODERN ARCHITECTURE," *price 6s.*

☞ *The ‡ indicates that these vols. may be had strongly bound at 6d. extra.*

Architecture, Building, etc., *continued.*

22. **THE ART OF BUILDING**, Rudiments of. General Principles of Construction, Strength and Use of Materials, Working Drawings, Specifications, &c. By EDWARD DOBSON, M.R.I.B.A., &c. . . . 2/0‡
"A good book for practical knowledge, and about the best to be obtained."—*Building News.*

25. **MASONRY AND STONECUTTING**: The Principles of Masonic Projection and their application to Construction. By E. DOBSON, M.R.I.B.A. . . . . 2/6‡

42. **COTTAGE BUILDING.** By C. BRUCE ALLEN. Eleventh Ed., with Chapter on Economic Cottages for Allotments, by E. E. ALLEN, C.E. 2/0

45. **LIMES, CEMENTS, MORTARS, CONCRETES, MASTICS, PLASTERING,** &c. By G. R. BURNELL, C.E. Thirteenth Edition 1/6

57. **WARMING AND VENTILATION** of Domestic and Public Buildings, Mines, Lighthouses, Ships, &c. By CHARLES TOMLINSON, F.R.S. 3/0

111. **ARCHES, PIERS, BUTTRESSES,** &c.: Experimental Essays on the Principles of Construction in. By WILLIAM BLAND . . 1/6

116. **ACOUSTICS IN RELATION TO ARCHITECTURE AND BUILDING**: The Laws of Sound as applied to the Arrangement of Buildings. By Professor T. ROGER SMITH. F.R.I.B.A. New Edition, Revised. With numerous Illustrations . . . [*Just published.* 1/6

127. **ARCHITECTURAL MODELLING IN PAPER**, The Art of. By T. A. RICHARDSON. With Illustrations, engraved by O. JEWITT . 1/6
"A valuable aid to the practice of architectural modelling."—*Builder's Weekly Reporter.*

128. **VITRUVIUS—THE ARCHITECTURE OF MARCUS VITRUVIUS POLLO.** In Ten Books. Translated from the Latin by JOSEPH GWILT, F.S.A., F.R.A.S. With 23 Plates 5/0
N.B.—This is the only Edition of VITRUVIUS procurable at a moderate price.

130. **GRECIAN ARCHITECTURE**, An Inquiry into the Principles of Beauty in ; with an Historical View of the Rise and Progress of the Art in Greece. By the EARL OF ABERDEEN. 1/0
_{*} *The two preceding Works in One handsome Vol., half bound, entitled* "ANCIENT ARCHITECTURE," *price 6s.*

132. **DWELLING-HOUSES**, The Erection of, Illustrated by a Perspective View, Plans, Elevations, and Sections of a Pair of Villas, with the Specification, Quantities, and Estimates. By S. H. BROOKS, Architect. 2/6‡

156. **QUANTITIES AND MEASUREMENTS**, in Bricklayers', Masons', Plasterers', Plumbers', Painters', Paperhangers', Gilders', Smiths', Carpenters' and Joiners' Work. By A. C. BEATON, Surveyor . . . 1/6
"This book is indispensable to builders and their quantity clerks."—*English Mechanic.*

175. **LOCKWOOD'S BUILDER'S PRICE BOOK FOR** 1896. A Comprehensive Handbook of the Latest Prices and Data for Builders, Architects, Engineers, and Contractors, Re-constructed, Re-written, and greatly Enlarged. By FRANCIS T. W. MILLER, A.R.I.B.A. 700 pages. . 4/0

182 **CARPENTRY AND JOINERY**—THE ELEMENTARY PRINCIPLES OF CARPENTRY. Chiefly composed from the Standard Work of THOMAS TREDGOLD, C.E. With Additions, and a TREATISE ON JOINERY by E. W. TARN, M.A. Fifth Edition, Revised and Extended . 3/6‡

182*. **CARPENTRY AND JOINERY. ATLAS** of 35 Plates to accompany and illustrate the foregoing book. With Descriptive Letterpress. 4to 6/0
"These two volumes form a complete treasury of carpentry and joinery, and should be in the hands of every carpenter and joiner in the empire."—*Iron.*

185. **THE COMPLETE MEASURER;** setting forth the Measurement of Boards, Glass, Timber and Stone. By R. HORTON. Fifth Edition . 4/0
_{*} *The above, strongly bound in leather, price 5s.*

187. **HINTS TO YOUNG ARCHITECTS.** By GEORGE WIGHTWICK, Architect, Author of "The Palace of Architecture," &c., &c. Fifth Edition, revised and enlarged by G. HUSKISSON GUILLAUME, Architect. 3/6‡
"A copy ought to be considered as necessary a purchase as a box of instruments."—*Architect.*

☞ *The* ‡ *indicates that these vols. may be had strongly bound at 6d. extra.*

Architecture, Building, etc., *continued.*

188 *HOUSE PAINTING, GRAINING, MARBLING, AND SIGN WRITING:* With a Course of Elementary Drawing, and a Collection of Useful Receipts. By ELLIS A. DAVIDSON. Sixth Edition. Coloured Plates 5/0
*** *The above in cloth boards, strongly bound,* 6s.
"A mass of information of use to the amateur and of value to the practical man."—*English Mechanic.*

189. *THE RUDIMENTS OF PRACTICAL BRICKLAYING.* General Principles of Bricklaying; Arch Drawing, Cutting, and Setting; Pointing; Paving, Tiling, &c. By ADAM HAMMOND. With 68 Woodcuts . 1/6
"The young bricklayer will find it infinitely valuable to him."—*Glasgow Herald.*

191. *PLUMBING:* A Text-Book to the Practice of the Art or Craft of the Plumber. With Chapters upon House Drainage and Ventilation. By WM. PATON BUCHAN, R.P., Sanitary Engineer. Sixth Edition, revised and enlarged, with 380 Illustrations . . . . 3/6‡
"A text-book which may be safely put into the hands of every young plumber, and which will also be found useful by architects and medical professors."—*Builder.*

192. *THE TIMBER IMPORTER'S, TIMBER MERCHANT'S, AND BUILDER'S STANDARD GUIDE.* By R. E. GRANDY . 2/0
"Everything it pretends to be: built up gradually, it leads one from a forest to a treenail, and throws in, as a makeweight, a host of material concerning bricks, columns, cisterns, &c."—*English Mechanic.*

206. *A BOOK ON BUILDING, Civil and Ecclesiastical.* By Sir EDMUND BECKETT, Bart., LL.D., Q.C., F.R.A.S., Author of "Clocks and Watches and Bells," &c. Second Edition, enlarged . . . . 4/6‡
"A book which is always amusing and nearly always instructive."—*Times.*

226. *THE JOINTS MADE AND USED BY BUILDERS.* By WYVILL J. CHRISTY, Architect. With 160 Woodcuts . . . . 3/0‡
"The work is deserving of high commendation."—*Builder.*

228. *THE CONSTRUCTION OF ROOFS, OF WOOD AND IRON:* Deduced chiefly from the Works of Robison, Tredgold, and Humber. By E. WYNDHAM TARN, M.A., Architect. Second Edition, revised . 1/6
"Mr. Tarn is so thoroughly master of his subject, that although the treatise is founded on the works of others, he has given it a distinct value of his own. It will be found valuable by all students."—*Builder.*

229. *ELEMENTARY DECORATION:* As applied to Dwelling Houses, &c. By JAMES W. FACEY. Illustrated 2/0
"The principles which ought to guide the decoration of dwelling-houses are clearly set forth, and elucidated by examples; while full instructions are given to the learner."—*Scotsman.*

257. *PRACTICAL HOUSE DECORATION.* A Guide to the Art of Ornamental Painting, the Arrangement of Colours in Apartments, and the Principles of Decorative Design. By JAMES W. FACEY . . . 2/6
*** Nos. 229 and 257 in One handsome Vol., half-bound, entitled "HOUSE DECORATION, ELEMENTARY AND PRACTICAL," price 5s.

230. *A PRACTICAL TREATISE ON HANDRAILING;* Showing New and Simple Methods. By GEO. COLLINGS. Second Edition. Revised, including a TREATISE ON STAIRBUILDING. With Plates . 2/6
"Will be found of practical utility in the execution of this difficult branch of joinery."—*Builder.*

247. *BUILDING ESTATES:* A Treatise on the Development, Sale, Purchase, and Management of Building Land. By F. MAITLAND. Second Edition, revised 2/0
"This book should undoubtedly be added to the library of every professional man dealing with building land."—*Land Agent's Record.*

248. *PORTLAND CEMENT FOR USERS.* By HENRY FAIJA, A.M. Inst. C.E. Third Edition, Corrected 2/0
"Supplies in a small compass all that is necessary to be known by users of cement."—*Building News.*

252. *BRICKWORK:* A Practical Treatise, embodying the General and Higher Principles of Bricklaying, Cutting and Setting; with the Application of Geometry to Roof Tiling, &c. By F. WALKER 1/6
"Contains all that a young tradesman or student needs to learn from books."—*Building News.*

259. *GAS FITTING:* A Practical Handbook. By JOHN BLACK. Second Edition, Enlarged. With 130 Illustrations . . . . 2/6‡
"Contains all the requisite information for the successful fitting of houses with a gas service, &c. It is written in a simple practical style, and we heartily recommend it."—*Plumber and Decorator.*

☞ *The* ‡ *indicates that these vols. may be had strongly bound at 6d. extra.*

Architecture, Building, etc., *continued.*

23, 189, 252. **THE PRACTICAL BRICK AND TILE BOOK.** Comprising: BRICK AND TILE MAKING, by E. DOBSON, A.I.C.E.; Practical BRICKLAYING, by A. HAMMOND; BRICKWORK, by F. WALKER. 550 pp. with 270 Illustrations, strongly half-bound 6/0

258. **CIRCULAR WORK IN CARPENTRY AND JOINERY.** A Practical Treatise on Circular Work of Single and Double Curvature. By GEORGE COLLINGS. Second Edition 2/6
"Cheap in price, clear in definition, and practical in the examples selected."—*Builder.*

261. **SHORING, and Its Application:** A Handbook for the Use of Students. By GEORGE H. BLAGROVE. With 31 Illustrations . . . 1/6
"We recommend this valuable treatise to all students."—*Building News.*

265. **THE ART OF PRACTICAL BRICK CUTTING AND SETTING.** By ADAM HAMMOND. With 90 Engravings . . . 1/6

267. **THE SCIENCE OF BUILDING:** An Elementary Treatise on the Principles of Construction. By E. WYNDHAM TARN, M.A. Lond. Third Edition, revised and enlarged . . . . 3/6‡

271. **VENTILATION:** A Text Book to the Practice of the Art of Ventilating Buildings. By W. P. BUCHAN, R. P., Author of "Plumbing," &c. With 170 Illustrations . . . . . . 3/6‡

272. **ROOF CARPENTRY;** Practical Lessons in the Framing of Wood Roofs. For the Use of Working Carpenters. By GEO. COLLINGS, Author of "Handrailing and Stairbuilding," &c. 2/-

273. **THE PRACTICAL PLASTERER:** A Compendium of Plain and Ornamental Plaster Work. By WILFRED KEMP . . . 2/-

SHIPBUILDING, NAVIGATION, etc.

51. **NAVAL ARCHITECTURE:** An Exposition of the Elementary Principles. By JAMES PEAKE, H.M. Dockyard, Portsmouth . . 3/6‡

53*. **SHIPS FOR OCEAN AND RIVER SERVICE,** Elementary and Practical Principles of the Construction of. By HAKON A. SOMMERFELDT. 1/6

53**. **AN ATLAS OF ENGRAVINGS** to Illustrate the above. Twelve large folding Plates. Royal 4to, cloth 7/6

54. **MASTING, MAST-MAKING, AND RIGGING OF SHIPS.** Also Tables of Spars, Rigging, Blocks; Chain, Wire, and Hemp Ropes, &c., relative to every class of vessels. By ROBERT KIPPING, N.A. 2/0

54*. **IRON SHIP-BUILDING.** With Practical Examples and Details. By JOHN GRANTHAM. Fifth Edition 4/0

55. **THE SAILOR'S SEA BOOK:** A Rudimentary Treatise on Navigation. By JAMES GREENWOOD, B.A. With numerous Woodcuts and Coloured Plates. New and enlarged Edition. By W. H. ROSSER . 2/6‡
"Is perhaps the best and simplest epitome of navigation ever compiled.—*Field.*

55 & 204. **PRACTICAL NAVIGATION.** Consisting of THE SAILOR'S SEA-BOOK, by JAMES GREENWOOD and W. H. ROSSER; together with Mathematical and Nautical Tables for the Working of the Problems, by HENRY LAW, C.E., and Prof. J. R. YOUNG. Half-bound in leather . 7/0
"A vast amount of information is contained in this volume, and we fancy in a very short time that it will be seen in the library of almost every ship or yacht afloat."—*Hunt's Yachting Magazine.*

80. **MARINE ENGINES AND STEAM VESSELS.** By R. MURRAY, C.E. Eighth Edition, thoroughly Revised, with Additions by the Author and by GEORGE CARLISLE, C.E. . . . . . 4/6‡
"An indispensable manual for the student of marine engineering."—*Liverpool Mercury.*

83*bis*. **THE FORMS OF SHIPS AND BOATS.** By W. BLAND. Seventh Edition, revised, with numerous Illustrations and Models . . 1/6

99. **NAVIGATION AND NAUTICAL ASTRONOMY,** in Theory and Practice. By Prof. J. R. YOUNG. New Edition. Illustrated . 2/6
"A very complete, thorough, and useful manual for the young navigator."—*Observatory.*

106. **SHIPS' ANCHORS,** a Treatise on. By GEORGE COTSELL. 1/6

149. **SAILS AND SAIL-MAKING.** With Draughting, and the Centre of Effort of the Sails. Also, Weights and Sizes of Ropes; Masting, Rigging, and Sails of Steam Vessels, &c. By ROBERT KIPPING, N.A. . 2/6‡

155. **THE ENGINEER'S GUIDE TO THE ROYAL AND MERCANTILE NAVIES.** By a PRACTICAL ENGINEER. Revised by D. F. M'CARTHY, late of the Ordnance Survey Office, Southampton . 3/0

☞ *The* ‡ *indicates that these vols. may be had strongly bound at 6d. extra.*

AGRICULTURE, GARDENING, etc.

61*. *A COMPLETE READY RECKONER FOR THE AD-MEASUREMENT OF LAND, &c.* By A. ARMAN. Third Edition, revised and extended by C. NORRIS, Surveyor, Valuer, &c. . . . 2/0
"A very useful book to all who have land to measure."—*Mark Lane Express.*
"Should be in the hands of all persons having any connection with land."—*Irish Farm.*

131. *MILLER'S, CORN MERCHANT'S, AND FARMER'S READY RECKONER.* Second Edition, revised, with a Price List of Modern Flour Mill Machinery, by W. S. HUTTON, C.E. 2/0
"Will prove an indispensable *vade mecum*. Nothing has been spared to make the book complete and perfectly adapted to its special purpose.'—*Miller.*

140. *SOILS, MANURES, AND CROPS.* (Vol. I. OUTLINES OF MODERN FARMING.) By R. SCOTT BURN. Woodcuts. 2/0

141. *FARMING AND FARMING ECONOMY*, Historical and Practical. (Vol. II. OUTLINES OF MODERN FARMING.) By R. SCOTT BURN. 3/0
"Eminently calculated to enlighten the agricultural community on the varied subjects of which it treats; hence it should find a place in every farmer's library."—*City Press.*

142. *STOCK; CATTLE, SHEEP, AND HORSES.* (Vol. III. OUTLINES OF MODERN FARMING.) By R. SCOTT BURN. Woodcuts. . 2/6
"The author's grasp of his subject is thorough, and his grouping of facts effective. . . . We commend this excellent treatise "—*Weekly Dispatch.*

145. *DAIRY, PIGS, AND POULTRY.* (Vol. IV. OUTLINES OF MODERN FARMING.) By R. SCOTT BURN. Woodcuts 2/0
"We can testify to the clearness and intelligibility of the matter, which has been compiled from the best authorities."—*London Review.*

146. *UTILIZATION OF SEWAGE, IRRIGATION, AND RECLAMATION OF WASTE LAND.* (Vol. V. OUTLINES OF MODERN FARMING.) By R. SCOTT BURN. Woodcuts 2/6
"A work containing valuable information, which will recommend itself to all interested in modern farming."—*Field.*

140.⎫
141.⎬ *OUTLINES OF MODERN FARMING.* By R. SCOTT
142.⎬ BURN, Author of "Landed Estates Management," "Farm Management,"
145.⎬ and Editor of "The Complete Grazier." Consisting of the above Five
146.⎭ Volumes in One, 1,250 pp., profusely Illustrated, half-bound . . 12/0
"The aim of the author has been to make his work at once comprehensive and trustworthy, and in this aim he has succeeded to a degree which entitles him to much credit."—*Morning Advertiser.*
"Should find a place in every farmer's library."—*City Press.*
"No farmer should be without it."—*Banbury Guardian.*

177. *FRUIT TREES*, The Scientific and Profitable Culture of. From the French of M. DU BREUIL. Fourth Edition, carefully Revised by GEORGE GLENNY. With 187 Woodcuts . . . . . . 3/6‡
"The book teaches how to prune and train fruit trees to perfection."—*Field.*

198. *SHEEP:* The History, Structure, Economy, and Diseases of. By W. C. SPOONER, M.R.V.C., &c. Fifth Edition, with fine Engravings, including Specimens of New and Improved Breeds. 366 pp. . . . 3/6‡
"The book is decidedly the best of the kind in our language."—*Scotsman.*

201. *KITCHEN GARDENING MADE EASY.* Showing the best means of Cultivating every known Vegetable and Herb, &c., with directions for management all the year round. By GEO. M. F. GLENNY. Illustrated 1/6‡
"This book will be found trustworthy and useful."—*North British Agriculturist.*

207. *OUTLINES OF FARM MANAGEMENT.* Treating of the General Work of the Farm; Stock; Contract Work; Labour, &c. By R. SCOTT BURN, Author of "Outlines of Modern Farming," &c. . . 2/6‡
"The book is eminently practical, and may be studied with advantage by beginners in agriculture, while it contains hints which will be useful to old and successful farmers."—*Scotsman.*

208. *OUTLINES OF LANDED ESTATES MANAGEMENT:* Treating of the Varieties of Lands, Methods of Farming, the Setting-out of Farms, &c.; Roads, Fences, Gates, Irrigation, Drainage, &c. By R. S. BURN. 2/6‡
"A complete and comprehensive outline of the duties appertaining to the management of landed estates."—*Journal of Forestry.*

⁎ *Nos. 207 & 208 in One Vol., handsomely half-bound, entitled* "OUTLINES OF LANDED ESTATES AND FARM MANAGEMENT." By ROBERT SCOTT BURN. *Price 6s.*

☞ The ‡ *indicates that these vols. may be had strongly bound at 6d. extra.*

Agriculture, Gardening, etc., *continued.*

209. *THE TREE PLANTER AND PLANT PROPAGATOR:*
With numerous Illustrations of Grafting, Layering, Budding, Implements, Houses, Pits, &c. By S. WOOD, Author of "Good Gardening," &c. . . 2/0
"Sound in its teaching and very comprehensive in its aim. It is a good book."—*Gardeners' Magazine.*
"The instructions are thoroughly practical and correct."—*North British Agriculturist.*

210. *THE TREE PRUNER*: Being a Practical Manual on the Pruning of Fruit Trees, including also their Training and Renovation, also treating of the Pruning of Shrubs, Climbers and Flowering Plants. With numerous Illustrations. By SAMUEL WOOD, Author of "Good Gardening," &c. 1/6
"A useful book, written by one who has had great experience."—*Mark Lane Express.*
"We recommend this treatise very highly."—*North British Agriculturist.*
** Nos. 209 & 210 *in One Vol., handsomely half-bound, entitled* "THE TREE PLANTER, PROPAGATOR AND PRUNER." By SAMUEL WOOD. *Price* 3s. 6d.

218. *THE HAY AND STRAW MEASURER*: New Tables for the Use of Auctioneers, Valuers, Farmers, Hay and Straw Dealers, &c., forming a complete Calculator and Ready Reckoner. By JOHN STEELE . 2/0
"A most useful handbook. It should be in every professional office where agricultural valuations are conducted."—*Land Agent's Record.*

231. *THE ART OF GRAFTING AND BUDDING.* By CHARLES BALTET. With Illustrations . . . . . . 2/6‡
"The one standard work on this subject."—*Scotsman.*

232. *COTTAGE GARDENING*; or, Flowers, Fruits, and Vegetables for Small Gardens. By E. HOBDAY 1/6
"Definite instructions as to the cultivation of small gardens."—*Scotsman.*
"Contains much useful information at a small charge."—*Glasgow Herald.*

233. *GARDEN RECEIPTS.* Edited by CHARLES W. QUIN. 1/6
"A singularly complete collection of the principal receipts needed by gardeners."—*Farmer.*
"A useful and handy book, containing a good deal of valuable information."—*Athenæum.*

234. *MARKET AND KITCHEN GARDENING.* By C. W. SHAW, late Editor of "Gardening Illustrated" . . . . . 3/0‡
"The most valuable compendium of kitchen and market-garden work published."—*Farmer.*
"A most comprehensive volume on market and kitchen-gardening."—*Mark Lane Express.*

239. *DRAINING AND EMBANKING.* A Practical Treatise. By JOHN SCOTT, late Professor of Agriculture and Rural Economy at the Royal Agricultural College, Cirencester. With 68 Illustrations . . . 1/6
"A valuable handbook to the engineer, as well as to the surveyor."—*Land.*
"This volume should be perused by all interested in this important branch of estate improvement."—*Land Agent's Record.*

240. *IRRIGATION AND WATER SUPPLY*: A Practical Treatise on Water Meadows, Sewage Irrigation, Warping, &c.; on the Construction of Wells, Ponds and Reservoirs, &c. By Prof. J. SCOTT. With 34 Illusts. 1/6
"A valuable and indispensable book for the estate manager and owner."—*Forestry.*
"Well worth the study of all farmers and landed proprietors."—*Building News.*

241. *FARM ROADS, FENCES, AND GATES*: A Practical Treatise on the Roads, Tramways, and Waterways of the Farm; the Principles of Enclosures; and the different kinds of Fences, Gates, and Stiles. By Professor JOHN SCOTT. With 75 Illustrations 1/6
"Mr. Scott's treatise will be welcomed as a concisely compiled handbook."—*Building News.*
"A useful practical work, which should be in the hands of every farmer."—*Farmer.*

242. *FARM BUILDINGS.* A Practical Treatise on the Buildings necessary for various kinds of Farms, their Arrangement and Construction, with Plans and Estimates. By Prof. JOHN SCOTT. With 105 Illustrations . 2/0
"No one who is called upon to design farm-buildings can afford to be without this work."—*Builder.*
"This book ought to be in the hands of every landowner and agent."—*Kelso Chronicle.*

243. *BARN IMPLEMENTS AND MACHINES.* Treating of the Application of Power to the Operations of Agriculture; and of the various Machines used in the Threshing-barn, in the Stock-yard, Dairy, &c. By Professor JOHN SCOTT. With 123 Illustrations 2/0

☞ *The* ‡ *indicates that these vols. may be had strongly bound at* 6d. *extra.*

Agriculture, Gardening, etc., *continued.*

244. **FIELD IMPLEMENTS AND MACHINES**: With Principles and Details of Construction and Points of Excellence, their Management, &c. By Prof. JOHN SCOTT. With 138 Illustrations 2/0
245. **AGRICULTURAL SURVEYING**: A Treatise on Land Surveying, Levelling, and Setting-out; with Directions for Valuing and Reporting on Farms and Estates. By Prof. J. SCOTT. With 62 Illustrations 1/6
239.⎫ **FARM ENGINEERING**: By Professor JOHN SCOTT. Comprising the above Seven Volumes in One, 1,150 pages, and over 600 Illustrations.
to ⎬
245.⎭ Half-bound 12/0
"A copy of this work should be treasured up in every library where the owner thereof is in any way connected with land."—*Farm and Home.*
250. **MEAT PRODUCTION**: A Manual for Producers, Distributors, and Consumers of Butchers' Meat. By JOHN EWART. . . . 2/6
"A compact and handy volume on the meat question."—*Meat and Provision Trades' Review.*
266. **BOOK-KEEPING FOR FARMERS AND ESTATE OWNERS.** A Practical Treatise, presenting, in Three Plans, a System adapted for all classes of Farms. By J. M. WOODMAN, Chartered Accountant. Third Edition, revised 2/6
⁎ *The above in cloth boards, strongly bound, 3s. 6d.*
"Will be found of great assistance by those who intend to commence a system of book-keeping, the author's examples being clear and explicit, and his explanations full and accurate."—*Live Stock Journal.*

MATHEMATICS, ARITHMETIC, etc.

32. **MATHEMATICAL INSTRUMENTS**, a Treatise on; Their Construction, Adjustment, Testing, and Use concisely Explained. By J. F. HEATHER, M.A., of the Royal Military Academy, Woolwich. Fourteenth Edition, Revised, with Additions, by A. T. WALMISLEY, M.I.C.E., Fellow of the Surveyors' Institution. Original Edition, in 1 vol., Illustrated . . 2/0‡
⁎ *In ordering the above, be careful to say "Original Edition," or give the number in the Series (32), to distinguish it from the Enlarged Edition in 3 vols. (Nos. 168-9-70).*
76. **DESCRIPTIVE GEOMETRY**, an Elementary Treatise on; with a Theory of Shadows and of Perspective, extracted from the French of G. MONGE. To which is added a Description of the Principles and Practice of Isometric Projection. By J. F. HEATHER, M.A. With 14 Plates . 2/0
78. **PRACTICAL PLANE GEOMETRY**: giving the Simplest Modes of Constructing Figures contained in one Plane and Geometrical Construction of the Ground. By J. F. HEATHER, M.A. With 215 Woodcuts . 2/0
"The author is well-known as an experienced professor, and the volume contains as complete a collection of problems as is likely to be required in ordinary practice."—*Architect.*
83. **COMMERCIAL BOOK-KEEPING.** With Commercial Phrases and Forms in English, French, Italian, and German. By JAMES HADDON, M.A., formerly Mathematical Master, King's College School . 1/6
84. **ARITHMETIC**, a Rudimentary Treatise on: with full Explanations of its Theoretical Principles, and numerous Examples for Practice. For the Use of Schools and for Self-Instruction. By J. R. YOUNG, late Professor of Mathematics in Belfast College. Eleventh Edition . . 1/6
84*. **A KEY TO THE ABOVE.** By J. R. YOUNG . . . 1/6
85. **EQUATIONAL ARITHMETIC**, applied to Questions of Interest, Annuities, Life Assurance, and General Commerce; with various Tables by which all Calculations may be greatly facilitated. By W. HIPSLEY. 2/0
86. **ALGEBRA**, the Elements of. By JAMES HADDON, M.A., formerly Mathematical Master of King's College School. With Appendix, containing Miscellaneous Investigations, and a collection of Problems . 2/0
86*. **A KEY AND COMPANION TO THE ABOVE.** An extensive repository of Solved Examples and Problems in Illustration of the various Expedients necessary in Algebraical Operations. By J. R. YOUNG.
88. ⎫ **EUCLID**, THE ELEMENTS OF: with many Additional Propositions and Explanatory Notes; to which is prefixed an Introductory Essay on
& ⎬
89. ⎭ Logic. By HENRY LAW, C.E. 2/6
⁎ *Sold also separately, viz.:—*
88. EUCLID, The First Three Books. By HENRY LAW, C.E. . . . 1/6
89. EUCLID, Books 4, 5, 6, 11, 12. By HENRY LAW, C.E. . . . 1/6

☞ *The* ‡ *indicates that these vols. may be had strongly bound a 6d. extra.*

Mathematics, Arithmetic, etc., *continued.*

90. **ANALYTICAL GEOMETRY AND CONIC SECTIONS**, a Rudimentary Treatise on. By JAMES HANN. A New Edition, re-written and enlarged by Professor J. R. YOUNG . . . . . **2/0‡**

"The author's style is exceedingly clear and simple, and the book is well adapted for the beginner and those who may be obliged to have recourse to self-tuition."—*Engineer.*

91. **PLANE TRIGONOMETRY, the Elements of.** By JAMES HANN, formerly Mathematical Master of King's College, London . . **1/6**

92. **SPHERICAL TRIGONOMETRY, the Elements of.** By JAMES HANN. Revised by CHARLES H. DOWLING, C.E. **1/0**

*** Or with "The Elements of Plane Trigonometry," in One Volume, 2s. 6d.

93. **MENSURATION AND MEASURING**, for Students and Practical Use. With the Mensuration and Levelling of Land for the purposes of Modern Engineering. By T. BAKER, C.E. New Ed. by E. NUGENT, C.E. **1/6**

101. **DIFFERENTIAL CALCULUS, Elements of the.** By W. S. B. WOOLHOUSE, F.R.A.S., &c. **1/6**

102. **INTEGRAL CALCULUS.** By HOMERSHAM COX, B.A. . **1/0**

136. **ARITHMETIC**, Rudimentary, for the Use of Schools and Self-Instruction. By JAMES HADDON, M.A. Revised by ABRAHAM ARMAN . **1/6**

137. **A KEY TO THE ABOVE.** By A. ARMAN . . . **1/6**

168. **DRAWING AND MEASURING INSTRUMENTS.** Including—I. Instruments employed in Geometrical and Mechanical Drawing, and in the Construction, Copying, and Measurement of Maps and Plans. II. Instruments used for the purposes of Accurate Measurement, and for Arithmetical Computations. By J. F. HEATHER, M.A. . . . **1/6**

"Valuable and instructive to all whose occupations require exceptional accuracy in measurements."—*Jeweller and Metal Worker.*

169. **OPTICAL INSTRUMENTS.** Including (more especially) Telescopes, Microscopes, and Apparatus for producing copies of Maps and Plans by Photography. By J. F. HEATHER, M.A. Illustrated . . **1/6**

"An excellent treatise."—*British Journal of Photography.*

170. **SURVEYING & ASTRONOMICAL INSTRUMENTS.** Including—I. Instruments used for Determining the Geometrical Features of a portion of Ground. II. Instruments employed in Astronomical Observations. By J. F. HEATHER, M.A. Illustrated **1/6**

"A good, sensible, useful book."—*School Board Chronicle.*

*** *The above three volumes form an enlargement of the Author's original work, "Mathematical Instruments": price 2s. (See No. 32 in the Series.)*

168. ⎫ **MATHEMATICAL INSTRUMENTS:** Their Construction,
169. ⎬ Adjustment, Testing and Use. Comprising Drawing, Measuring, Optical,
170. ⎭ Surveying, and Astronomical Instruments. By J. F. HEATHER, M.A. Enlarged Edition, for the most part entirely re-written. The Three Parts as above, in One thick Volume . . . . **4/6‡**

"An exhaustive treatise, belonging to the well-known Weale's Series. Mr. Heather's experience well qualifies him for the task he has so ably fulfilled."—*Engineering and Building Times.*

158. **THE SLIDE RULE, AND HOW TO USE IT.** Containing full, easy, and simple Instructions to perform all Business Calculations with unexampled rapidity and accuracy. By CHARLES HOARE, C.E. With a Slide Rule, in tuck of cover. Fifth Edition . . . . . **2/6‡**

196. **THEORY OF COMPOUND INTEREST AND ANNUITIES**; with Tables of Logarithms for the more Difficult Computations of Interest, Discount, Annuities, &c., in all their Applications and Uses for Mercantile and State Purposes. By FEDOR THOMAN, of the Société Crédit Mobilier, Paris. Fourth Edition, carefully revised and corrected . . **4/0**

"A very powerful work, and the author has a very remarkable command of his subject."—Professor A de MORGAN. "We recommend it to the notice of actuaries and accountants."—*Athenæum.*

☞ *The ‡ indicates that these vols. may be had strongly bound at 6d. extra.*

Mathematics, Arithmetic, etc., *continued.*

199. *THE COMPENDIOUS CALCULATOR (Intuitive Calculations);* or, Easy and Concise Methods of Performing the various Arithmetical Operations required in Commercial and Business Transactions; together with Useful Tables, &c. By DANIEL O'GORMAN. Twenty-seventh Edition, carefully revised by C. NORRIS 2/6
  ⁎⁎ *The above strongly half-bound, price 3s. 6d.*
  "It would be difficult to exaggerate the usefulness of this book to everyone engaged in commerce or manufacturing industry. It is crammed full with rules and formulæ for shortening and employing calculations in money, weights and measures, &c. of every sort and description."—*Knowledge.*

204. *MATHEMATICAL TABLES,* for Trigonometrical, Astronomical, and Nautical Calculations; to which is prefixed a Treatise on Logarithms. By H. LAW, C.E. Together with a Series of Tables for Navigation and Nautical Astronomy. By Professor J. R. YOUNG. New Edition 4/0

204.⁎ *LOGARITHMS.* With Mathematical Tables for Trigonometrical, Astronomical, and Nautical Calculations. By HENRY LAW, C.E. Revised Edition. (Forming part of the above work.) 3/0

221. *MEASURES, WEIGHTS, AND MONEYS OF ALL NATIONS,* and an Analysis of the Christian, Hebrew, and Mahometan Calendars. By W. S. B. WOOLHOUSE, F.R.A.S., F.S.S. Seventh Edition, 2/6‡
  "A work necessary for every mercantile office."—*Building Trades Journal.*

227. *A TREATISE ON MATHEMATICS,* as applied to the Constructive Arts. By FRANCIS CAMPIN, C.E., &c. Second Edition . . 3/0‡
  "Should be in the hands of everyone connected with building construction."—*Builder's Weekly Reporter.*

PHYSICAL SCIENCE, NATURAL PHILOSOPHY, etc.

1. *CHEMISTRY,* for the Use of Beginners. By Prof. GEO. FOWNES, F.R.S. With an Appendix on the Application of Chemistry to Agriculture. 1/0

2. *NATURAL PHILOSOPHY,* for the Use of Beginners. By CHARLES TOMLINSON, F.R.S. 1/6

6. *MECHANICS:* Being a concise Exposition of the General Principles of Mechanical Science, and their Applications. By CHARLES TOMLINSON, F.R.S. 1/6

7. *ELECTRICITY;* showing the General Principles of Electrical Science, and the Purposes to which it has been applied. By Sir W. SNOW HARRIS, F.R.S., &c. With considerable Additions by R. SABINE, C.E., F.S.A. 1/6

7⁎. *GALVANISM.* By Sir W. SNOW HARRIS. New Edition, revised, with considerable Additions, by ROBERT SABINE, C.E. . . . 1/6

8. *MAGNETISM.* By Sir W. SNOW HARRIS. New Edition, revised and enlarged by H. M. NOAD, Ph.D. With 165 Woodcuts . . 3/6‡
  "The best popular exposition of magnetism, its intricate relations and complicating effects, with which we are acquainted."—*School Board Chronicle.*

11. *THE ELECTRIC TELEGRAPH:* its History and Progress; with Descriptions of some of the Apparatus. By R. SABINE, C.E., F.S.A., &c. 3/0
  "Essentially a practical and instructive work."—*Daily Telegraph.*

12. *PNEUMATICS,* including Acoustics and the Phenomena of Wind Currents, for the Use of Beginners. By CHARLES TOMLINSON, F.R.S. Fourth Edition, enlarged. Illustrated. 1/6

72. *MANUAL OF THE MOLLUSCA:* A Treatise on Recent and Fossil Shells. By Dr. S. P. WOODWARD, A.L.S. With Appendix by RALPH TATE, A.L.S., F.G.S. With numerous Plates and 300 Woodcuts. cloth boards, gilt 7/6
  "A storehouse of conchological and geological information."—*Hardwicke's Science Gossip.*
  "An important work, with such additions as complete it to the present time."—*Land and Water.*

96. *ASTRONOMY.* By the late Rev. ROBERT MAIN, M.A., F.R.S., formerly Radcliffe Observer at Oxford. Third Edition, revised and corrected to the Present Time, by WILLIAM THYNNE LYNN, B.A., F.R.A.S. . . 2/0
  "A sound and simple treatise, very carefully edited, and a capital book for [beginners."—*Knowledge*

97. *STATICS AND DYNAMICS,* the Principles and Practice of; embracing also a clear development of Hydrostatics, Hydrodynamics, and Central Forces. By T. BAKER, C.E. Fourth Edition 1/6

☞ *The* ‡ *indicates that these vols. may be had strongly bound at 6d. extra.*

Physical Science, Natural Philosophy, etc., *continued.*

173. **PHYSICAL GEOLOGY**, partly based on Major-General PORT-
LOCK's "Rudiments of Geology." By RALPH TATE, A.L.S., &c. Woodcuts. 2/0

174. **HISTORICAL GEOLOGY**, partly based on Major-General
PORTLOCK's "Rudiments." By RALPH TATE, A.L.S., &c. Woodcuts. . 2/6

173. **GEOLOGY**, PHYSICAL and HISTORICAL. Consisting of
& "Physical Geology," which sets forth the Leading Principles of the Science;
174. and "Historical Geology," which treats of the Mineral and Organic Conditions
of the Earth at each successive epoch. By RALPH TATE, F.G.S., &c., &c.
With 250 Illustrations . . . . 4/6‡

"The fulness of the matter has elevated the book into a manual. Its information is exhaustive and well arranged, so that any subject may be opened upon at once."—*School Board Chronicle.*

183. **ANIMAL PHYSICS**, Handbook of. By DIONYSIUS LARD-
184. NER, D.C.L. With 520 Illustrations. In One Vol. (732 pages), cloth boards. 7/6
* *Sold also in Two Parts, as follows:—*

183. ANIMAL PHYSICS. By Dr. LARDNER. Part I., Chapters I.–VII. . . 4/0
184. ANIMAL PHYSICS. By Dr. LARDNER. Part II., Chapters VIII.–XVIII. . 3/0

"This book contains a great deal more than an introduction to human anatomy. In it will be found the elements of comparative anatomy, a complete treatise on the functions of the body, and a description of the phenomena of birth, growth, and decay."—*Educational Times.*

269. **LIGHT:** An Introduction to the Science of Optics. Designed
for the Use of Students of Architecture, Engineering, and other Applied
Sciences. By E. WYNDHAM TARN, M.A., Author of "The Science of
Building," &c. 1/6

FINE ARTS, etc.

20. **PERSPECTIVE FOR BEGINNERS.** Adapted to Young
Students and Amateurs in Architecture, Painting, &c. By GEORGE PYNE. 2/0

40. **GLASS STAINING, AND THE ART OF PAINTING**
ON GLASS. From the German of Dr. GESSERT and EMANUEL OTTO
FROMBERG. With an Appendix on THE ART OF ENAMELLING. . . 2/6

69. **MUSIC**, A Rudimentary and Practical Treatise on. With
numerous Examples. By CHARLES CHILD SPENCER 2/6

"Mr. Spencer has marshalled his information with much skill, and yet with a simplicity that must recommend his works to all who wish to thoroughly understand music."—*Weekly Times.*

71. **PIANOFORTE**, The Art of Playing the. With numerous
Exercises and Lessons. By CHARLES CHILD SPENCER 1/6

"A sound and excellent work, written with spirit, and calculated to inspire the pupil with a desire to aim at high accomplishment in the art."—*School Board Chronicle.*

69, 71. **MUSIC, AND THE PIANOFORTE.** One Vol. Half-bound. 5/0

181. **PAINTING POPULARLY EXPLAINED.** By THOMAS
JOHN GULLICK, Painter, and JOHN TIMBS, F.S.A. Including Fresco, Oil,
Mosaic, Water Colour, Water-Glass, Tempera, Encaustic, Miniature, Painting
on Ivory, Vellum, Pottery, Enamel, Glass, &c. Fifth Edition . . 5/0‡
* *Adopted as a Prize book at South Kensington.*

"Much may be learned, even by those who fancy they do not require to be taught, from the careful perusal of this unpretending but comprehensive treatise."—*Art Journal.*

186. **A GRAMMAR OF COLOURING.** Applied to Decorative
Painting and the Arts. By GEORGE FIELD. New Edition, revised and
enlarged by ELLIS A. DAVIDSON. With Coloured Plates . . . 3/0‡

"The book is a most useful *resumé* of the properties of pigments."—*Builder.*
"One of the most useful of students' books."—*Architect.*

246. **A DICTIONARY OF PAINTERS, AND HANDBOOK**
FOR PICTURE AMATEURS; being a Guide for Visitors to Public and
Private Picture Galleries, and for Art-Students, including Glossary of Terms,
Sketch of Principal Schools of Painting, &c. By PHILIPPE DARYL, B.A. . 2/6‡

"Considering its small compass, really admirable. We cordially recommend the book."—*Builder.*

☞ *The* ‡ *indicates that these vols. may be had strongly bound at 6d. extra.*

INDUSTRIAL AND USEFUL ARTS.

23. *BRICKS AND TILES,* Rudimentary Treatise on the Manufacture of; containing an Outline of the Principles of Brickmaking. By E. DOBSON, M.R.I.B.A. Additions by C. TOMLINSON, F.R.S. Illust. 3/0‡
"The best handbook on the subject. We can safely recommend it as a good investment."—*Builder*

67. *CLOCKS AND WATCHES, AND BELLS,* a Rudimentary Treatise on. By Sir EDMUND BECKETT, Bart. Q.C. Seventh Edition. . 4/6
*** *The above handsomely bound, cloth boards, 5s. 6d.*
"The best work on the subject probably extant. The treatise on bells is undoubtedly the best in the language."—*Engineering.* "The only modern treatise on clock-making."—*Horological Journal.*

83**. *CONSTRUCTION OF DOOR LOCKS.* From the Papers of A. C. HOBBS. Edited by CHARLES TOMLINSON, F.R.S. With a Note upon IRON SAFES by ROBERT MALLET. Illustrated 2/6

162. *THE BRASS FOUNDER'S MANUAL:* Instructions for Modelling, Pattern Making, Moulding, Turning, &c. By W. GRAHAM. . 2/0‡

205. *THE ART OF LETTER PAINTING MADE EASY.* By JAMES GREIG BADENOCH. With 12 full-page Engravings of Examples . 1/6
"Any intelligent lad who fails to turn out decent work after studying this system, has mistaken his vocation."—*English Mechanic.*

215. *THE GOLDSMITH'S HANDBOOK*, containing full Instructions in the Art of Alloying, Melting, Reducing, Colouring, Collecting and Refining. The processes of Manipulation, Recovery of Waste, Chemical and Physical Properties of Gold; Solders, Enamels and other useful Rules and Recipes, &c. By GEORGE E. GEE. Third Edition, considerably enlarged . 3/0‡
"A good, sound, technical educator."—*Horological Journal.*
"A standard book, which few will care to be without."—*Jeweller and Metalworker.*

225. *THE SILVERSMITH'S HANDBOOK*, on the same plan as the above. By GEORGE E. GEE. Second Edition, Revised . . . 3/0‡
"A valuable sequel to the author's 'Practical Goldworker.'"—*Silversmith's Trade Journal.*
"As a guide to workmen it will prove a good technical educator."—*Glasgow Herald.*
*** *The two preceding Works, in One handsome Vol., half-bound, entitled "* THE GOLDSMITH'S AND SILVERSMITH'S COMPLETE HANDBOOK," *7s.*

249. *THE HALL-MARKING OF JEWELLERY.* Comprising an account of all the different Assay Towns of the United Kingdom; with the Stamps at present employed; also the Laws relating to the Standards and Hall-Marks at the various Assay Offices. By GEORGE E. GEE . . 3/0‡
"Deals thoroughly with its subject from a manufacturer's and dealer's point of view."—*Jeweller.*
"A valuable and trustworthy guide."—*English Mechanic.*

224. *COACH-BUILDING:* A Practical Treatise, Historical and Descriptive. By JAMES W. BURGESS. With 57 Illustrations . . . 2/6‡
"This handbook will supply a long-felt want, not only to manufacturers themselves, but more particularly apprentices and others whose occupations may be in any way connected with the trade of coach-building."—*European Mail.*

235. *PRACTICAL ORGAN BUILDING.* By W. E. DICKSON, M.A., Precentor of Ely Cathedral. Second Edition, Revised, with Additions. 2/6‡
"The amateur builder will find in this book all that is necessary to enable him personally to construct a perfect organ with his own hands."—*Academy.*
"The best work on the subject that has yet appeared in book form."—*English Mechanic.*

262. *THE ART OF BOOT AND SHOEMAKING*, including Measurement, Last-fitting, Cutting-out, Closing and Making; with a Description of the most Approved Machinery employed. By JOHN BEDFORD LENO, late Editor of "St. Crispin" and "The Boot and Shoemaker." With numerous Illustrations. Third Edition . . . . 2/0‡
"This excellent treatise is by far the best work ever written on the subject. The chapter on clicking, which shows how waste may be prevented, will save fifty times the price of the book."—*Scottish Leather Trader.*

263. *MECHANICAL DENTISTRY:* A Practical Treatise on the Construction of the Various Kinds of Artificial Dentures, comprising also Useful Formulæ, Tables and Receipts for Gold Plate, Clasps, Solders, &c. By CHARLES HUNTER. Third Edition, revised, with additions . . 3/0‡
"We can strongly recommend Mr. Hunter's treatise to all students preparing for the profession of dentistry, as well as to every mechanical dentist."—*Dublin Journal of Medical Science.*

270. *WOOD ENGRAVING:* A Practical and Easy Introduction to the Study of the Art. By W. N. BROWN 1/6

☞ *The ‡ indicates that these vols. may be had strongly bound at 6d. extra.*

MISCELLANEOUS VOLUMES.

36. *A DICTIONARY OF TERMS used in ARCHITECTURE,
BUILDING, ENGINEERING, MINING, METALLURGY, ARCHÆ-
OLOGY, the FINE ARTS, &c.* By JOHN WEALE. Sixth Edition.
Edited by ROBT. HUNT, F.R.S., Keeper of Mining Records, Editor of
"Ure's Dictionary." Numerous Illustrations 5/0

*** *The above, strongly bound in cloth boards, price 6s.*

"The best small technological dictionary in the language."—*Architect.*
"The absolute accuracy of a work of this character can only be judged of after extensive consultation and from our examination it appears very correct and very complete."—*Mining Journal.*
"There is no need now to speak of the excellence of this work; it received the approval of the community long ago. Edited now by Mr Robert Hunt, and published in a cheap, handy form, it will be of the utmost service as a book of reference scarcely to be exceeded in value."—*Scotsman.*

50. *THE LAW OF CONTRACTS FOR WORKS AND
SERVICES.* By DAVID GIBBONS. Fourth Edition, with Appendix of
Statutes by T. F. UTTLEY, Solicitor. Cloth boards 3/6

"A very compendious, full and intelligible digest of the working and results of the law, in regard to all kinds of contracts between parties standing in the relation of employer and employed."—*Builder.*
"This exhaustive manual is written in a clear, terse, and pleasant style, and is just the work for masters and servants alike to depend upon for constant reference."—*Metropolitan.*

112. *MANUAL OF DOMESTIC MEDICINE.* By R. GOODING,
B.A., M.D. Intended as a Family Guide in all cases of Accident and Emer-
gency. Third Edition, carefully revised 2/0

"The author has, we think, performed a useful service by placing at the disposal of those situated, by unavoidable circumstances, at a distance from medical aid, a reliable and sensible work in which professional knowledge and accuracy have been well seconded by the ability to express himself in ordinary untechnical language."—*Public Health.*

112.* *MANAGEMENT OF HEALTH.* A Manual of Home
and Personal Hygiene. By the Rev. JAMES BAIRD, B.A. . . . 1/0

"The author gives sound instructions for the preservation of health."—*Athenæum.*
"It is wonderfully reliable, it is written with excellent taste, and there is instruction crowded into every page."—*English Mechanic.*

150. *LOGIC,* Pure and Applied. By S. H. EMMENS. Third Edition 1/6

"This admirable work should be a text-book not only for schools, students and philosophers, for all *literateurs* and men of science, but for those concerned in the practical affairs of life, &c."—*The News.*

153. *SELECTIONS FROM LOCKE'S ESSAYS ON THE
HUMAN UNDERSTANDING.* With Notes by S. H. EMMENS . . 1/6

154. *GENERAL HINTS TO EMIGRANTS.* Containing No-
tices of the various Fields for Emigration. With Hints on Preparation for
Emigrating, Outfits, &c., Useful Recipes, Map of the World, &c. . . 2/0

157. *THE EMIGRANT'S GUIDE TO NATAL.* By ROBERT
JAMES MANN, F.R.A.S., F.M.S. Second Edition, revised. Map . . 2/0

193. *HANDBOOK OF FIELD FORTIFICATION.* By Major
W. W. KNOLLYS, F.R.G.S. With 163 Woodcuts . . . . 3/0‡

"A well-timed and able contribution to our military literature. . . . The author supplies, in a clear business style, all the information likely to be practically useful."—*Chambers of Commerce Chronicle.*

194. *THE HOUSE MANAGER:* Being a Guide to Housekeep-
ing, Practical Cookery, Pickling and Preserving, Household Work, Dairy
Management, the Table and Dessert, Cellarage of Wines, Home-brewing and
Wine-making, the Boudoir and Dressing-room, Travelling, Stable Economy,
Gardening Operations, &c. By AN OLD HOUSEKEEPER . . . 3/6‡

"We find here directions to be discovered in no other book, tending to save expense to the pocket, as well as labour to the head."—*John Bull.*
"Quite an Encyclopædia of domestic matters. We have been greatly pleased with the neatness and lucidity of the explanatory details."—*Court Circular.*

194. } *HOUSE BOOK (The).* Comprising: I. THE HOUSE MANAGER.
112. } By an OLD HOUSEKEEPER. II. DOMESTIC MEDICINE. By RALPH GOODING,
&   } M.D. III. MANAGEMENT OF HEALTH. By JAMES BAIRD. In One Vol.,
112*. } strongly half-bound 6/0

☞ *The ‡ indicates that these vols. may be had strongly bound at 6d. extra.*

www.ingramcontent.com/pod-product-compliance
Lightning Source LLC
Chambersburg PA
CBHW022058230426
43672CB00008B/1207